W9-BLS-601

Bureaucratic Landscapes

Politics, Science, and the Environment
Peter M. Haas, Sheila Jasanoff, and Gene Rochlin, editors

Peter Dauvergne, *Shadows in the Forest: Japan and the Politics of Timber in Southeast Asia*

Peter Cebon, Urs Dahinden, Huw Davies, Dieter M. Imboden, and Carlo C. Jaeger, eds., *Views from the Alps: Regional Perspectives on Climate Change*

Clark C. Gibson, Margaret A. McKean, and Elinor Ostrom, eds., *People and Forests: Communities, Institutions, and Governance*

The Social Learning Group, *Learning to Manage Global Environmental Risks*. Volume 1: *A Comparative History of Social Responses to Climate Change, Ozone Depletion, and Acid Rain.* Volume 2: *A Functional Analysis of Social Responses to Climate Change, Ozone Depletion, and Acid Rain*

Clark Miller and Paul N. Edwards, eds., *Changing the Atmosphere: Expert Knowledge and Environmental Governance*

Craig W. Thomas, *Bureaucratic Landscapes: Interagency Cooperation and the Preservation of Biodiversity*

Bureaucratic Landscapes

Interagency Cooperation and the Preservation of Biodiversity

Craig W. Thomas

The MIT Press
Cambridge, Massachusetts
London, England

This book was set in Sabon by SNP Best-set Typesetter Ltd., Hong Kong and was printed and bound in the United States of America.

Library of Congress Cataloging-in-Publication Data

Thomas, Craig W.
Bureaucratic landscapes : interagency cooperation and the preservation of biodiversity / Craig W. Thomas.
p. cm.—(Politics, science, and the environment)
Includes bibliographical references and index.
ISBN 0-262-20141-0 (hc. : alk. paper)—ISBN 0-262-70089-1 (pbk. : alk. paper)
1. Biological diversity conservation—Government policy—California—Case studies. 2. Biological diversity conservation—Government policy—United States—Case studies. 3. Natural resources—California—Management—Case studies. 4. Natural resources—United States—Management—Case studies. 5. Interorganizational relations—California—Case studies. 6. Interorganizational relations—United States—Case studies. I. Title. II. Series.

QH76.5.C2 T52 2003
333.95′16′0976—dc21

2002071779

10 9 8 7 6 5 4 3 2 1

For Leanne, Ellie, and Wyatt

"Every scrap of biological diversity is priceless, to be learned and cherished, and never to be surrendered without a struggle."

—E. O. Wilson, *The Diversity of Life*

"Biodiversity conservation will not succeed if constrained by political boundaries and the disparate mandates of multiple agencies."

—Reed F. Noss and Allen Y. Cooperrider, *Saving Nature's Legacy: Protecting and Restoring Biodiversity*

"The only thing the public fears more than an uncoordinated bureaucracy is a coordinated bureaucracy."

—Douglas P. Wheeler, Secretary for Resources, State of California

Contents

Series Foreword

As our understanding of environmental threats deepens and broadens, it is increasingly clear that many environmental issues cannot be simply understood, analyzed, or acted upon. The multifaceted relationships between human beings, social and political institutions, and the physical environment in which they are situated extend across disciplinary as well as geopolitical confines, and cannot be analyzed or resolved in isolation.

The purpose of this series is to address the increasingly complex questions of how societies come to understand, confront, and cope with both the sources and the manifestations of present and potential environmental threats. Works in the series may focus on matters political, scientific, technical, social, or economic. What they share is attention to the intertwined roles of politics, science, and technology in the recognition, framing, analysis, and management of environmentally related contemporary issues, and a manifest relevance to the increasingly difficult problems of identifying and forging environmentally sound public policy.

Peter M. Haas
Sheila Jasanoff
Gene Rochlin

Preface

Government in the United States is extremely fragmented.

This statement is obvious and reasonable to political scientists familiar with U.S. politics, public policy, and public management, but it surprises and perplexes many natural scientists and resource management professionals, who wonder why the physical landscape is crisscrossed by multiple jurisdictions that bear little or no resemblance to ecological systems. Political scientists accept jurisdictional fragmentation as a well-known fact, while tracing its policy impacts and debating its democratic justifications. Natural scientists are familiar instead with habitat fragmentation and its deleterious effects on biodiversity. The U.S. landscape looks much different today than it did before European settlers moved west across the interior, transforming prairies into farmland, altering the flow of rivers, and cutting down forests to build houses, fences, and cities. The extent of this ecological transformation surprises many social scientists, who are much less familiar with the dramatic impacts of habitat fragmentation on biodiversity. This book addresses both types of fragmentation by focusing on the ways scientific knowledge has altered relationships among public agencies in our fragmented governmental system. In reading this book, I hope that social scientists, policy analysts, and public managers will better understand how natural scientists view the world, while natural scientists and resource managers will better see the order in our fragmented governmental system.

Political scientists have long been concerned with the tension between institutional fragmentation and policy coordination in the U.S. bureaucracy. While some have acknowledged that public agencies are able occasionally to coordinate their management plans, service delivery systems, and regulatory mechanisms, it remains unclear what factors lead individuals in different agencies to cooperate. On what kinds of tasks do agency officials seek to cooperate? What role do ideas, knowledge, and legal mandates play in shaping cooperative behavior? Are some types of agencies and officials more likely to cooperate than others? If so, what organizational characteristics facilitate or impede cooperation? In short, what are the organizational and institutional incentives that encourage individuals in different agencies and in different parts of agencies to cooperate rather than compete over turf or enhance their autonomy as traditionally expected?

The literature is rife with examples of agencies either competing with each other or asserting their independence. Cooperation appears to be relatively rare. So it is important to understand why cooperation occurs when it does, and why cooperative efforts fail when they do. Interagency cooperation is also important for addressing some policy issues. Biodiversity is a case in point because species, habitats, and ecosystems sprawl across agency jurisdictions. Thus, cooperative planning and management is an important component of efforts to preserve biodiversity.

This book analyzes several case studies of successful and unsuccessful efforts to preserve biodiversity in California through interagency cooperation. By "success," I mean whether cooperation emerged, grew, and became stable. I do not attempt to evaluate the policy impacts of these efforts. Instead, the focus is simply to understand why public officials tried to cooperate and what obstacles they faced. The case studies nevertheless provide some information about the policy impacts of these cooperative efforts. Moreover, if preserving biodiversity indeed requires coordinated action, knowing the extent to which cooperation occurred provides indirect evidence of policy impacts as well.

Audience

This book is written for several audiences in the social and natural sciences. For political scientists, the book bridges the fields of public policy and public management by examining the relationship between policy formulation and implementation. It also examines the role of courts in prompting agency action, the role of scientific knowledge in organizational learning, and the emergence of new institutions to resolve collective-action problems. In addition, it provides in-depth case studies of public managers working the seams of government to bring agencies together to accomplish new tasks. Rather than focusing solely on traditional missions, these public officials seek to create public value (Moore 1995) by developing new strategies to produce joint gains on a new mission: the preservation of biodiversity.

In the natural sciences, this book is pitched to those interested in biodiversity policy and management, and to those interested in how scientific knowledge is used in practice. In particular, it is pitched to conservation biologists, some of whom have called for increased integration with the social sciences and humanities. As Noss and Cooperrider (1994: 84) argued, "Geography, geology, sociology, education, philosophy, law, economics, and political science are just as important to the successful practice of conservation biology as are wildlife biology, forestry, ecology, zoology, botany, genetics, and other biological sciences." Political scientists can certainly contribute to an understanding of agency behavior. Given the prominent role of conservation biology in this book, my hope is that the book will advance this integration.

More generally, this book is situated in the emerging literature on collaboration (Gray 1989; Bardach 1998) and collaborative environmental management (Brick, Snow, and Van de Wetering 2001; Wondolleck and Yaffee 2000; John 1994). Yet it differs somewhat from this literature in two respects. First, much of this literature focuses on public-private partnerships. Though the theoretical framework in this book focuses on public-public partnerships, the case studies also include private stakeholders. Second, the book gives significant attention to why cooperation fails. Much of the existing literature focuses on cases of successful collaboration or how to build successful collaborative efforts. By contrast, this book critically assesses the causal roots of both success and failure. Indeed, from a methodological standpoint, if one seeks to understand the conditions that foster success, one must also understand failure.

The Endangered Species Act (ESA) plays a prominent role in this book, so it complements the growing literature on implementation of the ESA. This includes research on the role of science in the ESA (Czech and Krausman 2001; Kareiva et al. 1999; National Research Council 1995), habitat conservation planning (Beatley 1994; Yaffee et al. 1998; Noss, O'Connell, and Murphy 1997), and improving endangered species recovery programs (Clark 1997; Clark, Reading, and Clarke 1994).

For resource management practitioners, environmental advocates, and concerned citizens, this book offers lessons about why cooperation succeeds and fails, but it is not a guide on how to develop cooperative efforts. For the latter, I recommend Bardach 1998 and Wondolleck and Yaffee 2000. Yet there are certainly lessons in this book for those who want to understand why cooperative efforts succeed or fail, lessons that should help them map their own strategies. These readers should also note a second caveat. This book does not attempt to paint a current picture of public agencies, public laws, scientific knowledge, or ecosystems. Instead, it seeks to explain how local, state, and federal officials understood their world and their actions in the 1990s. Therefore, I present evidence that helps us understand the conditions they experienced at the time, not the conditions that might confront current practitioners, advocates, or local stakeholders.

Book Structure

Chapter 1 reviews the public agencies, public laws, and technical tasks associated with preserving biodiversity in the United States. It introduces social scientists to ecological principles for managing biodiversity, and offers natural scientists and resource managers a broad understanding of the political and administrative problems associated with managing natural landscapes in a fragmented governmental system. The chapter also frames the central question of the book: *Under what conditions do individuals in different public agencies cooperate with one another?*

Chapter 2 presents a framework for understanding the diverse motivations of agency officials, and how these motivations shape cooperative behavior. Specifically, I slice the agencies into three cross sections, representing line managers, professional staff, and field staff, and demonstrate how these individuals understand the world in fundamentally different ways. Line managers are primarily interested in providing stability and certainty for their organizational units, a goal best achieved by enhancing their autonomy. Professionals are less concerned with agency stability. Instead, they define problems, recommend solutions, and follow best management practices that have been socially constructed within their profession. For resource management professionals rooted in the ecological sciences, these solutions and practices necessarily include interagency coordination because ecological systems transcend agency jurisdictions. Field staff develop a third logic of interdependence because they live and work in local communities, where they identify with community needs. These community ties generate social capital, which binds public-private partnerships together at the local level. Management responsibilities, professional training, and community ties lead each set of actors to see a different logic of interdependence, which shapes how they perceive the potential costs and benefits of cooperation.

Chapters 3 through 7 present several case studies that explain why cooperation occurred in some cases but not in others. Because some cases progressed further than the existing literature led me to expect, I sometimes sound like a cheerleader celebrating the accomplishments of my favorite team. Yet these teams could have done much more. Accordingly, the case studies include a subplot that suggests what the agencies could have done if they had made a greater effort to cooperate and to protect biodiversity. Having conducted the research in pursuit of hypotheses, not to test hypotheses, I present the case studies primarily as narratives rather than statistical tables. (See appendix A for research methodology.)

Chapters 3 and 4 present a single case study that focuses on a pivotal interagency agreement known as the California Memorandum of Understanding on Biological Diversity (MOU on Biodiversity). (See appendix B for the full title and text of this agreement.) Originally signed in 1991 by political appointees and high-level line managers in six state agencies and four federal regional offices, this agreement soon gathered many more signatures. The agreement is remarkable because the signatory agencies had previously shown little concern for biodiversity. Chapter 3 explains the emergence of this agreement, focusing on why professional staff wrote it and why line managers signed it. Chapter 4 completes the case study by focusing on the Executive Council on Biological Diversity, which was formed to implement the agreement. This chapter analyzes what line managers and professional staff accomplished after the agreement was signed, and highlights the local backlash against the MOU and Executive Council, and the subsequent co-optation of county supervisors to assuage local concerns. The Executive Council is now called the California Biodiversity Council, and is still active in 2002.

Chapters 5 through 7 present three bioregional case studies. The MOU envisioned cooperative efforts at three organizational levels: at the state level, through the new Executive Council; at the bioregional level, through new bioregional councils; and within each bioregion, through local groups of public and private actors. I selected the Klamath, South Coast, and San Joaquin Valley Bioregions for analysis because they are biologically diverse; because the human communities within them have widely differing social, economic, and political characteristics; and because each of the bioregions is home to both successful and unsuccessful cooperative efforts. These differences help us to compare interagency cooperation in different parts of the state, and to compare cooperative dynamics at the local, regional, and state levels.

Chapter 5 examines the Klamath Bioregion, where vast expanses of evergreen forests predominate. Most of this forested landscape had been logged at least once, thereby depleting and fragmenting the habitat of numerous species that depend on the structural characteristics of old-growth forests. Here, the Executive Council sought to create a bioregional council from the ground up, and hired contractors to facilitate regionwide meetings of local stakeholders, local government officials, field staff, and regional line managers. Yet after only two bioregional meetings, participants split themselves into several subregional groups. This chapter focuses on the forces for integration at the local level and the forces for disintegration at the regional level, emphasizing the desire of field staff to participate in public-private partnerships at the local level despite lack of support from their line managers.

Chapter 6 looks at the South Coast Bioregion, which includes the Los Angeles and San Diego metropolitan areas and the diminishing open space around them. This bioregion is largely urban and suburban rather than rural. It differs from the Klamath Bioregion due to rapid suburbanization and the size and technical sophistication of local governments. Instead of timber harvests, bulldozers were scraping away much of the native vegetation, as developers replaced it with houses, roads, and shopping malls. Litigation over several endangered species was poised to impede developers, with whom Governor Pete Wilson maintained close ties. Wilson's political appointees accordingly played lead roles, working with local governments and developers to create a program known as Natural Communities Conservation Planning (NCCP). In contrast to the situation in the Klamath Bioregion, where the Executive Council sought to nurture the formation of a bioregional council from the bottom up, NCCP was a centralized program that offered participants an alternative means to meet regulatory requirements under the ESA. Line managers were also removed from the decision-making process, so that the agencies could speak and act with one voice.

Chapter 7 turns to the San Joaquin Valley Bioregion, where the socioeconomic profile was dominated by industrial agriculture and fossil fuel extraction. In the foothills surrounding the valley floor, the Bureau of Land Management (BLM) owned a large number of fragmented parcels. BLM line managers and staff led an interagency effort to create a bioregional council that would develop a regional plan to protect biodiversity, and they asked county supervisors to lead this effort. As in the Klamath, this bioregional effort floundered because it lacked the legal incentives for cooperation that were established in the South Coast. County supervisors also intentionally sabotaged the effort by leading state and federal officials to believe they would participate, while privately agreeing to pull out before anything could be accomplished.

Chapter 8 weaves the case studies together by exploring the organizational and institutional incentives for cooperation in preserving biodiversity. Though line managers generally sought autonomy, they pooled some agency resources in these cases to reduce the likelihood of lawsuits, particularly under the ESA. Line managers feared lawsuits that would diminish their management discretion on traditional tasks, so they turned to their professional staff for advice. It was here that they found consensual ecological knowledge that suggested coordinated habitat planning and management as a best management practice. These ideas and strategies also appealed to field staff, who sought to enhance the socioeconomic health of their local communities. While these incentives varied over time, other incentives varied across agencies. Variation in participation by agencies was largely determined by the likelihood of lawsuits. BLM managers led many of the cooperative efforts because environmentalists routinely sued the agency, and BLM land provided habitat for roughly half the listed species in California. Moreover, BLM's holdings were dispersed across the landscape in many small parcels, which increased the agency's dependence on activities in neighboring jurisdictions. In the National Park Service (NPS), by contrast, line managers did not actively participate in most cooperative efforts because environmental groups seldom sued the NPS, so park managers did not fear lawsuits—regardless of the number of listed species within the parks.

Acknowledgments

This book began life as a dissertation in political science at the University of California, Berkeley. It has undergone many revisions since then, but I want to thank those who helped me in the early years. I was particularly fortunate to work with an especially supportive dissertation committee. If there are good qualities in this book, to a large extent they can be traced to the thoughtful guidance of Todd La Porte, Bruce Cain, and Gene Bardach. Other Berkeley faculty also provided valuable commentary, including Chris Ansell, John Ellwood, Sally Fairfax, Judy Gruber, and Laura Stoker. I am particularly indebted to Ansell for reading my work closely and helping me situate the findings within numerous literatures.

A remarkable group of graduate students made the early years of this research experience a pleasure rather than a chore. In particular, I want to thank Taylor Dark, John Gerring, Marissa Martino Golden, Chris Muste, Paul Neibergs, and Rob Thompson for adding order to the chaos and fun to the process. I owe special thanks to Neibergs for helping me identify potential case studies, and to Golden and Thompson for repeated readings of several chapters.

Several wonderful organizations provided financial and administrative support. Resources for the Future provided a Joseph L. Fisher Dissertation Fellowship. At Berkeley, the Department of Political Science supported me through a Henry Robert Braden Dissertation Scholarship, and the Institute of Governmental Studies provided extensive administrative support, a terrific library, and a warm working environment.

At the University of Massachusetts, Amherst, my thanks go to the Center for Public Policy and Administration for providing research assistance and administrative support. Here, I was fortunate to work with four talented research assistants: Lori Schwarz, Jennifer Balkcom, Alice Napoleon, and Jessica Rajotte Wozniak. In the Department of Political Science, I also want to thank John Brigham, Peter Haas, John Hird, and M. J. Peterson for their feedback. I want to thank Peterson, in particular, for reading the entire manuscript closely and providing extensive editorial comments.

In the larger worlds of collaborative environmental management and deliberative democracy, I would like to thank Archon Fung, Bradley Karkkainen, Judy Layzer, Cassandra Moseley, and Charles Sabel for expanding my horizons. Of these, I need to thank Layzer for providing extensive and insightful commentary during the final revisions.

At MIT Press, I want to thank Clay Morgan, the production staff, and the anonymous referees. I also want to acknowledge three journals, in which pieces from this book previously appeared. Some of the background on habitat conservation planning in chapter 6 appeared in *Politics and Society* (Thomas 2001). Part of the Klamath Bioregion case in chapter 5 appeared in *Policy Studies Journal* (Thomas 1999). Excerpts from several chapters appeared in the *Journal of Public Administration Research and Theory* (Thomas 1997).

Nearly last but certainly not least, I must thank the public officials who took time from their busy schedules to talk with me. My field research depended on their goodwill, and they repeatedly surprised me by their candor, sincerity, and generosity. Although public employees are often maligned, I was greatly impressed by their hard work and dedication, and I apologize if the book suggests otherwise. Having promised anonymity to many, let me simply express a collective thanks and my continuing appreciation. In addition to those public officials with whom I conducted tape-recorded interviews, several individuals involved with resource management issues in California provided invaluable assistance in navigating me through the shoals of biodiversity policy and management. In particular, I want to thank Cecilia Danks, Yvonne Everett, Deborah Jensen, Jonathan Kusel, Mark Nechodom, Maria Rea, and Larry Ruth for their advice, encouragement, and hospitality. Nechodom deserves special thanks for his multifaceted support, for which I am deeply indebted.

Throughout it all and forever, my thanks and love go to my partner, Leanne Robertson, with whom life is joyful, serene, and sweet. Long projects like this often demarcate significant points in life. We were married as this research began, and expanded to three with the wonderful arrival of Ellie as it came to a close.

List of Abbreviations

ACE	U.S. Army Corps of Engineers
AMA	adaptive management area
BLM	U.S. Bureau of Land Management
BoR	U.S. Bureau of Reclamation
CARCD	California Association of Resource Conservation Districts
CDF	California Department of Forestry and Fire Protection
CDFG	California Department of Fish and Game
CDPR	California Department of Parks and Recreation
CDWR	California Department of Water Resources
CEC	California Energy Commission
CEQA	California Environmental Quality Act
CERES	California Environmental Resources Evaluation System
CERT	Community Economic Revitalization Team
CESA	California Endangered Species Act
CFA	California Forestry Association
COG	council of governments
CRMP	Coordinated Resource Management and Planning
CSLC	California State Lands Commission
CSS	coastal sage scrub
DoD	U.S. Department of Defense
EPA	U.S. Environmental Protection Agency
ESA	U.S. Endangered Species Act
FLPMA	Federal Land Policy and Management Act
FRRAP	Forest and Rangeland Resources Assessment Program

FS	U.S. Forest Service
FWS	U.S. Fish and Wildlife Service
GAO	U.S. General Accounting Office
GIS	geographic information system
HCP	habitat conservation plan
INACC	Interagency Natural Areas Coordinating Committee
MMBF	million board feet
MOU	memorandum of understanding
MVP	minimum viable population
NCCP	Natural Communities Conservation Planning
NEPA	National Environmental Policy Act
NFMA	National Forest Management Act
NMFS	U.S. National Marine Fisheries Service
NPS	U.S. National Park Service
NRCS	U.S. Natural Resources Conservation Service
OTA	U.S. Office of Technology Assessment
RCD	resource conservation district
RCHCA	Riverside County Habitat Conservation Agency
SANDAG	San Diego Association of Governments
SCAG	Southern California Association of Governments
SCS	U.S. Soil Conservation Service
SNAP	Significant Natural Areas Program
SOP	standard operating procedure
SRP	scientific review panel
SWRCB	State Water Resources Control Board
TBRG	Trinity Bio Region Group
TNC	The Nature Conservancy
TRRP	Trinity River Restoration Program
TTF	Timberland Task Force
TVA	Tennessee Valley Authority
UC	University of California
UC/DANR	UC Division of Agriculture and Natural Resources

1

Fragmented Jurisdictions, Fragmented Habitat

Fragmented Jurisdictions and the Possibility of Cooperation

The U.S. federal system was designed to prevent tyranny, not to achieve policy coherence. This was the essence of the debate between James Madison and Alexander Hamilton in the *Federalist Papers* more than two hundred years ago. Madison favored a system of fragmented power so that no single faction would be able to tyrannize over other factions. Hamilton favored unified power under a strong presidency. The Madisonians prevailed, and created a federal system of divided power with checks and balances, a system that necessarily entails negotiated policymaking among the seats of power.

Within this fragmented system, elected officials at the local, state, and national levels have created numerous public agencies to carry out public functions. These agencies are essentially political organizations because they are the primary means through which public policies are implemented. Not only are the political battles of interest groups and elected officials played out through the bureaucracy, the very structures and processes of these agencies are themselves reflections of past political conflicts. Public agencies are simply another political arena, providing an additional point of access for a wide variety of interests in a fragmented governmental system.

Given that public agencies are creatures of politics, political scientists have traditionally examined the relationships among agencies, elected officials, and clienteles.[1] They have also studied the mutually reinforcing relationships among all three sets of actors, using terms like *iron triangle* and *subgovernment* to characterize policy arenas in which the policymaking process is largely controlled by an agency, its clientele, and a congressional committee.[2] By comparison, political scientists gave little attention to relationships among public agencies before the 1970s, perhaps because most policy arenas did not overlap in significant ways, so there may have been little to study.

This situation changed dramatically when Pressman and Wildavsky published *Implementation* in 1973, thereby sparking interest in social policy implementation and the multiagency character of President Lyndon Johnson's "Great Society" programs. Relationships among local, state, and federal agencies took center stage in this new literature, but in a limited and negative sense. Implementation studies presented a limited view of interagency relationships because they focused on how agencies respond to formal mandates, while overlooking a broader array of agency activities. Moreover, in describing why policy outcomes usually differed from the intentions and formal mandates of elected officials, implementation studies tended to dwell on bureaucratic pathologies rather than success stories. Interagency relationships became a causal variable explaining implementation failure. Numerous case studies suggested that the inability of local, state, and federal agencies to work out their differences was simply one more reason public policies were not implemented as designed.[3]

Yet interagency conflicts are not always the problem. Moreover, agency officials sometimes resolve social, economic, and environmental problems in ways not specifically mandated. Despite popular and academic stereotypes, many unelected officials demonstrate great initiative. They can be creative, proactive, and cooperative in solving problems. In part, political scientists accentuate the negative characteristics of public agencies by the questions they address. If a researcher asks why a particular program was not implemented as intended, the answer will likely emerge as a list of pathologies, ranging from poor policy design to interest-group manipulation or an entrenched bureaucracy. On the other hand, if a researcher asks why public agencies pursued a particular course of action, the answer is likely to be less critical because the research does not begin with expectations about what the agencies should have done. This distinction is important, because even if agency officials have not done precisely what political principals mandated, it is nevertheless possible that they developed novel and acceptable approaches to meet the intent of the mandates.

I follow the latter approach in this book, and simply ask, "Under what conditions do individuals in different public agencies cooperate with one another?"[4] My working hypothesis is that individuals in different agencies can be expected to cooperate with one another under certain conditions. Yet these conditions are not obvious because they are not simple extensions of legislative mandates or standard operating procedures. Instead, cooperation depends largely on how public officials understand their world. While formal institutions such as laws and procedures shape cooperative behavior on the margin, informal institutions such as norms, worldviews, and scientific knowledge provide the cognitive foundation on which the gains from cooperation are constructed. For public officials, cooperation is a practical and rational solution to certain kinds of problems, but to understand why it is rational we must first understand how they interpret their world.

Before presenting this interpretive framework in chapter 2, it is important to note that a necessary condition for cooperation to occur among any set of actors is interdependence (Ouchi 1980: 130). Public officials and public agencies—like private individuals, firms, and nonprofit organizations—are interdependent to the extent that each has resources or performs services desired by the others (Thompson 1967; Emerson 1962). Interdependence can lead to other forms of interaction, such as competition, but cooperation necessarily requires interdependence. Thus, any study of interagency cooperation must begin with an issue in which public officials in two or more agencies are in some way interdependent, because interdependence gives them a reason to work together toward some common objective. Accordingly, Weiss (1987) examined public education administration, in which school districts could lower costs and improve effectiveness by sharing resources; Chisholm (1989) analyzed public transit systems, in which routes and riders overlapped; Gruber (1994) looked at regional growth management efforts, in which local governments experienced negative externalities from development occurring in neighboring jurisdictions; and Bardach (1996) studied the delivery of human services, in which service providers shared clients in common.

I examine the management of biological resources because species and ecosystems are affected by the activities of many public agencies. Migratory birds, for example, fly from one jurisdiction to another. Thus, at various times of the year, they may be within wildlife refuges managed by the U.S. Fish and Wildlife Service (FWS), national forests managed by the U.S. Forest Service (FS), or private farms responding to federal regulations and incentives. If the FWS lists a species as endangered (at imminent risk of extinction) or threatened (likely to become endangered in the foreseeable future) under the ESA, these agencies and private landowners are also regulated by the FWS, albeit under a separate division from the agency's National Wildlife Refuge System. This list of jurisdictions would be much longer if it included local, state, and international agencies.

While birds are obvious examples of species that move with ease from one jurisdiction to another, they are by no means the only examples. Plant species favor particular types of soil and climate, neither of which are confined within the boundaries of a single jurisdiction, except by happenstance. Insect species search out their preferred plants. Deer and other fauna roam to find their favorite browse or prey. Salmon are born upstream, migrate to the ocean, and return to the same stream to spawn. During this journey, their fate is affected by a wide variety of human activities, which vary across agency jurisdictions. Upstream, logging on federal lands managed by the FS causes erosion, filling streams with silt and smothering the gravel beds in which salmon spawn. Midstream, dams operated by the Bureau of Reclamation (BoR) impede adult salmon moving upstream, suck immature salmon moving downstream through spinning turbine blades, and raise the temperature of impounded water to lethal temperatures. All of these unintended impacts, and more, occur inland. The salmon must still contend with the fishing industry and sport fishing—both of which are regulated by state and federal agencies—on reaching the sea.

The case studies in this book focus on California because it contains a large number of endangered species and interlaced jurisdictions, which jointly create interdependence. Endangered species give agency officials an incentive to manage habitat in a way that benefits these species, while interlaced jurisdictions increase the likelihood that this habitat is interwoven among local, state, and federal jurisdictions. In the 1980s and 1990s, this interdependence increased dramatically for two reasons. First, successful lawsuits filed against some agencies compelled them to be much more concerned with protecting endangered species than they had been in the past. Consensual scientific knowledge also emerged regarding the causal links between the size of a species' population and habitat management practices. As agency officials confronted the legal consequences of having endangered species occur within their jurisdictions, they also learned about the collective-action problem of protecting species whose habitats sprawl across agency jurisdictions. In short, the more endangered species and the more jurisdictions, the more likely we are to find agencies cooperating to manage this habitat.

With regard to jurisdictions, California was carved up into 58 counties, 468 cities, and about 5000 special districts in the early 1990s, when the events described in the case studies unfolded.[5] State and federal agencies also managed thousands of distinct parcels of land overlapping these local jurisdictions. The California Department of Parks and Recreation (CDPR) managed 12 underwater parks, 35 natural preserves, 17 reserves, and 7 wilderness areas, while the California Department of Fish and Game (CDFG) managed 80 wildlife areas, 53 marine refuges, and 67 ecological reserves. Federal regional offices managed a similar number of diverse administrative units in California, albeit covering much more acreage. The FS managed 22 national forests; the FWS managed 34 wildlife refuges; and the National Park Service (NPS) managed 6 national parks, 7 national monuments, and 3 national recreation areas.[6] The Bureau of Land Management (BLM) owned so many scattered and fragmented parcels of land in California that the BLM did not even attempt to count them. In addition to land holdings, six federal agencies ran 32 programs, while eight state agencies ran 13 programs, to protect natural areas in California (Cochrane 1986).

Sprawling across this bureaucratic landscape lay a great storehouse of biodiversity. California's Mediterranean climate, varied topography, and diverse soil types provided habitat for more endemic species than found in any area of equivalent size in North America, including 5000 native plants, at least one-third of which were found nowhere else on earth (Schoenherr 1992: x). In the United States, California's biodiversity was rivaled only by Hawaii. Yet, as in Hawaii and elsewhere in the world, California's biodiversity was increasingly threatened by human developments on the landscape.

Fragmented Habitat and the Decline of Biodiversity

Preserving species is both a complex technical task and a highly politicized endeavor. Species tend to prefer specific types of soil, climate, and food, and may reproduce only under specific conditions. Populations of specialized species accordingly decline when the quantity and quality of their habitat is diminished. Therefore, the fundamental requirement for maintaining a viable, if not large, population of any species in the wild is to preserve its habitat, including functioning ecological processes.[7] Without sufficient habitat, a species' population gradually declines, regardless of whether humans are directly responsible for destroying individual members of the species. Setting aside habitat for species is politically controversial, however, because it entails limitations on human uses of land and aquatic resources. In practice, setting aside habitat means smaller housing developments, fewer roads, changes in logging, ranching, and farming practices, and numerous other trade-offs.

Technical uncertainty confounds the political problem. It is not clear, for example, how much habitat a given species requires to avoid extinction in the long run. In part, habitat requirements depend on the desired size of a species' population, which itself is open to debate. How large must a population be for a species to survive over the next 100 years, or 1000 years? The smaller the population, the more likely the species will be weakened by inbreeding, wiped out by an environmental disaster, or quickly overrun by competing or predatory species. Theoretically, there is a "minimum viable population" (MVP) below which a species is likely to become extinct. In common parlance, this idea is captured by the phrase "balancing on the brink of extinction." Phrased as a question, how big does a population of species *y* have to be so that it has a 95 percent probability of surviving for *x* number of years?

Empirically, this question is difficult to answer because extensive genetic studies may be required simply to understand the effects of inbreeding in small populations of a given species (Lacy 1992). Regardless of environmental catastrophes and other disturbances, such as predation by exotic species, each population must maintain genetic diversity to remain viable. While no single MVP applies to all species (Gilpin and Soulé 1986: 20), MVPs are probably measured in the thousands for most species (Soulé 1987). Natural scientists speak loosely of a 50–500 rule for genetic health: within a given population, at least 50 individuals capable of breeding are needed to avoid the deleterious effects of inbreeding in the short run, while 500 breeders are needed to avoid genetic drift (i.e., random changes in the genetic pool) in the long run (Wilson 1992: 236; Grumbine 1992: 34–35). Because only a fraction of the individuals in a population actually breed, the 50–500 rule means that thousands of individuals are needed to ensure the population's viability in the long run.

Yet, even if the MVP were known for a species, we would still need to know how many acres of habitat are necessary to sustain that population. Moreover, if the habitat is fragmented by clearcuts, housing developments, or farms, does the species need even more habitat to survive, or less? Fragmentation occurs when a habitat patch is transformed into several smaller patches, with each patch isolated from the others by a developed landscape. Some species benefit from habitat fragmentation because they thrive in the transformed habitat or along habitat edges. Deer, for example, find food in meadows and shelter in forests, so deer proliferated when harvesting fragmented forests. Other species are weakened by habitat fragmentation because they thrive only in the interior of large habitat patches.

Another type of problem arises with habitat fragmentation when individuals within a population are unable to move from one habitat patch to another. In these cases, the isolated populations within each habitat patch may individually succumb to inbreeding, predation, competition, or environmental disturbances. Unless humans physically transport these individuals from one patch to another, or design a system of habitat corridors to connect the patches, the populations in each patch slowly die out, ultimately culminating in the global extinction of the species. Extinction becomes more likely for these species as habitat patches become smaller and separated by greater distances.

These questions become increasingly important as humans develop species' habitats for other purposes. The leading cause of the decline and extinction of native species in California and the world is habitat transformation—including habitat fragmentation, degradation, and outright loss—caused by human population growth and resource consumption.[8] Some scientists believe the earth is currently experiencing a wave of extinctions unrivaled since the event(s) that wiped out the dinosaurs at the end of the Mesozoic era, 65 million years ago. At then-current rates of habitat destruction, E. O. Wilson (1992: 346) estimated that one-fifth of the earth's plant and animal species could be extinct by the year 2020. Unlike previous extinction spasms, this is the first thought to be caused by a single species: *Homo sapiens*.

Ecologists coined the term *biodiversity* (from *biological diversity*) to provide a political buzzword that would promote the importance of the diversity of life (Takacs 1996). The term has many definitions, because the idea itself is complex, but a simple definition will suffice for this book: "*Biological diversity means the full range of variety and variability within and among living organisms, and the ecological complexes in which they occur; it encompasses ecosystem or community diversity, species diversity, and genetic diversity.*"[9] While laypersons tend to think of biodiversity in terms of species diversity, ecologists stress the importance of genetic diversity and ecosystem diversity as well. Preserving biodiversity means more than simply saving as many species as possible from extinction; it also means preserving healthy gene pools by maintaining sufficiently large populations of each species, and preserving the ecosystems within which species thrive and evolve. Given that species have been naturally selected to fill a niche in which they are interdependent with other species and the abiotic elements in their environment, ecosystem stability depends on the preservation of species in their native habitats.

Regardless of whether society believes there is, or should be, a moral imperative to protect biodiversity, many species and ecosystems have utilitarian value for humans. Viewed simply in anthropocentric terms, biodiversity clearly provides innumerable health benefits, amenity values, and economic returns.[10] The costs of extinction may therefore be enormous to humans because most species have not even been named, let alone studied for their health benefits. The case of penicillin, a powerful antibiotic obtained from several species of fungi growing as green mold, is well known. Yet relatively few people know about more recent medical discoveries, such as taxol, a chemical found in the bark of the Pacific yew. Taxol appears to be a highly effective drug for treating ovarian cancer. In light of these examples, we might hope that a rationally self-interested society would perceive the value of preserving biodiversity, without recourse to ecocentric ethics. The problem is that the utilitarian value of biodiversity is largely unknown, dispersed, and long-term, while the benefits to development are largely known, concentrated, and short-term. In the absence of regulation, development pressures overwhelm efforts to preserve undeveloped habitats.

The Decline of California's Biodiversity

Little more than a century ago, the California landscape appeared limitless and threatening to newcomers. Yet the state's biodiversity was rapidly reduced to a fragment of its former self, beginning with the first large-scale wave of modern immigrants arriving after the discovery of gold in the American River near Sutter's Mill in 1848. Prior to the Gold Rush, thousands of grizzly bear roamed wild in the state. Today, there are none. The last California grizzly was shot in 1922 on the southwestern slope of the Sierra Nevada under a state-supported predator-control program. Ironically, the grizzly is California's official state animal, and its image appears prominently on the state flag, making California the only state to have as its symbol an animal that humans forced to extinction (Schoenherr 1992: 25, 386). Yet, by the 1990s, the grizzly was only one of 70 plant and animal species known to have been lost from California, of which 21 animal species are now globally extinct (Jensen, Torn, and Harte 1993: 49).

By the 1990s, California contained more species listed as threatened or endangered under the ESA than any state other than Hawaii, and the numbers were growing. In 1994, California contained 138 listed species of plants and animals, or roughly 10 percent of all listed species. A year later, the number reached 156—a 13 percent increase.[11] Yet listed species represented only the tip of the iceberg. Approximately 600 plants and 300 vertebrate species in California were documented to be declining or seriously at risk of extinction (Jensen, Torn and Harte 1993, 53). In this regard, California contained more species proposed for listing than any other state—including Hawaii. In 1995, California contained 119 species proposed for listing by the FWS; its closest competitor was Arizona, with just 9 species proposed for listing. These figures dramatically depict the quandary in which public officials in California found themselves. Indeed, the predicament was unrelenting during the 1990s. By 2001, California contained 290 listed species.

For agency officials, it was not simply the number of listed species that mattered. They were also concerned with species likely to be listed in the near future, including those not yet proposed for listing. While listed species provided immediate avenues for lawsuits under the ESA, species queued up for listing provided avenues for future lawsuits. This is why the listing process has been so politically charged—despite the legal requirement that the FWS make listing decisions "solely on the basis of the best scientific and commercial data available."[12] One strategy agency officials could pursue to manage the uncertainty of potential lawsuits was to slow the listing process through administrative and political machinations. Another strategy was to manage species' habitats to maintain viable populations. While more difficult, the latter strategy had several potential benefits. For example, it could ward off lawsuits over listed species, obviate the need for the FWS to list other species, and perhaps even limit lawsuits under other federal and state environmental laws.

Yet proactive habitat planning and management to maintain viable populations is difficult because listed and unlisted species seldom heed the jurisdictional boundaries that humans impose on the landscape. While fences around military bases may limit the movement of some faunal species, and elk herds may stay within the unfenced boundaries of national parks if they become savvy about the presence of hunters outside the parks, most animal species ignore the boundaries we have superimposed on their preferred habitat. Hence, habitat management is often a collective-action problem, one made all the more difficult because public agencies pursue widely varying missions, each of which generates specific tasks that affect species in different ways. Local planning agencies approve housing developments and new roads, state and federal water agencies divert stream flows, and federal multiple-use agencies sell timber and grazing allotments. Because many species suffer from the cumulative effects of activities promoted by local, state, and federal agencies, habitat preservation is a collective-action problem. If a species would benefit from compact blocks of habitat and a system of corridors connecting them, habitat management is more effective if agencies coordinate their planning and management processes to preserve blocks of habitat straddling agency boundaries and allow habitat corridors to meet at the boundaries.

As a technical task, species preservation promotes interdependence because joint action is often necessary for maintaining viable populations. Agency managers must nevertheless commit agency resources to this task. Without management support, staff will likely make little progress when attempting to develop and implement coordinated plans. Gathering the support of agency managers is not a simple matter because agency missions have contributed to the decline of biodiversity. Not only have these agencies pursued traditional missions with little attention to the long-term viability of the vast majority of species living within or straddling their jurisdictions, but the agencies were not known for cooperation. Prior to lawsuits, species preservation was a low priority in most agencies, and cooperative planning and management was limited, at best.

Resource Management Agencies Rarely Cooperated in the Past

Prior to the 1990s, resource management agencies, like most public agencies, rarely cooperated with one another or with other agencies. Sometimes they competed, as when the BoR and the U.S. Army Corps of Engineers (ACE) competed feverishly for congressional appropriations to build dams in the West, exaggerating potential benefits and underestimating costs to construct dams of limited economic value (Reisner 1993; Maass 1951). Usually, public agencies simply avoided competition and conflict. Research suggests that the FS and NPS, for example, preferred to enhance their autonomy from one another rather than compete over new tasks (Kunioka and Rothenberg 1993). The FWS was also reluctant to pressure state and federal agencies to comply with federal wildlife laws (Tobin 1990).[13] In the NPS, park managers were unable or unwilling to negotiate with individuals in neighboring jurisdictions regarding external impacts on park resources, even when they had the authority to do so (Sax and Keiter 1987; Freemuth 1991). Some park managers even limited the access of university researchers whose findings contradicted park policies (Chase 1987: 60, 253; Wagner et al. 1995: 101–102).

Interagency cooperation is nearly absent in the established literature on these agencies. For example, Kaufman (1960) noted the existence of cooperative programs with state agencies and private landowners in several passing references, but he did not pursue this angle. Instead, he focused on the administrative processes through which professional foresters are socialized within the agency. Culhane (1981) later analyzed the FS and BLM, finding historical conflicts and competition between the agencies (1981: 188, 191), but not cooperation. In sum, the traditional literature on federal resource management agencies suggests that they rarely attempted to develop—let alone implement—cooperative management plans to achieve common goals, and there is no evidence to suggest that their state counterparts behaved differently, at least not in California (Hammond 1979).

Cooperation is an important problem in managing biological resources because local, state, and federal boundaries were not drawn to encompass ecosystems. Major John Wesley Powell, director of the U.S. Geological Survey in the late nineteenth century, clearly failed in his attempt to convince Congress and the newly forming Western states to draw public boundaries and to write settlement laws that conformed to watershed boundaries in the arid West; instead, Congress relied on the grid system, which had worked well for farming in the wetter Midwest (Stegner 1954). As Powell feared, the result was ecological ruin and economic chaos throughout much of the West. Nor were public lands plotted with the habitat requirements of native species in mind. National forests were created at the turn of the century to counter unsustainable logging practices on private lands, but these boundaries were drawn before detailed studies of plant communities and species distribution were available to guide decisions about boundary placement. Therefore, national forest boundaries do not reflect the careful thought of planners thinking about ecosystems. BLM holdings basically represent the lands not withdrawn from the public domain for other federal purposes or turned over to settlers and robber barons by the infamously corrupt and incompetent General Land Office. Hence, BLM lands are highly fragmented and scattered throughout the West. Although BLM lands long held little commercial value, ecologists increasingly prize them as a repository of biodiversity. Even national park boundaries do not conform to ecosystem boundaries. Most national parks encompass monumental natural wonders, such as canyons and mountains, with boundaries carefully tailored to exclude areas thought profitable for resource extraction (Runte 1987). Everglades National Park was the first—and remains one of the few—national parks set aside for ecological preservation rather than scenery, but even it "was hamstrung with artificial rather than natural boundaries" (Runte 1987: 229).

Some state and federal wildlife refuges and ecological reserves were designed with ecosystems in mind, but many of them are too small or widely scattered to maintain the long-term viability of species therein, and some do not even provide habitat for endangered species. Many refuges and reserves are so small that ecologists, with a touch of black humor, refer to them as "postage-stamp" parcels. Given the size and distribution of such parcels, buffer zones around them and habitat corridors between them are probably necessary to provide adequate protection for species living within the protected habitat cores (Noss and Harris 1986). Even large parcels may be insufficient for species preservation. In 1994, the National Wildlife Refuge System included about 500 refuges, totaling 91 million acres. Yet only 24 percent of federally listed species occurred on these refuges, and only 66 refuges—just 13 percent—provided a significant portion of the habitat needs of listed species (U.S. General Accounting Office 1995). Put another way, 76 percent of all listed species did not occur anywhere in the Refuge System, and 87 percent of those that did found significant portions of their habitat outside the Refuge System. Thus, even if species are nominally protected within administrative parcels scattered across the landscape, agencies still need to coordinate their activities if they want to ensure habitat integrity.

In the 1980s, scientists and resource managers increasingly recognized the disjunction between ecosystems and land ownership patterns. Research indicated that national parks were not big enough to maintain viable populations of large carnivores within their boundaries (Newmark 1985; Salwasser, Schonewald-Cox, and Baker 1987). Even designated wilderness areas, where human impacts are greatly diminished, did little to preserve biodiversity because most wilderness areas are situated on the top of mountain ranges, where biodiversity is low. Biodiversity tends to be concentrated in valleys, along streams, and in forests and wetlands, where human impacts have been greatest. Most ecosystems types are not even represented within the wilderness system in sizes large enough to matter (Noss and Cooperrider 1994: 172–174). Thus, scientists and professional resource managers increasingly recommended that public agencies work cooperatively with one another, and with private landowners, to preserve biodiversity.[14] Yet these recommendations seemed to fall on deaf ears because there were few examples of coordinated habitat planning and management efforts during the 1980s. This situation changed remarkably in the 1990s. Interagency cooperation on resource management issues suddenly sprouted everywhere in the United States, under the rubrics of ecosystem management (Yaffee et al. 1996; Grumbine 1994), watershed management (Kenney 1997; Natural Resources Law Center 1996), collaboration (Wondolleck and Yaffee 2000; Brick, Snow, and Van de Wetering 2001), stewardship (Knight and Landres 1998), and bioregionalism (Jensen 1994).

The Traditional Missions of Resource Management Agencies

The absence of cooperation is not surprising given that most public agencies have long been disinterested in protecting biodiversity. Federal land management agencies focused on their primary missions and constituencies, such as the logging, mining, ranching, and tourism industries. Although vague references to biodiversity in some agency statutes seemingly required them to manage natural resources in ways that would maintain biodiversity, there is little evidence that these agencies earnestly sought to meet these mandates until lawsuits compelled them to do so. The National Forest Management Act (NFMA) of 1976, for example, requires FS managers in each national forest to "provide for diversity of plant and animal communities based on the suitability and capability of the specific land area to meet overall multiple-use objectives."[15] Similarly, the Federal Land Policy and Management Act (FLPMA) of 1976 declares that BLM's "public lands be managed in a manner that will protect the quality of scientific, scenic, historical, ecological, environmental, air and atmospheric, water resource, and archeological values; that, where appropriate, will preserve and protect certain public lands in their natural condition; that will provide food and habitat for fish and wildlife and domestic animals; and that will provide for outdoor recreation and human occupancy and use."[16] Both agencies have been sued repeatedly for not meeting these provisions. Indeed, it was primarily through NFMA—not the ESA—that litigants won the infamous injunction against logging on national forests within the range of the northern spotted owl.

Yet, to be fair, agency managers find it difficult to meet conflicting congressional mandates. Multiple-use acts, like NFMA and FLPMA, do not set explicit standards specifying the proper balance between wildlife needs and competing resource demands by humans. Therefore, concludes a U.S. General Accounting Office (1991: 2–3) report, "No definitive basis exists to judge whether these agencies are appropriately considering wildlife in their land use planning and resource management decisions." Nevertheless, "GAO found that wildlife management receives only a small percentage of available staffing and funding. Further, while wildlife needs were uniformly considered during land use planning at the locations we visited, in some cases, the agencies' choice of consumptive interests in land use decisions adversely affected wildlife. Moreover, when actions to benefit wildlife are included in land use plans, they are frequently not performed." While GAO found no single reason why the BLM and FS gave little attention to wildlife protection and enhancement, it cited the agencies' "traditional deference to consumptive uses of the land" as a key factor. In a similar vein, the FWS has long permitted competing uses within many wildlife refuges to generate political support for its Refuge System; this includes grazing and recreation, which are harmful to the very wildlife these refuges were established to protect (Curtin 1993).

Even the NPS, long considered an ally of the environmental movement (or at least that part of the movement oriented toward tourism), has increasingly come under fire for poorly managing natural resources. Some critics have been relatively kind. One GAO study, citing several internal threats to parks, such as increased visitation, noted that the NPS does not maintain a national inventory of such threats (U.S. General Accounting Office 1996). Other critics have been scathing. Alston Chase (1987) extensively documented the manipulation of natural systems in Yellowstone National Park by NPS employees bent on satisfying tourists, even when that meant sacrificing the agency's mandate to preserve natural resources. Karl Hess (1993) found similar behavior at Rocky Mountain National Park, where park managers ignored the advice of staff ecologists regarding the ecologically insidious effects of park policies. One park manager even eliminated two of the park's three research ecology positions in 1991, and "changed the third beyond recognition," thereby "ridding himself and the park of an embarrassing reminder of three-quarters of a century of failed policy and bad management" (Hess 1993: 78). Although many environmentalists find it difficult to believe that their sacred cow, the NPS, might be a villain, and readily dismiss Chase because his politics lean to the right, several notable scientists accept his conclusions and provide additional evidence that this type of behavior is widespread within the agency (Wagner et al. 1995). Edward Abbey's (1968) once-radical idea that the NPS is wedded to "industrial tourism" as a management philosophy has been increasingly validated.

Like the BLM and FS, which must interpret inherently contradictory multiple-use mandates, the NPS also faces a contradiction in its two-pronged mandate. As stated in the agency's Organic Act, the NPS mission is "to conserve the scenery and the natural and historic objects and the wildlife therein and to provide for the enjoyment of the same in such manner and by such means as will leave them unimpaired for the enjoyment of future generations."[17] Providing increased access for humans, however, necessarily entails habitat degradation for some species. Building roads within national parks to provide access for visitors leads to habitat fragmentation, which constrains the ability of some national parks to protect sensitive species (Schonewald-Cox and Buechner 1992). Far from being "islands of preservation" as depicted by some authors, threatened only by pollution and development pressures emanating from outside the parks, species have long been threatened by the management practices of the very custodians charged with protecting them.[18]

Not only have the NPS, FS, and BLM not met the species-management requirements embedded within their multiple-use mandates, but these and other agencies often acted as if the ESA did not apply to them. The best evidence for this lies in court records. Environmental activists have filed numerous lawsuits against local, state, and federal agencies, charging failure to comply with the ESA. Because lawsuits filed under the ESA have played such a crucial role in motivating agency behavior, it is necessary to explain how the law works in practice.

The Blunt Hammer of the Endangered Species Act

In practice, the ESA focuses primarily on species and their habitat requirements, not gene pools or ecosystems. Although Section 2(b) states that one of the ESA's purposes is "to provide a means whereby the ecosystems upon which endangered species and threatened species depend may be conserved," the mechanics of the law do not operate at the level of ecosystems, and thus neglect an important aspect of biodiversity. Moreover, as Noss and Cooperrider (1994: 27) argue, "the agencies have never taken this ecosystem protection mandate seriously, and Congress has never told them how they might do so." Nevertheless, lawsuits filed under the ESA have indirectly led some agencies to consider all scales of biodiversity, including ecosystems. To understand how this happens, we need to understand the ESA's prohibitions.

These prohibitions are of two types: the Section 9 prohibition on take, which applies to all persons subject to U.S. jurisdictions, and the Section 7 jeopardy standard, which applies only to federal agencies. Neither prohibition requires cooperation; they only limit what is permissible. Section 9 prohibits any person or organization from taking fish or wildlife species listed as endangered by the FWS, with *take* defined broadly in Section 3 to include "harass, harm, pursue, hunt, shoot, wound, kill, trap, capture, or collect, or to attempt to engage in any such conduct." The FWS subsequently expanded this definition by issuing a rule that defines *harm* to include "significant habitat modification or degradation where it actually kills or injures wildlife by significantly impairing essential behavioral patterns, including breeding, feeding or sheltering."[19] Therefore, environmental activists can successfully sue a private landowner for altering habitat (e.g., through logging, farming, or housing developments), and they can sue a local or state agency for either engaging in such activities or permitting them to occur. If a federal court rules in favor of the plaintiff, it can prohibit these activities, or fine and even jail those committing the offense. Property owners have felt sufficiently threatened by the Section 9 prohibition on take that they have attempted (so far unsuccessfully) to reverse the charges, by claiming that the federal government is "taking" their private property without just compensation, as guaranteed by the Fifth Amendment.

Yet the prohibition on take is not as strict as it may seem. For example, it applies only to fish and wildlife species, not to plants. While plant species can be listed, they are only covered indirectly by the prohibition on take, in that plant species (such as trees) provide habitat for wildlife (such as owls). The FWS can also issue a permit allowing "incidental take" of listed fish and wildlife species, provided that permit applicants submit a habitat conservation plan (HCP) for the species in question. (See chapter 6 for more background on HCPs.)

The ESA holds federal agencies to a different standard. Specifically, Section 7 requires federal agencies to "consult" with the FWS before undertaking activities to "insure that any action authorized, funded, or carried out by such agency . . . is not likely to jeopardize the continued existence of any endangered species or threatened species or result in the destruction or adverse modification of habitat of such species."[20] The bilateral Section 7 consultation process formally culminates with a biological opinion, in which FWS staff present factual details about a listed species and its habitat, and discuss possible effects of the proposed agency action. If the FWS issues a "no jeopardy" opinion, the agency is in compliance with Section 7 and can lawfully proceed with the action. If the FWS issues a "jeopardy" opinion, the agency cannot proceed with the action, unless the FWS recommends alternatives to mitigate the effects of the action.

Many federal agencies once routinely ignored Section 7 consultation requirements, for at least two reasons. First, it was the responsibility of each agency to decide whether an adverse impact was likely from their activities, and therefore whether to consult with the FWS. With thousands of federal actions occurring every year, and with the understaffed FWS consumed by ongoing consultations, unreported actions typically went unnoticed by FWS staff, unless pursued by environmentalists. Second, even if agencies went through the consultation process, they could still ignore recommended alternatives in the biological opinion, and possibly even ignore a jeopardy decision, because the FWS did not have enough staff to monitor compliance. The FWS did not even have the authority to require federal agencies to report actions taken following a jeopardy opinion (Tobin 1990: 174).

Even the Environmental Protection Agency (EPA), which one might assume worked closely with the FWS because both agencies have environmental goals, has a checkered history with regard to Section 7 consultation (Serfis 1991). One EPA ecologist in the San Francisco regional office, for example, reported that the "EPA basically was ignoring . . . Section 7 requirements for years and years and years."[21] In part, the agency's recalcitrance was due to EPA managers, who led staff to believe that Section 7 compliance was a low priority (Tobin 1990: 195–199). It was not until the EPA was sued, and the Department of Justice refused to defend it, that EPA officials began to take Section 7 consultation seriously. If the EPA did not comply with Section 7, it is difficult to believe that other federal agencies did so in the absence of lawsuits.

In sum, the FWS did not make the ESA a regulatory burden. Citizen lawsuits did. For federal agencies, the first hint of the ESA's power in the courts came in the 1970s, when a lawsuit filed under Section 7 almost killed the Tellico Dam project, which the Tennessee Valley Authority (TVA) had nearly completed. By itself, the FWS lacked the administrative capability and political clout to compel federal agencies to fulfill their Section 7 requirements. Moreover, consultations themselves rarely impeded the activities of federal agencies (U.S. General Accounting Office 1987). This did not mean that the FWS was instead targeting private landowners or local and state agencies under Section 9. Rather, the FWS was woefully understaffed and buried in paperwork. FWS staff also faced political obstacles to enforcement during the 1980s because the Reagan and Bush administrations opposed new regulations, and the Secretary of Interior was the final arbiter of listing decisions.[22] In short, the ESA's language and FWS regulations were incredibly far-reaching, but the agency was barely enforcing them. Therefore, environmental activists stepped into the void by bringing cases to court.

It is also important to note that the ESA does not mandate coordinated action. Other than consulting with the FWS on a bilateral basis under Section 7, federal agencies are not obligated to coordinate with one another under the ESA. Nor, for that matter, are local and state agencies. More generally, Congress has given federal agencies little direction as to how they should work out their differences regarding transjurisdictional issues such as species management (Sax and Keiter 1987: 209). Therefore, to the extent that coordination occurs, it is because agency officials depend on one another for other reasons, and choose to cooperate for these reasons.

Staff ecologists, for example, quickly realized they were unable to carry out their tasks independently of one another because ecosystems are not coterminous with agency jurisdictions. For them, interagency planning and management made so much sense that noncooperation bordered on nonsense. For agency managers, however, an additional impetus was needed, because they were steeped in traditional missions. They had to confront external demands for changes in agency priorities that elevated the cause of biodiversity. This occurred in the 1980s and 1990s, as environmental activists successfully sued local, state, and federal agencies under the ESA and other environmental laws. Agency managers then began to see benefits to cooperative planning and management. Moreover, they undertook and supported staff efforts to coordinate agency activities to protect the long-term viability of species—not just listed species, but entire natural communities of species sprawled across the fragmented bureaucratic landscape.

The implications of this transition are profound because biodiversity has never been a priority for these agencies. Even the wildlife agencies charged with enforcing the state and federal ESAs have historically catered to sports enthusiasts. For the BLM, long derided by environmentalists as the "Bureau of Livestock and Mining," this change reflected broader transformations within the agency. Though academics have traditionally focused on the BLM's weakness and resulting capture by grazing interests (Clarke and McCool 1996; Foss 1960; Calef 1960), the BLM has increasingly shown itself willing and able to preserve natural resources (Fairfax 1984). Even military bases incorporated ecological protection into their missions.[23] What is surprising about all of this activity is that the agencies were not confronted with new legislative mandates to protect species or to coordinate their habitat planning and management practices. Of course, top-down mandates have never been a particularly good means for encouraging interagency coordination.

Why Top-Down Mandates Seldom Generate Coordination among Public Agencies

Academics, popular pundits, and public officials have long wrung their hands over the perceived lack of coordination within American government, and the inefficiency, ineffectiveness, and incoherence that supposedly result from it. Seidman and Gilmour (1986: 219) captured such apolitical thinking in a humorous and oft-quoted passage:

> In ancient times alchemists believed implicitly in the existence of a philosopher's stone, which would provide the key to the universe and, in effect, solve all of the problems of humankind. The quest for coordination is in many respects the twentieth-century equivalent of the medieval search for the philosopher's stone. If only we can find the right formula for coordination, we can reconcile the irreconcilable, harmonize competing and wholly divergent interests, overcome irrationalities in our government structures, and make hard policy choices to which no one will dissent.

Traditional public administration theorists argued that coordination should be forced on public agencies through reorganizations and interagency committees. The president, it was argued, is in the best position to lead because of being a single actor; thus, reorganizers rearranged organizational boxes to centralize authority structures, such that no one would have more than one boss and the president would be the ultimate boss of the bureaucrats (Emmerich 1971; Gulick 1938). These theorists believed that coordination imposed from the top would promote effectiveness and efficiency, not simply the concentration of political power. Today, top-down attempts to consolidate the bureaucracy tend to be seen as symbolic actions or struggles for political control rather than as instrumentally effective means to coordinate agency activities (March and Olsen 1983; Thomas 1993). Evidence suggests that merging agencies does not lead to increased coordination unless the fundamental tasks, financial resources, and career rewards within the agencies are changed (Wilson 1989: 266).

Interagency committees are another traditional means used to coordinate agency activities from the top down. Yet politically imposed interagency committees typically provide little more than the appearance of coordination. As Seidman and Gilmour (1986: 226) noted, "Interagency committees are the crabgrass in the garden of government institutions. Nobody wants them, but everyone has them. Committees seem to thrive on scorn and ridicule and multiply so rapidly that attempts to weed them out appear futile." Requiring agency officials to sit on such committees is like leading the proverbial horse to water: you can require them to sit at the table, but that does not mean they will cooperate. Maass (1951) depicted the futility of relying on such politically imposed interagency committees in a study of river-basin planning. President Franklin

Roosevelt intended the Columbia Basin Inter-Agency Committee to be a vehicle for coordinating agency planning processes, but instead it became a venue for the BoR and ACE to divvy up construction projects. The committee did not participate in the planning process of either agency. This suggests that externally imposed interagency committees may be less likely to promote coordination than to provide agency officials with a means to enhance their autonomy.

In part, top-down reorganizations and externally imposed interagency committees have not been successful means for inducing interagency coordination because agencies have demanding clients, both within Congress and outside government. Some have argued that a powerful system of incentives binds congressional committees, federal agencies, and clienteles together into subgovernment alliances (Knott and Miller 1987). Within these subsystems, agencies increase their power by expanding their clientele base rather than by working with other agencies (Rourke 1969). Political appointees are unable to break these ties because their tenure in office is relatively short. Lacking the means and the incentives to forge interagency relationships, appointees often "go native," working with civil servants to get things done (Heclo 1977).

Yet agency-centered subgovernments are certainly not ubiquitous. Since at least the 1960s, public agencies have been linked by complex mandates and regulatory relationships, as described in the literatures on policy implementation (Pressman and Wildavsky 1984), intergovernmental regulation (Wilson and Rachal 1977; Durant 1985; Hamilton 1990), and intergovernmental management (Agranoff 1986). In the implementation literature, interdependence is portrayed as a direct result of political mandates, which stipulate—sometimes in great detail—the relationships among agencies. Pressman and Wildavsky (1984) likened such politically mandated interagency relationships to a Rube Goldberg machine, in which the design is so excessively complex it is bound to fail. Public agencies are also not unitary actors. Their very organizational structure itself may reflect political compromise, with competing interests lodged in different units within the same agency. As Terry Moe (1989: 267) sums up the situation, "American public bureaucracy is not designed to be effective."

As new policy problems and solutions arose, they were superimposed on preexisting agency jurisdictions. These jurisdictions were initially designed for other purposes, with many agencies serving as monopoly providers. Interagency relationships emerged as the original organizational designs became less appropriate for addressing the problems at hand. As Yates (1982: 79) puts it,

> A simpler world may once have existed in which the jurisdictions of cabinet departments were relatively distinct, but today any major policy problem affects various bureaucracies. Problems like inflation, energy, international economic policy, urban development, and environmental policy all call on the imagined prerogatives and the programs of many different agencies. It follows that the level of interbureaucratic conflict over jurisdictions is likely to be increasing.

Yet conflict need not be the only type of interagency relationship on the rise. When complex problems and solutions are imposed on a greater number of agencies, conflict or competition may indeed result, but under what conditions might cooperation arise instead?

In spite of the disarray depicted by academics and popular pundits, some agency officials go to great lengths to make the system work. The empirical evidence presented in this book suggests that, even if multiple political principals have not collectively designed the bureaucracy to be effective, agency officials nevertheless try to make things work. Rather than consistently butting heads, competing for resources, or asserting their autonomy, individuals in different public agencies sometimes negotiate among themselves the definitions of the problems they face and the range of alternatives they will consider, and then work the seams of the formal organizational structures of government to manage transjurisdictional issues. They may complain about the political milieu within which they operate, but few throw up their hands and simply give up trying. After all, many individuals became bureaucrats in the first place because they wanted to make a difference, not because they could not get jobs in the private sector. Sometimes organizations pull things off in practice, if not in theory (La Porte and Consolini 1991; La Porte and Thomas 1995).

Making Things Work through Cooperation

The common problem with top-down methods for inducing coordination is insufficient attention to the incentives for cooperation. Interagency committees become the crabgrass of government because the individuals assigned to sit on them do not want them. Merging agencies by shuffling boxes on organizational charts is similarly problematic because it does not address the individual incentives—rooted in tasks, resources, and career rewards—that promote either autonomous, competitive, or coordinated action (Wilson 1989). Rather than trying to force coordination from the top down, political principals should think instead about the incentives that lead individuals to work cooperatively.

I define *interagency cooperation* as an unmandated effort by public officials in at least two local, state, or federal agencies to coordinate their activities or share resources.[24] I emphasize "unmandated" because top-down tactics to coordinate agency activities have proven relatively ineffective. Therefore, it is useful to know how coordination can be achieved through cooperation, what incentives encourage or discourage cooperation, and how these incentives vary within and among agencies. In a cooperative relationship, individuals work together because they want to achieve a common goal, not because they are told to work together by their bosses or elected officials. Nor do they simply agree to disagree, or agree to stay out of each other's affairs. Instead, they "accommodate their actions in accord with the desires of their partners" (P. Haas 1990: 33).

In cooperative relationships, individuals work together to achieve things they cannot achieve individually (Barnard 1938: 23). They share resources and inform one another of their planned behaviors to coordinate their activities and pursue mutually consistent strategies (Simon 1976: 72, 139). Individuals must therefore share a common goal (Simon 1976: 72) or purpose (Barnard 1938: 82), or at least have congruous goals. Goal congruity requires only that goals be complementary, but not necessarily identical. According to Seidman and Gilmour (1986: 223), "If agencies are to work together harmoniously, they must share at least some community of interests about basic goals." Cooperation can thus occur if individuals have different, but not mutually exclusive, goals.

Cooperative behavior differs from competitive behavior in the willingness of individuals to share resources with one another, including truthful information about their intentions. In competitive relationships, individual goals are incompatible, perhaps even zero-sum. Individuals therefore hoard resources, deceive rivals, and design their strategies on the basis of whatever information is available regarding the intentions of others. Game theorists model such behavior in the stylized form of noncooperative games, using a variety of mathematical techniques to find equilibria for hypothetical situations. Cooperative outcomes may result from noncooperative games, as Axelrod (1984) has shown with tit-for-tat strategies, which are based on the threat of mutual punishment. Yet nonregulatory agencies are seldom able to punish one another for defection. If punishment is not germane, then interagency cooperation is largely shaped by the potential for joint gains. In other words, rather than simply dividing an existing pie, they see the possibility of making the pie bigger through joint gains (Lax and Sebenius 1986).

Cooperation is a complex variable because it includes individual, agency, and temporal components. Cooperation at the agency level can be measured by counting agreements, plans, or treaties, but measuring cooperation as a dichotomous variable oversimplifies the behavior to be explained. Cooperation is a dynamic process, often without discernible points at which it begins or ends. Viewed in this way, interagency agreements are simply epiphenomena that emerge from ongoing cooperative efforts. Therefore, an empirical study of interagency cooperation should focus on a series of activities, some of which express themselves at the individual level, such as regular attendance at interagency meetings.

At the agency level, specific cooperative agreements should be measured in terms of their scope, strength, and duration, not simply their existence (P. Haas 1990: 33–34). *Scope* refers to the range of issues covered, varying from a narrow focus on a herd of elk in the Owens Valley of eastern California, to a broader focus on the preservation of biodiversity in California, the United States, or the entire world. *Strength* refers to the binding nature of the agreements, ranging from verbal or tacit agreements to legally binding documents. *Duration* refers to the endurance of an agreement. How long did it remain intact? How long did the signatories adhere to its principles? The duration of agreements differs from the temporal component of cooperation discussed above because formal agreements are parts of much larger cooperative processes, which give birth to these agreements, and which usually continue after the agreements are suspended, superseded, or dissolved.

For individuals and agencies, cooperation varies along a continuum measuring the voluntary accommodations that each makes.

Demonstrating the existence and extent of cooperation also requires a comparative baseline against which the behavior is measured. The baseline throughout this book is the historical pattern of agency behavior identified in this chapter and the case studies.

As the following chapters demonstrate, agency officials face a wide variety of incentives, some of which lead them to cooperate in the absence of specific mandates to do so. Chapter 8 summarizes these incentives and reviews how they vary across agencies in the case studies. Chapter 2 sets up the case studies by framing how agency officials understand the potential benefits and costs of cooperation. Interagency coordination, the "philosopher's stone" of public administration, may not be as difficult to comprehend as many have suggested. If we examine agencies comprehensively from the bottom up, rather than with a narrow eye from the top, we will better understand why some individuals choose to cooperate while others do not. Encouraging coordinated action requires thinking strategically about a logic of interdependence that entices agency officials to cooperate. Absent a logic of interdependence, agency officials will not understand how coordinated action can produce synergistic benefits—other than simply appeasing those political principals demanding coordinated action. The logic of interagency cooperation is much different from the logic of command and control.

2

What Cooperation Means to Agency Officials

Public agencies coexist in an ever-changing environment. Laws, budgets, technology, scientific knowledge, and public support are constantly in flux. These changes sometimes present opportunities for joint gains through cooperation. Yet agency officials do not always perceive these opportunities or choose to act on them. As I conducted interviews for this project, it became apparent that some agency officials demonstrated a greater propensity to cooperate than others because they saw the possibility of achieving joint gains. Later, while analyzing more than a hundred tape-recorded interviews and writing the case studies, it became clear that three sets of agency actors—line managers, professionals, and field staff—each perceived a different logic of interdependence that shaped their understanding of joint gains, and hence their willingness to pursue cooperative strategies.

In brief, line managers seek autonomy to provide stability and certainty for their organizational units. This leads them to limit their dependence on other agencies. Professionals split their loyalty between agency and profession. Though employed by agencies, professionals define problems and promote solutions that have been socially constructed within their profession, and that drive their view of agency interdependence. While not all professions develop a task-based logic of interdependence, professions rooted in the natural sciences have increasingly constructed worldviews in which events and tasks are embedded within webs of ecological relationships that transcend agency jurisdictions. Field staff develop a third logic of agency interdependence. The longer they live and work in a local community, the more they identify with that community, thereby blurring distinctions between agency, profession, and place. The following three sections examine the logic of agency interdependence from each of these perspectives. The remaining sections of this chapter tie these actors together into a multilayered model of interagency relationships.

Line Managers: The Pursuit of Autonomy

Line managers are responsible for the functional units of public agencies. They manage people who operate the agency's core technology. In agencies with traditional hierarchies, line managers are easy to find on formal organizational charts. In the FS, for example, they include district rangers, national forest supervisors, and regional foresters. Thompson (1967) argues that managers in all organizations strive for rationality in their technical core by buffering it from uncertainties, such as fluctuations in the supply of inputs or the demand for outputs. This desire for technical rationality leads managers to reduce the number of variables operating on the technical core and to smooth fluctuations in the remaining variables. For line managers in public agencies, this means enhancing their autonomy from other organizational units, including units within their own agencies. It also means that line managers will sacrifice other agency resources—including higher budgets—if those resources conflict with buffering the core technology from uncertainties, such as changes in missions or tasks.

While not a universal law, the managerial pursuit of autonomy is a sufficiently robust assumption to use as a foundation for constructing theory. It also has tremendous implications for the evolution of interagency relationships. Before delving into the idea of managerial autonomy and its implications, I first want to dispel the notion that there might be other things line managers desire more than autonomy. In this regard, the two prominent challengers are budgets and turf.

In the case of budgets, Niskanen (1971: 38) provided the basic rationale, arguing that budgets can serve as a proxy for nearly all agency preferences. He posited—without providing evidence—that bureaucrats seek to maximize their budgets because larger budgets lead to increases in many things they and their subordinates desire, including salary, perquisites, reputation, power, patronage, and policy outputs. In Niskanen's theory, it does not matter whether bureaucrats care more about some of these things than others because it is assumed that all are positively related to the size of an agency's budget. Therefore, the budget-maximization hypothesis does not preclude the possibility of altruism because bureaucrats may care most about the public interest and accordingly desire larger budgets to provide additional public services. In this sense, the budget-maximization hypothesis is similar to the hypothesized reelection imperative of members of Congress, because elected officials must first hold their offices if they are to continue providing legislative services (Mayhew 1974; Fiorina 1989). Yet Niskanen concluded that agencies oversupply services, which is a popular refrain among those advocating privatization and small government. This led many to employ the hypothesis to explain perceived bureaucratic failures, and to recommend reforms for specific agencies.[1]

Despite its popularity in some circles, the budget-maximization hypothesis is not supported by empirical evidence.[2] This deficiency is particularly pronounced in a collection titled *The Budget-Maximizing Bureaucrat: Appraisals and Evidence* (Blais and Dion 1991). While the editors clearly intended to praise Niskanen's contributions, the chapters offer only weak or contradictory evidence at best. Yet, despite the dearth of evidence to support the hypothesis, the editors surprisingly concluded that it "does make sense to assume that bureaucrats attempt to maximize their budgets and to make that assumption the central proposition of a theory of bureaucratic behavior" (1991: 359). This conclusion suggests that lack of evidence does not dissuade the faithful—particularly given that they had already noted the evidence was "rather thin" on a weaker version of the hypothesis, which claims only that bureaucrats systematically request larger budgets (1991: 355).[3]

Evidence supporting the strong and weak versions of the budget-maximization hypothesis is thin for several reasons. First, budgets are not positively associated with all other things bureaucrats want. As Niskanen (1971: 38) himself briefly noted, and subsequently ignored, managers also want to increase the ease of managing their agency—a goal that is not necessarily supported by higher budgets. Larger budgets may lead to bigger salaries, more perquisites, and greater prestige, but larger budgets may also lead to a loss of control. As Halperin (1974: 51) argued in his study of defense agencies, career officials sought primarily to maintain discretionary use of their budgets rather than increase the size of their budgets; therefore, they were "often prepared to accept less money with greater control rather than more money with less control." This suggests that larger budgets may be forced on line managers by members of Congress, who are themselves responding to the demands of voters and interest groups.

Second, Niskanen applied the budget-maximization hypothesis to the motivation and behavior of high-level officials with respect to their "sponsors" (or political principals). Though he used the term *bureaucrat* throughout his book, he noted at the outset that "the term will be used to define the senior official of any bureau with a separate identifiable budget" (1971: 22). Yet for most individuals within an agency, including lower-level line managers, the budget is a rather distant concept, one they affect only indirectly. While they may be pleased to know their agency's budget is increasing, they probably have little influence on that increase, and may not themselves benefit from it. Therefore, the budget-maximization hypothesis, even if empirically accurate for high-level line managers like agency directors, may tell us little about the motivations of lower-level line managers.

Third, even if line managers seek to increase their budgets, they face external constraints in doing so, particularly when local, state, or federal budgets are declining due to a weak economy or public hostility toward government spending. These constraints are usually overlooked by academics testing the budget-maximization hypothesis.[4] Given that this hypothesis emerged during a period in which the size of government was expanding rapidly, academics may have assumed that budgetary expansion resulted from imperialistic bureaucrats, rather than external sources such as Congress. In sum, budget-based hypotheses provide a poor empirical basis for developing a theory of managerial behavior within public agencies. There is very little direct evidence that supports the budget-maximization hypothesis, whether in its strong or weak forms, and there are several logical reasons why we should look for other explanations for what motivates line managers.

"Turf" is another popular concept that has been used to explain the behavior of line managers, though it is analyzed differently than budgets are. Turf is not associated with a specific indicator, such as budgets. Turf can include budgets, but it may also include landholdings, missions, tasks, staff, or other agency resources. Turf-related behavior is also not posited to be maximizing, or even necessarily growth oriented, but is instead contingent. Under certain circumstances, agency managers are imperialists, moving into turf controlled by others; under other circumstances, managers are defensive, protecting their turf from encroachment. Therefore, turf-related hypotheses provide a more dynamic depiction of interagency relationships than the budget-maximization hypothesis does. While turf-related arguments come in various forms, what most share is an underlying concern with managerial autonomy. Hence, as will be demonstrated, we can also dispense with turf, but for a different reason. While the budget-maximization hypothesis is shackled by lack of evidence, turf-related hypotheses mask the fundamental motivation of line managers to enhance their autonomy.

In popular usage, turf is a slang expression for "one's own territory or domain," such as the territory staked out by inner-city gangs (with graffiti), wildcats (with scents), and suburban homeowners (with fences).[5] In the bureaucratic politics literature, the concept is used in a similar fashion, though often denoted by less colloquial synonyms such as *domain*, *territory*, or *jurisdiction*. Downs (1967: 215), for example, draws an analogy between the territorial behavior of animals and public agencies, arguing that "the basic nature of all such struggles is the same—each combatant needs to establish a large enough territory to guarantee his own survival." For Downs (1967: 212), an agency's territory is composed of social functions, which he calls *policy space*. Because these territorial boundaries are ambiguous, Downs (1967: 213–216) argues that agencies constantly seek to defend or expand their territory depending on the disposition of their employees. A young agency tends to be dominated by climbers, zealots, and advocates, who stake out the agency's territory, while conservers, who predominate in later years, focus primarily on protecting the agency's existing territory from incursions by interlopers. In both cases, however, the fundamental motivation is autonomy. As Downs (1967: 215) states, "the struggle for territorial control among bureaus is another form of the struggle for autonomy."

This theme appears repeatedly in the bureaucratic politics literature. Ellison (1995) argues that agency managers generally seek to maintain their autonomy, but will engage in competition when their core tasks are threatened. Yates (1982: 74) argues that agencies try to avoid conflict "by creating a domain in which they have as much autonomy as possible and thus are free of jurisdictional disputes with other agencies and levels of government." Therefore, "we might reasonably suppose that the bureaucratic policymaker will seek to maximize the autonomy of his unit and its control over its own policy processes and agenda" (Yates 1982: 75). Dimock (1952: 282) similarly claims that the "only way an executive can be certain that any failure to perform effectively is his own is to assure himself in the first place that all the elements necessary for a unified administration are in his hands." Territorial aggressiveness is limited, however, because public managers must please their agency's existing constituency (Holden 1966; Rourke 1969). Therefore, managers in new agencies are more likely to be imperialists than managers in established agencies because the latter are more likely to be oriented toward maintenance rather than expansion (Holden 1966: 951; Downs 1967).

The contingent nature of turf-related imperialism is crucial. Because few agencies are young, we should expect to find few line managers with imperial designs. Instead, we should expect to find most line managers nurturing some tasks while shedding others. According to Wilson (1989: 371–372),

> No single organization . . . can perform well a wide variety of tasks; inevitably some will be neglected. In this case, the wise executive will arrange to devolve the slighted tasks onto another agency, or to a wholly new organization created for the purpose. . . . The turf-conscious executive who stoutly refuses to surrender any tasks, no matter how neglected, to another agency is courting disaster; in time the failure of his or her agency to perform some orphan task will lead to a political or organizational crisis.

Wilson thus takes strong exception to claims that managers seek to maximize their budgets (Niskanen 1971) or the number of people they supervise (Tullock 1965).

Hammond (1979) similarly argues that state agencies have "jurisdictional preferences." In his comparative study of state fish and game agencies in California and Wyoming, Hammond found that agency officials avoided new tasks and expanded jurisdictions when they believed these new responsibilities would be accompanied by increased conflict in their agency's environment or if the new tasks would not mesh well with their agency's current mission and professional competence. In laying out this argument, Hammond followed Rourke's (1969: 85) observations about bureaucratic power:

> While bureaucracies are often pictured as being implacably imperialist in their desire to expand their jurisdiction, in actual fact there are occasions when an agency may increase its power by narrowing, or refusing to expand, the scope of its legal authority. . . . Thus, in the quest for power an agency's strategy has to be one of optimizing rather than maximizing its jurisdiction. Activities that have weak political support, are inordinately expensive, or that divert an agency from its essential purposes represent liabilities rather than assets from the point of view of an agency's power balance.

Imperialism is therefore an anomaly, which we should expect only during an agency's young and expansive years.

In sum, turf is a category of agency resources that managers believe they need to carry out the programs and missions for which they are responsible with as much ease as possible. Public agencies do not possess turf; rather, agencies provide managers with the formal accoutrements around which they develop their own sense of turf. As Bardach (1996: 177) says, "From the manager's point of view, all other things equal, more turf is better." Instead of worrying about which indicators of turf matter, we should assume that all matter under certain circumstances, and focus on the underlying motivation of agency managers to increase their autonomy on the margin. It is the desire for managerial autonomy that has produced the American bureaucracy with which we are familiar: a hodgepodge of relatively independent agencies, each jostling cautiously against one another as they try to secure their jurisdictions, tasks, resources, and political support.

Because political scientists have largely identified agencies with their directors, only high-level agency officials appear to protect their turf as a means for enhancing their autonomy. Authors differ in their use of jargon, speaking either of executives (Wilson 1989), administrative politicians (Holden 1966), or dominant groups (Halperin 1974), but they are primarily concerned with the highest agency officials. According to Wilson (1989: 181), the daily business of public executives "involves performing agency tasks in a way that minimizes the effort needed to maintain the organization." Therefore, they seek out tasks not performed by other agencies, compete with others attempting to perform their tasks, and are wary of joint or cooperative ventures (1989: 189–190). Because executives view interdependence with other agencies as a threat to their autonomy, "many agencies that must cooperate (or at least appear to cooperate) enter into agreements designed to protect each other from any loss of autonomy" (1989: 192).

Yet Wilson discusses the relationship between turf and autonomy only at the executive level. Lower-level line managers do not appear to have turf-related concerns of their own. It seems unlikely, however, that turf is only the concern of high-level agency officials. There is no compelling reason why the same logic does not apply to lower-level line managers as well. Turf is simply a category of agency resources (such as budgets, landholdings, or staff), from which line managers select as needed to enhance their autonomy.

While authors tend to use different indicators of turf, perhaps for ease of exposition or because a particular indicator fits within a broader research agenda, there is no compelling reason why one indicator should matter more than others. It is likely that line managers vary in their desire for particular types of turf depending on their position, responsibilities, and mission. Therefore, rather than choosing among indicators, it is more important to identify the motivations underlying turf-related behavior and to look for commonalities within and among agencies. Turf is a category of agency resources; it is not a motivation.[6] The common motivation is the desire for managerial autonomy.

Land management agencies certainly fit this mold. Turf includes physical territory, as well as budgets, tasks, and other agency resources, and land managers are no less concerned about enhancing their autonomy than the managers of other agencies. Gifford Pinchot, the founder and first Chief of the FS, argued in his autobiography that line managers routinely sought autonomy from one another and thus seldom cooperated:[7]

> Every Bureau chief was for himself and his own work, and the devil take all the others. Everyone operated inside his own fence, and few were big enough to see over it. They were all fighting each other for place and credit and funds and jurisdiction. What little co-operation there was between them was an accidental, voluntary, and personal matter between men who happened to be friends.

More recently, Kunioka and Rothenberg (1993: 704) argued that the FS and NPS avoid competing with one another: "Each goes along grudgingly, at best, when opportunities for expansion are presented. They shun task diversification and are even passive in defending their own turf. Agency decision makers exhibit a strong predisposition to favor autonomy over competition." Even more germane to the issue of biodiversity preservation, Yaffee (1994: 272–273) detailed the extreme lengths to which FS managers went in attempting to maintain their autonomy when confronted by external demands to protect the northern spotted owl and its habitat:

> What motivated any response to the issue at all was largely a fear of losing control over it, and with it, losing control over the direction of national forest management. Listing the owl as a federally recognized threatened or endangered species was consistently seen as the ultimate horror. . . . The fear of listing was partly a fear of having the FWS involved in FS decision making, which at minimum meant more red tape in national forest management. Just as concerning were the new levers that a listing decision would give to nongovernmental groups, enabling them to use the courts to contest management actions.

As will be seen in the case studies, FS line managers were by no means the only ones concerned about threats to their autonomy posed by listed species and lawsuits.

The managerial desire for autonomy stems from several sources. First, autonomy has policy implications. Many line managers are convinced they know best, and therefore should decide how to carry out agency tasks, develop programs, and achieve agency missions. Second, autonomy is closely related to the issue of control. We all like to feel as if we are in control of events in our daily lives, whether at home or work. As Yaffee (1994: 265) argues,

> Organizations and individuals seek autonomy as both a means of carrying out their everyday tasks, and an end in and of itself. The ability to control one's own life is a powerful motivator of human behavior, and individuals translate their needs into organizational needs. To be out of control means a loss of predictability and stability, and requires the expenditure of a lot of energy to regain balance and momentum. Organizations that are out of control rapidly become inefficient and ineffective, and highly stressful places to be.

Moreover, psychological studies indicate that people place a greater value on losses than gains, a trait Tversky and Kahneman (1991) call *loss aversion*. This suggests that managers are willing to make forays into new territory, but only if in doing so they do not risk control of their current turf or confront other agencies on their turf. As Yates (1982: 104) states, "A minimaxer will cautiously and defensively inch ahead, being highly protective of his domain at the same time as he tests the possibilities of extending the limits of his expenditures and authority."

Third, autonomy reduces uncertainty. In organization theory, such diverse strands of the literature as contingency theory (Thompson 1967), resource dependence theory (Pfeffer and Salancik (1978), and transaction cost economics (Williamson 1985) are all based on the assumption that public and private managers seek to reduce uncertainty in their organizational environments. Yet enhancing autonomy is not the only means for coping with uncertainty. Private-sector managers, for example, may attempt to reduce uncertainty through hierarchical expansion (Williamson 1985) or cooperative strategies (Thompson 1967: 34–36; Pfeffer and Salancik 1978: 143–184). Public managers, however, usually lack the authority to expand hierarchically through mergers. Therefore, they are more likely to pursue cooperative strategies as a means for coping with uncertainty. Such cooperative strategies might include contractual relationships, nonbinding agreements, coordinating councils, joint ventures, and co-optation.

If line managers indeed desire autonomy more than anything else, agency relationships should reflect this motivation. Cooperative agreements, for example, might simply delineate turf boundaries rather than lead to the coordinated delivery of public services (Maass 1951; Wilson 1989). After all, the easiest way for line managers to maintain their autonomy is to agree to stay out of one another's turf. In doing so, they would avoid entering into agreements that would increase their dependence on other agencies; they would avoid mission dilution, which may result from intermixing tasks with other agencies; and they would not contribute limited staff time and resources to interagency efforts. Wilson (1989: 269) stated the autonomy problem succinctly: "No agency head is willing to subordinate his or her organization to a procedure that allows other agencies to define its tasks or allocate its resources."

Such claims about the pursuit of managerial autonomy have led many to believe that truly productive, value-creating cooperation is unlikely. Yet it is conceivable that line managers can enhance their autonomy over core tasks through cooperation on other tasks. By doing so, they may be able to secure more autonomy in the long run by working together than if they assert their independence. Accordingly, if we find line managers cooperating to achieve joint gains rather than delineating turf boundaries, we should find that these cooperative efforts enhance their autonomy rather than impinge on it. While this hypothesis may initially seem far-fetched, it is important to recall from chapter 1 that new demands—such as protecting the habitats of endangered species—are regularly imposed on agencies designed for other purposes. Line managers may be willing to cooperate on newly imposed tasks if these new tasks hamper agency performance on traditional core tasks. Within this context, line managers would be particularly open to professional advice they might otherwise ignore if agency professionals can show them how cooperating on newly imposed tasks can produce joint gains in managerial autonomy on traditional core tasks.

Professionals: Network Ties and Epistemic Influence

Unlike line managers, who are primarily autonomy seeking and agency centered, professionals belong to social communities that transcend agency boundaries. Wilensky (1964: 138) characterizes professionals by two criteria: (1) their jobs are based on technical knowledge acquired only through long prescribed training, and (2) they adhere to a set of professional norms. In other words, "professionals are not merely skilled technicians, they are committed to certain substantive values" (Bell 1985: 22). Accordingly, doctors apply their medical training to improve the health of their patients, lawyers apply their legal training to defend their clients, and ecologists apply their scientific training to ensure the survival of species and the integrity of ecosystems.

Each profession attempts to claim exclusive jurisdiction over the technical application of a body of knowledge by establishing schools to teach specific skills and a credential-awarding system to provide the public with a guarantee of trustworthiness (Bella 1987; Wilensky 1964). Some professions (e.g., medicine and law) have been very successful in carving out a jurisdiction of expertise and protecting it from encroachment by other professions, or by occupations seeking to achieve professional status (e.g., nursing and paralegal work). Yates (1982: 131) labels this behavior "guild professionalism," arguing that professions seek "to protect their own established norms of procedure and training and to defend their occupational turf against incursions by outsiders." In this way, professionals have collectively become a new class, competing for power with workers and capitalists by controlling knowledge and expertise, rather than labor or capital (Derber, Schwartz, and Magrass 1990).

Because professions seek to control the content of their work, they represent a form of "organized autonomy" (Freidson 1970). Yet professional autonomy differs from managerial autonomy in both motivation and behavior. While professionals and line managers each seek to define the content of their work, professionals do so to apply their technical expertise to solving problems while line managers do so to provide stability and predictability for their organizational units. Professionals also seek autonomy as a group, while line managers seek autonomy as individuals. Therefore, we should expect to find that a profession dispersed across several agencies provides a means for integrating these agencies, while line managers push the agencies apart.

Public-sector professionals maintain two allegiances—one to their agency, and one to their profession. Thus, agency socialization is tempered by continued fealty to competing norms and practices. Professional identity is instilled in undergraduate and graduate schools, which mold a common worldview and provide members with a common set of problem-solving tools. Professional associations further bind individuals together through conferences, meetings, newsletters, journals, and peer-review processes. Regardless of where professionals work, their primary loyalty is to their profession, which shapes their attitudes on many issues (Rourke 1969: 95–99). Professional training and commitment to an outside group give professionals "strength to resist the demands" of the organizations that employ them (Wilensky 1964: 151).

Wilson (1989: 60) argues that professionals also differ from other agency officials in that they "receive some significant portion of their incentives from organized groups of fellow practitioners located outside the agency. Thus, the behavior of a professional in a bureaucracy is not wholly determined by incentives controlled by the agency." Professional rewards and esteem, for example, often come from colleagues outside the agency. Professionals also develop interagency networks "to protect their professional perquisites and standards" (Heclo 1977: 128). Professionals are more willing to jump from one agency to another because they are not as committed to their agency for a career (Downs 1967: 95–96). Given these characteristics, professionals are more likely than line managers to look beyond the formal jurisdictions of their agencies and communicate with others in their professional community. Hodges and Durant (1989) found professional networking among foresters to be a "pronouncedly more powerful and significant predictor" of foresters' decisions to change their general approach to forest management than bureaucratic factors.

Professional views and commitments may clash with an agency's mission. For agencies recruiting primarily from a single profession, the dual-allegiance problem presents few conflicts, particularly if there is little distinction between the profession and the agency. The FS routinely hired professional foresters until the 1970s, and the agency's early leaders had themselves established the forestry profession and guided the formation of professional forestry schools. Observing the FS during the 1950s, Kaufman (1960: 166) noted that

> foresters have a common set of technical tools and techniques, a common lore and body of knowledge, so the Forest Service can take for granted many things about the way they would handle property under their jurisdiction. Over 90 percent of the more than 4000 professional employees of the Forest Service are foresters; the existence of a widespread consensus on technical matters within the agency is therefore not surprising.

Most agencies, however, including the contemporary FS, hire from several different professions.

The presence of multiple professions has important consequences for relationships both within and among agencies. On the one hand, it leads to rifts within agencies, as each profession "seeks to mold and shape the decision-making process so that issues will be presented and resolved in accordance with its professional standards" (Seidman and Gilmour 1986: 180). Agency policies and practices are then shaped by the profession that controls particular facets of the agency (Mendeloff 1979; Bell 1985). On the other hand, if several agencies hire individuals from the same professions, professional networks may bridge the chasms between public agencies.

Interagency professional networks operate at both the micro and macro levels. At the micro level, network structures are based on interpersonal relationships. At the macro level, networks encompass entire professional communities. The micro-macro distinction suggests two ways in which professions draw agencies together. At the micro level, individuals resolve uncertainty by seeking guidance from others who have better knowledge or higher status (Galaskiewicz 1985: 655). In the process of exchanging information for mutual gains, they develop interpersonal relationships governed by norms of reciprocity and trust (Chisholm 1989; Larson 1992). At the macro level, these informal exchanges disperse information, values, and beliefs throughout the professional community. Patterns of dispersal are not uniform, however, because individuals tend to seek information from those with equal or higher status (Galaskiewicz and Burt 1991; Galaskiewicz 1985). Nonetheless, professional networks draw agencies together by providing "the infrastructure upon which professional subcultures are built" (Galaskiewicz 1985: 640).

Even within a profession, individuals do not always share the same school of thought. Some professions welcome heterodoxy—they accept individuals with diverse backgrounds and beliefs. While heterodoxy opens a profession to wide-ranging and lively debates, such professions pay a cost in policymaking arenas because their internal diversity does not allow them to present a unified and cogent set of problem definitions and alternatives. If policymakers hear a broad range of advice from a profession, they have more options from which to proceed. Professional influence is thereby attenuated. If the political arena is divided on the issue at hand, differing professional opinions may be used to justify differing political positions, which undermines the credibility of the profession.

If a profession exhibits great uniformity in thinking, it is in a position to have a directed impact on the policymaking process. An "epistemic community" is a group of like-minded individuals that exhibits such uniformity. As defined by Peter Haas (1992: 3),

> An epistemic community is a network of professionals with recognized expertise and competence in a particular domain and an authoritative claim to policy relevant knowledge within that domain or issue-area. Although an epistemic community may consist of professionals from a variety of disciplines and backgrounds, they have (1) a shared set of normative and principled beliefs, which provide a value-based rationale for the social action of community members; (2) shared causal beliefs, which are derived from their analysis of practices leading or contributing to a central set of problems in their domain and which then serve as the basis for elucidating the multiple linkages between possible policy actions and desired outcomes; (3) shared notions of validity—that is, intersubjective, internally defined criteria for weighing and validating knowledge in the domain of their expertise; and (4) a common policy enterprise—that is, a set of common practices associated with a set of problems to which their professional competence is directed, presumably out of the conviction that human welfare will be enhanced as a consequence.

In short, the members of an epistemic community have similar normative values, believe in the same causal relationships, and have a common methodology for validating knowledge, all of which shape their formulation of best management practices.

An epistemic community produces consensual knowledge, regardless of whether this knowledge reveals truth or is socially constructed. Therefore, as Peter Haas (1990: 55) argued,

> Presented with incomplete or ambiguous evidence, members of an epistemic community would draw similar interpretations and make similar policy conclusions. If consulted or placed in a policymaking position, they would offer similar advice. Individuals who were not members of the same epistemic community would be much more likely to disagree in their interpretations. Unlike an interest group, confronted with anomalous data they would retract their advice or suspend judgment.

Suspending judgment in the face of anomalous data maintains the scientific legitimacy and authority of the community, thereby shielding its value premises from political conflict. This is important because an "epistemic community's power resource, domestically and internationally, is its authoritative claim to knowledge" (P. Haas 1990: 55).

Even if an epistemic community constitutes a relatively small portion of an agency or profession, it will likely have a disproportionate effect on organizational learning and behavior because the authority of its members in decision-making processes is based on consensual knowledge, not simply the positions of power they occupy or their strength in numbers. Unlike contested knowledge, consensual knowledge cannot be used to support opposing positions in a policy debate. Therefore, depending on the implications of this knowledge for interagency relationships, epistemic communities are well situated to provide a driving logic for cooperation.[8]

In sum, professional staff differ significantly from line managers in having dual loyalties. Whereas line managers tend to tie their careers to their agencies and focus their attention on maintaining the organizational units for which they are responsible, professionals look beyond agency boundaries for social esteem, career incentives, and technical assistance. Therefore, professionals are much more likely to develop interagency networks than line managers. This does not mean that the cooperative efforts professionals initiate will necessarily be supported by line managers. Yet the more professionals approximate an epistemic community by defining problems similarly and promoting similar solutions, the more their recommendations will be supported by line managers.

Field Staff: Living in and Working for a Place

Field staff develop a third logic of agency interdependence when they live and work in local communities for long periods of time. Unlike line managers, who must tend to the needs of their organizational units, and professionals, who are trained to step back from social situations and to view the environment from an analytic perspective, field staff sometimes immerse themselves in the social dynamics of local communities. In doing so, they develop interpersonal relationships with community members and a deep appreciation of the landscape within which a community is situated. This trichotomy is not absolute, however, because line managers and professionals who live and work in local communities for long periods may also become concerned with community well-being, thereby blurring the distinction between line managers, professionals, and field staff.

Tensions often arise between field offices and agency headquarters because duties to agency and profession become tangled with local concerns. Field staff may "express frustration at the arbitrariness of central policy directives, and complain about lack of concern at headquarters for the needs of their particular programs in local communities" (Yates 1982: 78). Agencies accordingly develop personnel rotation systems to move field staff periodically from one location to another so they do not become attached to local communities. Kaufman's (1960) classic depiction of the FS in the 1950s demonstrates how personnel systems can be used to centralize far-flung organizations through extensive socialization and reporting procedures. The FS personnel system rotated forest rangers routinely from one field office to another before they could become integrated into local communities. This personnel system was considered highly functional for the FS at the time, contributing greatly to its perceived success. In Kaufman's (1960: 177–178) words,

> The impact of rapid transfer is more profound than training alone; it also builds identifications with the Forest Service as a whole. For during each man's early years, he never has time to sink roots in the communities in which he sojourns so briefly. He gets to know the local people who do the manual work in the

woods, but not very well in the short time he spends with them. He barely becomes familiar with an area before he is moved again. Only one thing gives any continuity, any structure, to his otherwise fluid world: the Service.

Other agencies encourage rotation through indirect means. Management positions may require broad experience, which is gained through voluntary rotation. Thus, if an individual wants to move up the ranks, she may want to move from one unit to another or from one program to another to gain this experience.

Agency cultures vary with regard to tenure in local offices, in some cases encouraging longevity rather than discouraging it. The BLM long relied on extensive local input to determine the best uses of BLM land. Unlike their FS counterparts, BLM line managers actively encourage field staff to meet with local stakeholders and to address local concerns when developing resource management plans. They do so, in part, because the BLM does not emphasize a specific use of the land, and is therefore more open to local input. The Federal Land Policy and Management Act (FLPMA), which legitimated the agency's gradual move toward multiple-use management in the 1960s and 1970s (Fairfax 1984: 81), allowed BLM line managers and field staff great discretion in choosing among multiple uses at the local level, depending on local input from local stakeholders.

Rural ties have also been the hallmark of county-based extension offices in land-grant colleges (Selznick 1949) and the U.S. Soil Conservation Service (SCS), which was renamed the Natural Resources Conservation Service (NRCS) in 1994. These agencies rely on field staff, who have accumulated a great deal of trust in local communities, as a means for delivering technical assistance to private landowners. Given that state and federal agencies are generally distrusted in rural communities, the professional advice of extension advisors and soil conservationists might have been ignored had these individuals not spent long periods of time cultivating trust at the local level.

Some readers might wonder if field staff are simply "street-level bureaucrats" (Lipsky 1980). While similarities exist, particularly with regard to the use of discretion by service-delivery agents in face-to-face encounters with members of the public, there are significant differences. At one level, it is simply an issue of semantic connotation. "Street-level bureaucracy," with its image of police officers and social workers, connotes an urban rather than rural setting. As will become apparent in chapters 5 and 7, many field staff in natural resource agencies live and work in rural communities far from urban centers. Without being derogatory, a more appropriate term for these rural field staff might be "soil-level bureaucrats." Unlike line managers and professionals, who are more likely to wear ties, slacks, and skirts to work, rural field staff are more often garbed in jeans and boots. If they wear cowboy boots, it is for a practical purpose rather than an urban fad, and the boots are dirty and worn from working outside with ranchers, farmers, or loggers. These images are not intended to convey the idea that rural field staff are somehow country bumpkins or hicks. Many have extensive professional training in soil science, range management, silviculture, wildlife management, and agriculture. The earthy images should convey only a sense of their physical proximity to the natural resources for which they are responsible and the rural communities associated with these resources. Rural field staff lead a very different lifestyle from the professionals, line managers, and street-level bureaucrats who work in large office buildings and live in suburbs.

From a theoretical perspective, street-level bureaucracy is also not entirely apt. Weatherley and Lipsky (1977), for example, explored the coping methods street-level bureaucrats use to manage work-related overload in delivering public services. They found these coping methods to be dysfunctional for the education program they studied, and suggested these were "typical of the coping behaviors of street-level bureaucrats" (1977: 194). It is an empirical issue, however, as to whether particular coping behaviors are widespread or relatively isolated, and I do not want to imply that field staff discretion is necessarily functional or dysfunctional for a given program or agency. It is important simply to recognize that field staff may have a great deal of discretion, and that, depending on how long they have lived and worked in a particular community, their discretion may favor community concerns rather than the concerns of their agency or profession.

In sum, field staff tend to see a different logic of agency interdependence than line managers and professionals. By working in small offices and living in local communities, sometimes far removed from the headquarters of state agencies or federal regional offices, they develop close relationships with other community members and a deep appreciation for the natural resources on which these communities depend. While these three sets of agency actors are not mutually exclusive, because some individuals belong to more than one set, it is important to note that individuals develop a different logic of agency interdependence depending on whether they manage organizational units, participate in professional activities, or live and work in places far removed from agency headquarters or professional networks.

Three Types of Interagency Communities

While public agencies may appear to have little in common in terms of their missions and culture, thereby giving us little reason to expect them to work together toward common goals, we must remember that individuals—not agencies—develop cooperative relationships. Therefore, we should look for commonalities among individuals to see where interagency communities are likely to form. For this reason, line managers, professionals, and field staff provide useful constructs for analyzing interagency relationships because their internal dynamics lead to different types of interagency communities, which in turn provide different capacities for building cooperative relationships. This is not to say that cooperation occurs only within these three sets of agency actors, but there are compelling reasons to expect interagency cooperation to arise more frequently within one or two of them.

Recall from chapter 1 that cooperative relationships are characterized by individuals working together to produce what they cannot achieve individually (Barnard 1938: 23). For cooperation to occur, individuals need not share a common goal or purpose, but their goals should not be mutually exclusive. Yet even under these conditions, cooperation is not frictionless; it requires information, knowledge, and personal effort. Minimally, individuals must know of each other's existence and have some understanding of each other's interests. The more opportunities individuals have to interact with one another, to learn about one another, and to trust others' intentions, the more likely they are to work together toward a common goal. In these respects, some agency officials are more likely than others to perceive their common interests and to act on them. Therefore, they have a greater capacity for interagency cooperation.

Bardach (1998, 1996) provides a useful framework for analyzing this potential for cooperation, or what he calls *interorganizational collaborative capacity.* Specifically, he discusses three types of collaborative capacity: (1) the operational capacity of a system to produce synergistic benefits for those involved, (2) the resource-raising capacity of institutions to leverage resources outside the system for operational purposes, and (3) the constituent capacity of individuals to make constructive use of the system's operational and resource-raising capacities (1996: 180–181). In chapter 1, I discussed operational capacity, with regard to the utility of interagency coordination in maintaining viable populations of species. Resource-raising capacity will be discussed in the case studies. The remainder of this chapter focuses specifically on constituent capacity, or the willingness and ability of agency officials to work together toward a common goal.

Constituent capacity varies among line managers, professionals, and field staff because each set of actors coheres in different ways. Field staff tend to cohere around the geographic places in which they live and work, thereby forming place-based communities rich in social capital. Professionals tend to cohere around the technical application of a body of knowledge, thereby forming knowledge-based communities rich in intellectual capital. Line managers, on the other hand, usually have no preexisting source of coherence, such as social or intellectual capital. Nevertheless, line managers may form position-based communities, based on their similar positions in agency hierarchies, if leaders step forward within their ranks to organize a cooperative effort.

Place-based communities emerge in local communities and the natural landscapes of which they are a part (Lipschutz 1996; Kemmis 1990; Wondolleck and Yaffee 2000: 73–76). While our common image of bureaucrats usually places them in impersonal buildings located in large cities, many agency officials work and live in small communities most of their careers. By living and working in small communities, they amass social capital, a concept that includes interpersonal trust, reciprocity, and civic engagement. Like physical capital, social capital is a productive resource, but it is used primarily to facilitate cooperative action rather than to produce tangible economic goods. As Putnam (1993: 167) argued in his study of local and regional government in Italy, "Voluntary cooperation is easier in a community that has inherited a substantial stock of social capital, in the form of norms of reciprocity and networks of civic engagement." Social capital arises over time as individuals participate in community affairs, develop norms of reciprocity, and trust one another to follow through on commitments. For the most part, this is not a conscious exercise on their part; instead, social capital is a by-product of other activities. Because social capital accumulates through repeated exchanges, it is more likely found among individuals who interact often.

Ostrom (1990: 206), for example, depicted the importance of norms and shared experiences in common-pool resource conflicts between large-scale trawlers and local fishing villages:

> The reason for the general hostility of inshore, small-boat fishers toward large-scale trawlers is not simply that the appropriation technology used by the trawlers is so much more powerful than theirs. Often the operators of trawlers live elsewhere, belong to different ethnic or racial groups, and share few of the local norms of behavior. They do not drink in the same bars, their families do not live in the nearby fishing villages, and they are not involved in the network of relationships that depend on the establishment of a reputation for keeping promises and accepting the norms of the local community regarding behavior.

Place-based communities, such as this fishing village, are built on deep reservoirs of social capital.

Place-based communities may also include agency officials among their members. For this reason, interagency relationships at the local level may be subsumed within or closely tied to public-private partnerships. Field staff are the most likely agency officials to belong to place-based communities, but line managers and professionals may also become members if they live in these communities for long periods of time.

Knowledge-based communities differ markedly from place-based communities because they cohere around ideas rather than geographic locations. Instead of developing a sense of community through town meetings or watershed restoration projects, members of knowledge-based communities are amorphous groups of like-minded individuals who may share few, if any, personal ties. Therefore, knowledge-based communities are built primarily on intellectual capital rather than social capital. Gruber (1994: 4) argues that groups amass intellectual capital when participants bring technical or scientific information, which is shared in conversation among group members:

> Merely making information available to the group, however, is not enough to create intellectual capital. Information does not become intellectual capital until group members share it and accept it as valid. Thus, the creation of intellectual capital is a collective process in which group members learn about their environment and one another, and then construct a collective understanding of the tasks they face. Such an understanding then serves to define the problem, the universe of possible alternatives, and the criteria for evaluating those alternatives.

Like social capital, intellectual capital is amassed by groups, not by individuals.

Professions are the most obvious examples of knowledge-based communities, but they are not the purest form because some professions are riven by disputes over goals and management practices. These professions may lean heavily on social capital as an alternative means for cohesion. Epistemic communities are, by definition, the purest form of knowledge-based community because intellectual disputes among members are small or muted. Later in this chapter and in the case studies, I focus on a tightly knit epistemic community of conservation biologists, a knowledge-based community concerned specifically with the decline of biodiversity.

Position-based communities are a third type of community within which interagency cooperation may arise. Membership in position-based communities is assumed through one's hierarchical position rather than where one lives or what one believes. Line management duties confer on them a common outlook regarding their responsibilities for organizational units, such as state parks and national forests. The irony, of course, is that line managers tend not to form interagency communities because of their common desire for autonomy. Line managers generally seek to establish their independence from one another because, all else equal, greater autonomy means more stability for themselves and the organizational units for which they are responsible. As individuals, line managers may belong to place-based or knowledge-based communities, but they seldom develop position-based communities of their own.

Given that individuals within professional networks and local communities are much more likely to cooperate across agency jurisdictions than are line managers, we should give particular attention to the conditions that prompt line managers to participate in interagency processes. This situation might occur if a line manager belongs to a place-based or knowledge-based community. While helpful to specific interagency efforts initiated by professionals or field staff, such idiosyncrasies mean that other line managers will likely opt out, thereby stifling input from other agencies or other units within the same agency. Moreover, line managers cannot easily ignore their agency-based responsibilities. Therefore, membership in a place-based or knowledge-based community may increase the commitment of some line managers to particular interagency processes, but it is unlikely to override their primary responsibilities within their agencies.

Line managers may also commit themselves to interagency processes when external pressures increase their interdependence. Public agencies are open systems, which means that line managers must respond to changes in their organizational environments. For most agencies, these external pressures have been increasing in number and magnitude over time, much to the chagrin of line managers. These changes include responsibility for increasingly technical tasks with less public tolerance for error, new regulatory demands whittling away at internal line authority, and increasing volatility in policymaking processes (La Porte 1994: 8). In this context, cooperation on newly imposed tasks may be a means for maintaining or enhancing their control over traditional tasks. Yet, should line managers desire to establish a position-based community to cope with changing contextual circumstances, they would not be able to draw on the social and intellectual capital available within place-based and knowledge-based communities. Instead, they must rely to a much greater extent on individual leadership to pull the new group together.

Leadership is a set of focus-giving or unity-enhancing behaviors that help a collectivity accomplish useful work (Bardach 1998: 223). It is measured in individual behavior, but underlying this behavior must be the willingness and ability to persuade others to create and maintain cooperative arrangements. Thus, leaders must be highly motivated individuals who are willing to expend political capital to achieve a particular purpose.[9] Without leadership by one or more line managers, position-based communities are unlikely to form. Once formed, leadership becomes less important for sustaining them because social and intellectual capital become more prominent. For line managers, leadership in pursuit of cooperation might seem oxymoronic. Yet, as will be seen in the case studies (particularly chapters 3 and 4), leadership was crucial for line managers to form position-based communities on the issue of biodiversity, and it came from a seemingly unlikely agency: the BLM.

Merging the Three Types of Interagency Communities into a Cooperative System: The Crucial Role of Contextual Factors and Epistemic Knowledge

It is not difficult to find cases of interagency cooperation among individuals in the same profession or among field staff living in the same community. It is rare to find cases of interagency cooperation among line managers or among all three sets of actors. In chapters 3 through 7, I present several case studies embodying these permutations, all linked by a common argument. As exogenous forces constrained traditional agency tasks and missions, a previously marginalized epistemic community assumed prominence within the agencies by providing multijurisdiction solutions to problems vexing all three sets of actors. Specifically, lawsuits filed by environmental activists compelled some line managers to address ecological problems caused by the traditional missions of their agencies, at which point they turned to professional staff for advice on how to solve these problems. Yet they did not turn to the traditional professionals within their agencies, such as foresters, game managers, and landscape architects, who carried out traditional tasks, because these professionals did not have solutions for the new problems vexing line managers. Therefore, they turned to ecologists—conservation biologists in particular. As an epistemic community, their advice was consensual, and, as ecologists, their recommendations necessarily entailed coordinated action among agencies because habitats and ecosystems transcend agency jurisdictions. Given the dire straits in which some line managers found themselves due to environmental litigation and the presence of endangered species within their jurisdictions, cooperative planning and management emerged as the preferred solution.

Peter Haas (1990: 57–58) offers several hypotheses linking epistemic communities to the scope, strength, and duration of cooperation. He argues that the *scope* of cooperation depends primarily on the comprehensiveness of the epistemic community's beliefs; the *strength* of cooperative arrangements depends on the power amassed by the epistemic community's members within their respective governments, where power is a function of the community's authoritative claim to knowledge and the position of its members within public agencies; and the *duration* of cooperation is largely determined by the epistemic community's continued power. Context also matters greatly in shaping the scope, strength, and duration of cooperation because public officials search out and heed new advice when confronted by anomalous situations and uncertainty, particularly in the face of crises (Haas 1990: 54). Therefore, we need to understand the conditions under which an epistemic community's advice will be sought and its recommendations accepted.

I argue that the mere presence and knowledge-based authority of an epistemic community are insufficient conditions for cooperative efforts to expand beyond the membership of the epistemic community itself. After all, it would be a relatively simple exercise to claim that a large number of epistemic communities permeate the American bureaucracy, but it would be highly speculative to suggest that each such community significantly changes agency behavior. Public agencies will not change their missions simply because of the emergence of epistemic knowledge within their professional ranks. To the contrary, exogenous factors are definitive. Line managers and field staff are unlikely to adopt an epistemic community's logic of agency interdependence unless that logic presents a means for alleviating uncertainties that threaten managerial autonomy and the socioeconomic stability of local communities.

If an epistemic community operates at the margins of public agencies, it can claim few line managers as members or allies. Yet line managers command the legal, administrative, and budgetary resources crucial for interagency planning and management efforts to succeed. Therefore, an epistemic community must find some hook to gain their support, because line managers would otherwise seek to maintain their autonomy while pursuing traditional missions. It must provide a logic for cooperation that overrides or complements this centrifugal tendency, perhaps even solving an autonomy-threatening problem. The case studies will demonstrate how this occurred in California, where field staff also played important roles in developing and implementing multijurisdiction management plans, particularly in working with private landowners and other local stakeholders. Yet without support from line managers the efforts of field staff faltered.

Contextual factors—particularly court interpretations of the ESA and other environmental laws—produced immense uncertainty for some agencies, leading line managers in those agencies to seek new strategies to reduce this uncertainty. As James D. Thompson (1967: 159) argues, uncertainty is the fundamental problem for complex organizations, and coping with uncertainty is the essence of the administrative process. In this regard, agency officials confronted two basic types of uncertainty in their organizational environments: uncertainty about cause-effect relationships and uncertainty about the contingent probability that others will set these cause-effect relationships in motion. Uncertainty about cause-effect relationships is "the worst problem," according to Thompson (1967: 160), because it impedes managers from pursuing organizational goals: "Purpose without cause/effect understanding provides no basis for recognizing alternatives, no grounds for claiming credit for success or escaping blame for failure, no pattern for self-control." If line managers oversaw closed organizational systems, both types of uncertainty would be greatly reduced. Public agencies, however, are increasingly buffeted by many factors in their organizational environments, each posing numerous uncertainties (La Porte 1994).

Line managers therefore search out particular types of information to reduce this uncertainty. As stated by Peter Haas (1992: 4) in the context of international relations,

> The information needed does not consist of guesses about others' intentions, about the probability of discrete events occurring, or about a state's own ability to pursue unilaterally attainable goals that are amenable to treatment by various political rules of thumb. Rather, it consists of depictions of social or physical processes, their interrelation with other processes, and the likely consequences of actions that require application of considerable scientific or technical expertise. The information is thus neither guesses nor "raw" data; it is the product of human interpretations of social and physical phenomena.

This information is a form of intellectual capital. As Gruber (1994: 14) argues, "Intellectual capital reduces uncertainty within a group in at least three ways: it helps document the existence of a problem or set of problems that need to be addressed; it helps define the nature of those problems; and it builds the criteria to evaluate alternative courses of action."

Epistemic communities are strategically situated to reduce this uncertainty. They may also be a cause of this uncertainty. The intrusion of ecological knowledge into public agencies increased the level of uncertainty for line managers, field staff, and the traditional professions by suggesting innumerable contingencies regarding cause-effect relationships and by implying that an agency's technical core could not be demarcated from the activities of other agencies. The traditional resource management professions employed models based on closed ecological and organizational systems, which implied that line managers needed only to supply organizational resources and political support. Ecological knowledge fundamentally changed this situation by opening the technical core of agencies to environmental uncertainties resulting from activities in neighboring jurisdictions. Unless agency officials denied the validity of this increasingly consensual knowledge, they could not escape its implications. With ecological knowledge came interdependence.

Given that the agencies still had their traditional missions to fulfill, this emergent ecological interdependence appeared messy, complicated, and scattered. Environmental litigation and ecological knowledge combined to make traditional agency tasks increasingly difficult. Line managers were buffeted by this uncertainty, but did not have a common goal or objective to guide them in developing long-term, multijurisdictional arrangements to manage the uncertainty. They also lacked a preexisting community within which to work, and social and intellectual capital on which to draw. Instead, they found themselves fighting brush fires—stamping out isolated problems unilaterally without addressing the underlying issues. These multijurisdiction brush fires—on issues such as water quality, air quality, and habitat management—were nuisances that impeded the agencies' traditional missions. At a societal scale, ecologists created these problems by exposing this information and knowledge, which in turn positioned them to provide solutions at the organizational level.

Peter Haas (1992: 34) argues that researchers should address five points when demonstrating the impact of an epistemic community on organizational decision making: (1) identify membership in the community, (2) determine the community's principled and causal beliefs, (3) trace the activities of its members and demonstrate their influence on decision makers at various points in time, (4) identify alternative outcomes foreclosed as a result of their influence, and (5) explore alternative explanations for the actions of decision makers. I pursue the first two points in the next section, leaving the other points for subsequent chapters. In doing so, I emphasize epistemic beliefs rather than membership because consensual knowledge provided the logic for agency interdependence. Given that my argument is based on the existence of consensual knowledge within a particular context, rather than the magnitude of an epistemic community's presence within a set of agencies, it is not necessary to identify membership with precision.

Two Epistemic Communities: Conservation Biologists and Their Ecological Allies

Two epistemic communities, one nested within the other, play prominent roles in the case studies. The ecological epistemic community encompasses many subfields and specializations. Some ecologists focus on the cycling of nutrients in the biosphere; others focus on global warming, marine pollution, or the loss of biodiversity. Within the subset of ecologists concerned about biodiversity, a new discipline gained prominence in the 1980s. Known as *conservation biology*, its academic and professional adherents rapidly emerged as a distinct epistemic community.[10] Conservation biologists share with other ecologists many common values, causal beliefs, methods, and practices, but they pursue narrowly tailored ends. Specifically, conservation biologists seek to protect and enhance biodiversity by maintaining viable populations of species within healthy ecosystems and evolutionary processes. In pursuit of this overarching goal, conservation biologists draw from many ecological disciplines, including population genetics (the study of gene pools within a population), biogeography (the study of species distribution and dispersal), and landscape ecology (the study of habitat mosaics and disturbance regimes).

The Ecological Epistemic Community

The ecological epistemic community differs from other professions in the physical and life sciences because its members study ecosystems rather than analyzing biotic and abiotic parts in isolation from the whole. Ecological knowledge is constructed on the idea of interdependence.

Ecologists accordingly rebel against the reductive tendencies of modern scientific methodology because they believe that "scientists today are in danger of ignoring the complex whole of nature, the quality of organic interrelatedness that defies analysis by the physicist or chemist" (Worster 1994: 21). The ecological epistemic community's holistic view of ecological health and functioning binds its members together even though they have been trained in different disciplines and belong to different professional associations (P. Haas 1990: 75). For ecologists, interdependence is the object of interest. It is intellectually interesting and aesthetically pleasing; it is not a problem to be broken down or minimized.

Peter Haas (1990) examined the role of the ecological epistemic community in developing and promoting the Mediterranean Action Plan, an international agreement to clean up the Mediterranean Sea from industrial, oil, and sewage pollution. Marine pollution is a collective-action problem because marine currents transport pollutants from one country to another. Ecologists offered political leaders new attitudes toward the environment, and decision-making procedures to cope with international environmental problems. They also assumed prominence in international organizations (particularly the United Nations Environment Programme), and in the ministries of countries bordering the Mediterranean Sea. Their power, in terms of knowledge-based authority and the positions they occupied in each country, gave them the opportunity to have a large impact on the reformulation of national objectives, for which they sought to increase concern for the environment and to "reorganize political arrangements to better recognize the interlinkages between ecosystems" (P. Haas 1990: 76).

Similar transformations occurred within professional communities in the United States during the latter half of the twentieth century. Professional ideas about the proper role of humans with respect to nature gradually evolved toward a new management paradigm, spurred on by Aldo Leopold's "land ethic." As Leopold (1949: 203) wrote in *A Sand County Almanac*, "There is as yet no ethic dealing with man's relation to land and to the animals and plants which grow upon it. Land, like Odysseus' slave girls, is still property. The land-relation is still strictly economic, entailing privileges but not obligations." By contrast, "a land ethic changes the role of *Homo sapiens* from conqueror of the land-community to plain member and citizen of it" (Leopold 1949: 204). Largely unnoticed when first published, *A Sand County Almanac* has since been widely recognized for its influential role in sparking the emergence of a new resource management paradigm.[11]

Paradigm is an often-used and slippery term, one that Thomas Kuhn himself did not clearly define in the first edition of *The Structure of Scientific Revolutions.* In the second edition, Kuhn (1970) responded to his critics by discussing two different conceptions of the term, one of which he labeled "sociological." A sociological paradigm consists of "the entire constellation of beliefs, values, techniques, and so on shared by the members of a given community" (Kuhn 1970: 175). If we ignore Kuhn's careless use of the words "and so on," the community about which he writes is an epistemic community. Stated differently, a Kuhnian paradigm in the sociological sense constitutes the common understanding among individuals in an epistemic community. In Kuhn's (1970: 176) words, "A paradigm is what the members of a scientific community share, *and*, conversely, a scientific community consists of men who share a paradigm."[12]

During the first half of the twentieth century, the dominant paradigm in natural resource management was "sustained yield," which reflected the utilitarian values prevalent during the Progressive Era. As Cortner and Moote (1994: 168) argued, "These values state that the best use of resources is human consumption, and the purpose of resource management should therefore be to provide a continuous supply of market-oriented goods. In land management, this has meant an emphasis on maximizing production of a single resource or use, such as timber, livestock, game species, or aesthetics." Sustained yield is a utilitarian paradigm because it views the natural world as a bundle of commodities for human consumption.[13] The sustained-yield paradigm recognized that natural resources could be depleted, whereas Americans previously believed that natural resources were without limit. This paradigm also embodied the Progressive idea that conservation should be practiced by a professional cadre of politically insulated civil servants using scientific methods to develop plans for the efficient, long-term production of natural resources. Even Frederick Winslow Taylor (1911: 5), the icon of scientific management, acknowledged the efforts and foresight of the Progressive conservationists.

In practice, sustained yield embodied three management principles:

1. Give priority to species with the greatest utilitarian value (e.g., ducks, deer, trout, and marketable trees).[14]
2. Develop management plans for individual populations of these species.
3. Focus management efforts within the jurisdiction of particular agencies.

These principles were largely unstated and unquestioned while the sustained-yield paradigm was dominant during the first half of the twentieth century. It simply did not occur to most resource managers that public agencies should be concerned about species with no readily apparent value to humans. Instead, the consumption-driven "gospel of efficiency" provided the moral ethic of the Progressive conservation movement (Hays 1959). Moreover, given that resource agencies were created to manage species with utilitarian value, it made sense for agency officials to focus on populations of those species within their agency's jurisdiction. Thus, for example, professionally trained foresters developed management plans to harvest Douglas fir in the Gifford Pinchot National Forest in Washington State. Other plant species lacking economic value were cut down and burned to prepare the land for monoculture crops of marketable Douglas fir. These clearcuts sometimes extended right up to national forest boundaries, creating well-defined ecological edges between administrative jurisdictions.

Support for the sustained-yield paradigm gradually eroded during the latter half of the twentieth century as resource management professionals became increasingly sensitive to ecological issues. In part, Leopold's land ethic suggested an alternative ethic to the anthropocentric utilitarianism embodied by Progressive conservationism. Having previously authored a classic text on game management (Leopold 1933), which laid out methods for manipulating habitats and controlling predators to enhance game populations for human consumption, Leopold was not an obvious candidate to lead resource management professionals into a new era. Worster (1994: 271) called Leopold's *Game Management* (1933) "the bible of the wildlife profession," representing "the culmination of the entire Progressive environmental philosophy." Yet, after years in the field, Leopold came to believe that species were interdependent in ways biologists had only begun to fathom, and that species did not exist solely for human consumption. Therefore, Leopold (1949: 224–225) exhorted resource managers to think about the ethical, not simply the economic, consequences of their actions: "A thing is right when it tends to preserve the integrity, stability, and beauty of the biotic community. It is wrong when it tends otherwise." Scientific knowledge also expanded, providing increased evidence of the interdependent relationships among species and the ecological costs of managing land solely on the basis of the market value of a few species. The new ethic, combined with increasing ecological knowledge, led to new thinking about agency relationships. Leopold (1949: 198) himself recognized the implications for interagency coordination when he observed that national parks were too small to perpetuate large carnivores, and that national forests surrounding the parks offered the most feasible means to provide additional habitat for these species.

Ecology, as an academic discipline, also experienced a paradigm shift, moving from a belief in stable equilibria, as embodied in the popular idea of "the balance of nature," to a new paradigm emphasizing processes rather than end points (Pickett, Parker, and Fiedler 1992; Botkin 1990). In the traditional paradigm, ecologists assumed that ecosystems were closed, were self-regulating, and followed a standard path toward a unique equilibrium (or climax community). These assumptions implied that nature could be preserved, and would restore itself following any disturbance, simply by enclosing areas within legally designated wilderness areas or parks where human impacts would be greatly reduced. The new ecological paradigm rejects these beliefs because it assumes that ecosystems are open to, and contingent on, surrounding influences. The new paradigm has important implications for resource management because, if wilderness areas, wildlife refuges, and parks are open systems, they are affected by human impacts on surrounding lands. This implies that restricting development and resource extraction within some administrative jurisdictions does not mean that ecosystems will remain undisturbed therein or return to some idealized prehuman condition. Rather than simply leaving areas unmanaged, the new paradigm implies that resource managers should understand landscape-level processes and develop management plans at all scales, not simply within the boundaries of parks and wilderness areas (Agee and Johnson 1988).

Table 2.1
Professional diversification within the U.S. Forest Service

Discipline	1986	1994	Percent change
Foresters	5319	4959	−7
Civil engineers	1102	940	−15
Hydrologists	207	281	+36
Fishery biologists	122	353	+189
Botanists	28	117	+318
Wildlife biologists	490	900	+84
Ecologists	45	177	+293

Source: U.S. Forest Service, as compiled by Hagstrom (1994: 31).

As this new generation of resource managers grew in number, professional conflicts within agencies increased. In part, the influx of ecological thinking was driven by new laws mandating environmental analyses, such as the National Environmental Policy Act (Taylor 1984) and its state counterparts, including the California Environmental Quality Act. Table 2.1 shows a pronounced shift in FS hiring as the agency looked beyond traditional forestry schools to diversify its professional base. This transformation spawned a new literature on the attitudes of FS employees, with researchers attempting to establish the extent to which FS employees harbored a pro-timber bias (Twight, Lyden, and Tuchmann 1990) or had become environmentally conscious (Culhane 1981; Brown and Harris 1992a, 1992b, 1993).[15] Although empirical evidence demonstrates that the professional and demographic makeup of the FS has broadened, and that this shift has been accompanied by the emergence of ecologically sensitive attitudes, it is not yet clear whether and how these changes altered policy outcomes (Sabatier, Loomis, and McCarthy 1995). Therefore, one cannot simply claim that new thinking leads to new policies.

At the state level, the California Department of Fish and Game (CDFG) also increasingly hired wildlife biologists rather than traditional game wardens, leading to professional conflicts within the agency. Game wardens were trained, for example, to increase the size of deer herds for hunters by protecting female deer. By the 1950s, however, the deer herds in California were larger than they had ever been, and the agency's new biologists believed the herds "had to be brought into balance with the available vegetation" by allowing the hunting of female deer, a policy game wardens resisted (Hammond 1979: 169). In part, this simply reflected a difference of opinion regarding the validity of hypothesized cause-effect relationships among wildlife professionals. Philosophically, however, the new generation of ecological professionals generally disapproved of the agency's traditional focus on nurturing game animals, believing that greater attention should be given to California's diverse nongame species (Pister 1987; Hammond 1979). Game wardens continued to have strong allies outside the agency because hunting and fishing licenses funded wildlife management programs, and consumptive users of game animals preferred that their license fees enhance game species rather than songbirds, plants, and reptiles (Dasmann 1965: 52–54).

The federal and state ESAs also drew increased attention to nongame species. But these laws troubled ecologists because nongame species did not receive attention until they neared extinction, at which point recovery became increasingly difficult because their habitat was largely depleted. Moreover, these laws perpetuated the traditional species-by-species management approach by focusing attention on individual species rather than ecosystems. Some even argued that the federal ESA and accompanying regulations "contain significant biological deficiencies" (Rohlf 1991: 281) that preclude effective biodiversity protection (Doremus 1991: 265; Grumbine 1992: 95–100). The species-by-species approach was also administratively inefficient. As Jensen (1994: 274) argued, "California's resource managers are veterans of many endangered species battles. Several of these efforts have cost the parties involved years of negotiation, millions of dollars, and enormous amounts of political wrangling. . . . With more than 900 species in serious decline in California (Jensen, Torn, and Harte 1993), the prospect of species-by-species conservation was appalling." In light of this situation, agency professionals increasingly believed that multispecies planning and management was preferable, and, because species' habitats were interwoven with multiple jurisdictions, multispecies planning and management necessarily entailed looking beyond agency boundaries. As one federal ecologist put it,[16]

Agencies, for whatever reason, over the years, as they've evolved, internalize things. And I'm sure it has to do with turf—you know, "This is what my career is tied to; this is what I'm responsible for; and if you, as another agency, step over the line you got no business there." We found that, if we continue working like that, we're gonna find ourselves in the same dilemma we're in right now. Species are dying; ecosystems are going to hell; and we're failing not only in our responsibilities for managing natural resources, we're failing in our moral obligation to manage natural resources. And only by working cooperatively . . . are we going to make a difference.

In sum, the newer generation of ecological professionals sought to shift the prevailing paradigm from the traditional intra-agency focus on commodity species, to a new approach emphasizing noncommodity species and ecosystems. Arguably, this transition began before World War II (Dunlap 1988), but it gathered much more momentum in the 1970s and 1980s as a new wave of ecologically oriented resource managers entered the agencies. By the 1990s, the new paradigm had become so ingrained within the minds of some agency professionals that the boundaries drawn on maps decades earlier made little or no sense to them.

The Conservation Biology Epistemic Community

Conservation biology emerged within this changing context in the 1980s. While conservation biologists are themselves ecologists, they constitute a smaller and more tightly woven epistemic community because of their mission to protect biodiversity. More so than other ecologists, they have developed and espouse a set of best management practices for preserving gene pools, species, and ecosystems. This epistemic community similarly includes research academics and practitioners, but there is no clear demarcation between the two. As scientists, conservation biologists seek to understand the causal mechanisms of extinction. While they do not agree on all the fine details, they nevertheless constitute an epistemic community because they tend to suspend public debate on these details to offer a unified front to blunt the impending spasm of mass extinction they foresee. Rather than focus on points of academic disagreement regarding the specific causal mechanisms of extinction, conservation biologists focus on points of agreement, translating currently accepted hypotheses into management principles for practitioners to develop plans to protect biodiversity.

Michael Soulé (1985: 727), who founded the Society for Conservation Biology, argued that conservation biology, like medicine, is a "crisis discipline" because its practitioners must act without complete knowledge.[17]

Surgeons do not deny patients treatment simply because physiologists do not know all the facts about the human body. Conservation biologists similarly make recommendations and develop plans to protect biodiversity even though scientists do not completely agree about the mechanisms of human-caused extinction.

In laying out his manifesto for action, Soulé (1985) presented several normative and functional postulates. The normative postulates provide an ethic of appropriate attitudes toward other forms of life. Briefly stated, the normative postulates (in italics) and their corollaries are:

1. *Diversity of organisms is good.* Anthropogenic extinction of populations and species is bad; natural extinction is either value free or good because it is part of the evolutionary process of replacing less adapted gene pools with better adapted ones.
2. *Ecological complexity is good.* Wilderness is preferred to gardens and zoos.
3. *Evolution is good.* Continuity of evolutionary potential, particularly the continuation of evolutionary processes in undisturbed ecosystems, is good.
4. *Biodiversity has intrinsic value, irrespective of the instrumental or utilitarian value it provides to humans.* Species have rights, which spring from their evolutionary heritage and potential, or even their mere existence.

Soulé (1985: 730) believed these normative postulates are shared by "most conservationists and many biologists." Yet these postulates are closely aligned with the school of environmental ethics known as "deep ecology" (Devall and Sessions 1985; Naess 1973). Deep ecologists reject anthropocentric views of nature, believing that humans are simply citizens, like other species, within a larger ecological community. Though not all conservation biologists share this strong biocentric view, as a community conservation biologists do not represent the mainstream environmental movement, which remains largely anthropocentric and utilitarian—albeit with an increasingly aesthetic component.

Soulé's functional postulates distill generally accepted ecological knowledge about the causes of extinction and have direct implications for planning. The list of functional postulates and their corollaries is long, and some of them are not easily understood without a background in the natural sciences. Nevertheless, I list Soulé's postulates (in italics), and briefly discuss some of their corollaries and implications, to impart some idea of how conservation biologists link science-based knowledge with social planning processes.

1. *Many of the species that constitute natural communities are the products of coevolutionary processes.* In other words, species are genetically interdependent. Many species depend on a particular host, which means that the coattails of endangered host species can be very long. Moreover, the extinction of "keystone species" (e.g., predators and herbivores, which control the populations of other species, and plants, which provide breeding and feeding sites for animals) may initiate sequences of causally related events that ultimately lead to further extinctions. In light of this uncertainty, resource managers should strive to save all species and gene pools.[18]

2. *Many, if not all, ecological processes have thresholds below and above which they become discontinuous, chaotic, or suspended.* This esoteric statement basically implies that the rate of extinction within an area is inversely related to the size of the area. The smaller the island, preserve, park, or refuge, the more likely random events will trigger extinctions therein. Therefore, all else equal, resource managers should develop larger reserve systems rather than smaller ones.[19]

3. *Genetic and demographic processes have thresholds below which nonadaptive, random forces begin to prevail over adaptive, deterministic forces within populations.* In other words, small populations are more likely to suffer from inbreeding and genetic drift, which reduce the effectiveness of natural selection. Therefore, viable populations of species should be maintained to avoid inbreeding and genetic drift.

4. *Nature reserves are inherently disequilibrial for large, rare organisms.* Therefore, artificial gene flow should be undertaken for some species to survive within reserves, or reserve systems should be designed with buffer zones around the reserves and corridors between them to facilitate natural gene flow.

Soulé's functional postulates were simply a starting point. They have since been refined, updated, and expanded on by others.[20] Rather than present a concise and up-to-date list, the important point here is to characterize the links between ethical values, hypothesized causal mechanisms, and prescriptions for policy and planning that emanated from this epistemic community during the period in which the case studies unfolded.

To summarize, conservation biology is mission oriented and interdisciplinary. It is an epistemic community because its members share common values and believe in the same cause-effect relationships and truth tests to assess them. Although disagreement exists on some hypothesized causal relationships, the community's strong sense of mission and belief in an impending extinction crisis overshadow these concerns. As two prominent members of this epistemic community once stated, while exhorting an audience to action at the 1988 Annual Meeting of the Society for Conservation Biology (Thomas and Salwasser 1989: 127), "Don't be shy and don't hold back. Whatever you have to say, say it now. If you want to influence what happens on the ground, don't wait. Be prepared to do the best you can with what you've got—right now. And, please, hurry." One of these individuals, Jack Ward Thomas, was later appointed Chief of the FS by President Bill Clinton, but when he made these comments he was an FS research scientist in Oregon.

The Emergence of Interagency Cooperation in California

In the following case studies, conservation biologists similarly occupied technical or advisory positions in California, as did Jack Ward Thomas in Oregon. Some managed research units responsible for gathering and analyzing ecological data but did not have line authority within their agencies. This meant they were able to float ideas within and among agencies but did not have line authority over staff in headquarters or field offices. Unless they gained the support and participation of line managers, they were not in a position to have a direct or significant effect on agency activities. As it turned out, the policies recommended by conservation biologists, and supported by their ecological allies, provided a means both to protect biodiversity and for line managers to maintain their authority over traditional tasks in the face of external pressures that increasingly impinged on their autonomy. Foremost among these external pressures was enforcement of the federal ESA in the courts, which made habitat conservation mandatory. Within this context, the functional postulates of conservation biology offered the potential for joint gains through cooperation on this newly imposed task if agencies coordinated their planning and management processes.

On the margin, the potential for joint gains depended on the shape of agency jurisdictions, the presence of listed species, the likelihood of lawsuits, and other factors, which will be summarized in chapter 8. For now, it is important to note that these factors weighed heavily on the BLM because BLM landholdings were highly fragmented, BLM land provided habitat for numerous listed species, and the BLM was the target of numerous environmental lawsuits. This meant that BLM line managers had the most to gain from cooperation, which explains why the leaders of cooperative efforts among line managers emerged from the BLM.

As line managers in the BLM and other agencies learned about the benefits of cooperation in warding off lawsuits, they developed their own interagency community to spread the new gospel of ecological interdependence. Initially, however, these line managers did not have their own forums within which to interact and develop a sense of community. Agency professionals belonged to professional associations, such as the older Wildlife Society and the newer Society for Conservation Biology, both of which held meetings and disseminated journals and newsletters to facilitate the exchange of ideas. Professionals also worked together in a large number of interagency groups on technical issues ranging from fire ecology to wildlife corridors to data management. Line managers, who had to start from scratch, were led by Ed Hastey, the highest-ranking BLM line manager in California.

The following chapter focuses on a pivotal interagency agreement, the Memorandum of Understanding on Biological Diversity, signed by the directors of six state agencies and four federal regional offices in California. Having long maintained their independence from one another, these high-level career line managers and political appointees were suddenly confronted by new conditions in their organizational environments that challenged traditional missions and eroded managerial autonomy. Buffeted by external pressures that imposed problematic new tasks, they turned to their professional staff for advice. The traditional resource management professions were unable to provide solutions to these new problems. Numerically outnumbered and once marginalized within the agencies, ecologists gained power in this context. Conservation biologists, in particular, provided a knowledge-based logic of interdependence that allowed line managers to reduce the uncertainty that accompanied these newly imposed tasks while maintaining some semblance of their former autonomy on traditional tasks. Cooperative planning and management became a means to fulfill legal obligations to protect species, and thereby protect management autonomy on traditional tasks.

3

The Emergence of Cooperation among Agency Directors

On September 19, 1991, the directors of six state agencies and four federal regional offices in California signed the Memorandum of Understanding on Biological Diversity (or MOU on Biodiversity). This interagency agreement was a milestone because, as discussed in chapter 1, these agencies had previously shown little concern for biodiversity. By contrast, the MOU on Biodiversity boldly stated that the "signatory parties agree to make the maintenance and enhancement of biological diversity a preeminent goal in their protection and management policies." (See appendix B.) The MOU also stated that this newly preeminent goal would be achieved through "improved coordination, information exchange, conflict resolution, and collaboration among the signatory parties." In other words, rather than working independently in pursuit of traditional missions, the MOU indicated that each agency's director—the highest-level line manager or political appointee—now intended to cooperate for the purpose of maintaining and enhancing biodiversity.

This chapter traces the history of the MOU on Biodiversity, from its roots in earlier cooperative efforts among agency professionals and field staff to the emergence of cooperation among agency directors. Because these earlier cooperative efforts shaped the text and implementation of the MOU on Biodiversity, they provide important insights into the dynamics of interagency cooperation among line managers, professionals, and field staff. This chapter also analyzes agency relationships from the perspectives of the participants themselves because the MOU on Biodiversity cannot be understood apart from the experiences and motivations of the public officials who drafted, championed, and signed the document.

Chapter 4 completes the case study by examining what agency officials did after the MOU on Biodiversity was signed. Did the signatories actually contribute agency resources to interagency planning and management processes? Did they exert pressure on lower-level line managers to support agency professionals and field staff for these purposes? To what extent did the agencies actually implement the principles in the MOU? How and why did cooperation vary among the agencies? As will be seen in these two chapters, one of the surprising findings is that leadership emanated primarily from the BLM, an agency not traditionally known for environmental protection.

Three Interagency Efforts Give Rise to the MOU on Biodiversity

Professional staff in California's state agencies and federal regional offices routinely cooperated during the 1980s in both formal and informal groups, particularly on technical issues such as wildlife habitat and data management.[1] Line managers lacked similar forums. Thus, while state-level cooperation occurred among professionals, it was absent among line managers, including agency directors. By the late 1980s, however, three interagency efforts—each following distinct tracks—gradually brought agency directors into cooperative relationships with one another. These interagency efforts were known as

1. Coordinated Resource Management and Planning (CRMP)
2. The Interagency Natural Areas Coordinating Committee (INACC)
3. The Timberland Task Force (TTF)

In table 3.1, the first four columns identify formal agency membership in CRMP (pronounced "crimp"), INACC (pronounced "eye-knack"), and TTF. The last two columns indicate whether and when the first agency directors signed the MOU on Biodiversity, either initially in 1991 or in 1992. The overlapping membership of these interagency efforts suggests the formative roles of CRMP, INACC, and TTF in the evolution of the MOU on Biodiversity. Table 3.2 identifies four nonprofit organizations that also participated.

The following sections discuss these three interagency efforts to flesh out the reasons why line managers, professionals, and field staff participated in some but not in others. Though CRMP, INACC, and TTF are

Table 3.1
State and federal signatories to interagency efforts for managing natural resources in California

Agency	CRMP 1987	INACC 1983/ 1989	TTF 1989–1993	MOU 1991	MOU 1992
U.S. Bureau of Land Management (State Office)	X	X	X	X	X
U.S. National Park Service (Western Region)		X	X	X	X
U.S. Fish and Wildlife Service (Pacific Region)	X	X	X	X	X
U.S. Forest Service (Pacific Southwest Region)	X	X	X	X	X
U.S. Forest Service (Pacific Southwest Experiment Station)		X			
U.S. Soil Conservation Service (State Office)	X				X
U.S. Agricultural Stabilization and Conservation Service (State Office)	X				X
U.S. Bureau of Reclamation (Mid-Pacific Region)	X				X
California Resources Agency			X	X	X
California Department of Fish and Game	X	X	X	X	X
California Department of Forestry and Fire Protection	X		X	X	X
California Department of Parks and Recreation		X	X	X	X
California State Lands Commission	X	X		X	X
California Department of Water Resources	X				X
California Department of Conservation	X				X
California Department of Food and Agriculture	X				X
University of California (Division of Agriculture and Natural Resources)	X	X	X	X	X

Sources: Memorandum of Understanding for Coordinated Resource Management and Planning in California (1987); Memorandum of Understanding among Public and Private Organizations Relating to Natural Area Program Coordination in California (1989); Master Agreement Regarding the Preservation of Natural Areas through the California Significant Natural Areas Program (1983); *Report of the California Timberland Task Force* (Timberland Task Force 1993); Memorandum of Understanding on Biological Diversity (1991), plus additional signatories in 1992.

Table 3.2
Other signatories to the interagency efforts in table 3.1

Nonprofit organization	CRMP 1987	INACC 1983/ 1989	TTF 1989– 1993	MOU 1991	MOU 1992
California Association of Resource Conservation Districts	X				X
The Nature Conservancy (California Regional Office)		X			
California Forestry Association			X		
National Audubon Society (Western Regional Office)			X		

Sources: Memorandum of Understanding for Coordinated Resource Management and Planning in California (1987); Memorandum of Understanding among Public and Private Organizations Relating to Natural Area Program Coordination in California (1989); Master Agreement Regarding the Preservation of Natural Areas through the California Significant Natural Areas Program (1983); *Report of the California Timberland Task Force* (Timberland Task Force 1993); Memorandum of Understanding on Biological Diversity (1991), plus additional signatories in 1992.

presented chronologically, the intended image is "three tracks merging into one" rather than a logical or historical progression from one interagency effort to the next.

Coordinated Resource Management and Planning (CRMP): Cooperation among Field Staff

CRMP is a method for resolving local resource problems that has been thriving for several decades. It was developed in the 1950s by U.S. Soil Conservation Service (SCS) field staff in Oregon and Nevada who sought a means for resolving multijurisdiction problems on Western rangelands (Coordinated Resource Management and Planning, 1990: 3). CRMP subsequently spread throughout the American West because local participants were generally satisfied with the process.

CRMP is a voluntary, public-private, multijurisdiction process for making decisions and implementing solutions at the local level. It is supported at the state and federal levels by formal interagency agreements and a modest organizational structure. As a planning process, CRMP is very flexible because state and federal agencies establish a short list of procedural guidelines, which local groups are expected to follow.[2] In practice, participants define a planning area based on resource issues rather than jurisdictions, register their group with the state-level CRMP organization, and then recruit as many interested stakeholders as possible to define problems and develop consensus-based solutions. Because CRMPs are self-organized at the local level, they consist of and are defined by those who take the time to participate. The resource management plans developed by each CRMP must be consistent with local, state, and federal laws, but CRMPs are not themselves subject to any additional regulations or standards.

CRMP is based upon several principles. First, many resource problems are not confined within the boundaries of a private parcel or administrative jurisdiction; therefore, coordinated solutions are necessary to resolve resource-related problems. Second, local stakeholders should resolve local problems themselves. This is a common belief among advocates of local control. Some professional ecologists who are convinced that public agencies will never have the financial, administrative, or political resources to protect biological resources without local participation also hold this belief. Not only do private landowners manage significant tracts of habitat, many have an experiential-based understanding of natural resources that complements agency data and knowledge. Third, CRMP is based on the assumption that face-to-face communication among all interested parties is the best way to develop solutions because participants learn about each other's interests through deliberation and repeated encounters, thereby finding common ground from which to work. If local stakeholders become personally involved and jointly commit themselves to a plan they develop, they are more likely to implement the plan than if it is imposed on them by others.

Within public agencies, individual CRMPs are primarily the work of field staff—rural field staff, in particular. At higher levels of agency hierarchies, state and federal line managers have signed several interagency agreements to demonstrate their symbolic support for CRMP and to guide CRMP efforts at the local level. These memoranda of understanding lay out policy and guidelines for agency staff to follow in local and regional offices.[3] California, for example, has its own MOU on CRMP, signed in 1987 by the directors of seven state agencies and six federal regional offices, as indicated in table 3.1. Other than affixing their signatures to these MOUs, most agency directors have played negligible roles.

In California, CRMP was loosely orchestrated by an Executive Council, nominally composed of the signatories to California's MOU on CRMP. The CRMP Executive Council met annually to provide overall direction and review the progress of local CRMPs, but the meetings were poorly attended by agency directors.[4] The CRMP Executive Council was supported by a Technical Advisory Council, composed of a representative from each of the member agencies. The state-level Technical Advisory Council met more frequently than the Executive Council, with about half of the agencies represented at a typical meeting.[5] These technical representatives promoted CRMP by reviewing local plans, monitoring the effectiveness of local CRMPs, and training and assisting field staff. Local CRMPs chose a lead agency, usually based on which agency had primary planning responsibility in the area, but the lead agency primarily played a titular role because it was not supposed to impose its will on other members.

The SCS and BLM were the strongest advocates of CRMP. For the SCS, which provided technical assistance to private landowners and local governments, CRMP offered an efficient outreach mechanism. The SCS did not manage land or regulate public or private actors. Moreover, with only a few hundred employees spread across the entire state, interagency cooperation and public-private partnerships provided a sensible strategy for achieving the agency's mission to offer technical assistance when requested by private landowners and public agencies.[6] Understaffed, unable to compel compliance through regulations, and having no land of its own, the SCS depended on cooperative strategies to be effective.

The BLM, on the other hand, managed 17.1 million acres of land in California, or 17 percent of the nonaquatic surface area of the entire state. Most of this land was rangeland, meaning that it was not forestland, farmland, or urbanized. Federal rangelands in California and throughout the West (including FS rangelands) were leased to private ranchers as grazing allotments. The first CRMPs arose in Oregon and Nevada in the 1950s primarily to deal with problems related to overgrazing. In these disputes, SCS field staff acted as mediators, proposing a consensus-based planning process that brought local stakeholders to the table. Because participants believed the CRMP process was effective, BLM and SCS field staff used it to resolve other rangeland issues, including prescribed burning to control wildfires.

The FS relied less on the CRMP process for several reasons. First, the FS had a large staff with a dominant profession. The BLM, like the SCS, was relatively understaffed, and was thus more reliant on outsiders for management assistance. The BLM also did not have a dominant profession that defined best management practices, as did the FS with its large cadre of professional foresters. Therefore, it was sensible for BLM field staff to rely on local users of the land to resolve resource issues. This reliance, of course, led to repeated charges that the BLM was captured by local stakeholders, primarily ranchers.[7] Regardless of one's affinity for the capture metaphor, it was nevertheless true that BLM and SCS field offices were much more permeable than those of the FS.

By the late 1980s, CRMP was widely used throughout California. More than 80 local CRMPs, covering over 6.6 million acres, were underway (Coordinated Resource Management and Planning, 1990: 4). While these figures demonstrate the vitality of CRMP at the local level, line managers were seldom involved with individual CRMPs, and few agency directors ever attended CRMP Executive Council meetings. CRMP was created in and driven from the field, not state agency headquarters or federal regional offices. According to Leonard Jolley, an SCS range conservationist and CRMP spokesperson, individual CRMPs were "driven more by the Levi-types in the field" than by state-level officials.[8]

> Often, I think, the level of cooperation is greater at the local level, with local [agency staff] who begin to buy into the community and begin to get a sense of how to manage their particular resource or agency mandate within either their county, their town, their forest, or whatever local unit they have. I think the local people were striving for a way to get people to cooperate even when the powers-that-be in Sacramento, Portland, or San Francisco were loathe to do so or else had a number of meetings that yielded no result.

Agency directors provided symbolic support by signing the various MOUs on CRMP, and could claim credit for the perceived successes of local CRMPs. Yet rural field staff, driven largely by their concern for the communities in which they lived and the resources on which these communities depended, did most of the work.

Despite the enthusiasm of field staff in some agencies, their participation was occasionally stifled in seemingly innocuous ways by agency structures and decision-making processes. Unlike private landowners and other local stakeholders, field staff often lacked the authority to commit their agencies to a particular course of action. Moreover, local landowners generally had a stronger motivation to participate in long-term planning processes than field staff did because they lived in a particular area for longer periods of time and had a financial stake in their property. Despite the antigovernment rhetoric emanating from the Sagebrush Rebellion in the 1980s and the Wise Use Movement in the 1990s, several agency officials reported that rural landowners were usually more cooperative in CRMPs than government officials. As noted by a state biologist, who strongly supported CRMP, "There are a lot of private owners of land who are very interested in maintaining the resource base of that land, but don't really understand what's needed to do that. . . . People had no idea what they were doing to the land."[9] For these reasons, Jolley argued, "CRMP seems to work best when you cross the boundary between public and private land."

These comments by state and federal staff in California echo the experiences of Anderson and Baum (1987: 164) in Oregon, who noted that difficulties arise with the CRMP process when agency representatives are not authorized to make decisions and when agency personnel turn over during the planning process. More important, they argue, "the most critical difficulty is the need for agency executives to be committed personally to the CRMP process for it to be successful; experience has proved that the degree to which an agency participates effectively in the CRMP process at the field level is directly related to the agency executive's commitment to CRMP." This is because agency directors and lower-level line managers control agency resources necessary for planning and management purposes.

In sum, CRMP provided a multijurisdiction method for resolving natural resource problems at the local level. CRMP was not developed to protect biodiversity. Individual CRMPs addressed issues that spanned private parcels and public jurisdictions, but participants did not give priority to biodiversity. CRMP was simply a decision-making process. If a particular CRMP happened to focus on cattle grazing within a watershed, for example, native species might benefit from reduced impacts along riparian corridors, but such benefits were typically an incidental outcome, not a specific objective of that CRMP. Individual CRMPs protected some species and ecosystems directly, and indirectly protected others, but CRMPs generally overlooked biodiversity within the planning area. Moreover, individual CRMPs arose piecemeal at the local level; thus, CRMP was not an appropriate tool for addressing biodiversity loss at larger scales, such as landscapes and bioregions. In short, regardless of the effectiveness of individual CRMPs at resolving local resource problems, CRMP did not provide a comprehensive biodiversity strategy. This problem was tackled instead during the 1980s by a small group of state and federal professional ecologists.

The Interagency Natural Areas Coordinating Committee (INACC): Cooperation among Ecological Professionals

Developing an effective statewide or regional biodiversity strategy requires information on the status, location, and distribution of species, habitats, and natural areas. Without such information, it is difficult to establish that a species is headed toward extinction, let alone to identify sites that provide useful habitat for the species. Without detailed ecological information, protection is haphazard, regardless of whether agency activities are coordinated or not. If decision makers know where species at risk are located, it is possible to devise site-specific protection strategies and establish priorities by focusing on sites harboring multiple species.

State agencies across the country generally lacked such information in the 1960s. Either the data had not been collected, or it lay scattered in various repositories waiting to be organized into a standardized format. Interest groups accordingly lobbied state governments in the 1970s to create natural heritage programs, which would gather and maintain inventories on biological resources. This nationwide effort was led by The Nature Conservancy (TNC), a nonprofit environmental organization that focuses on protecting biodiversity in natural areas. In 1974,

South Carolina became the first state with a TNC-sponsored natural heritage data center, and nearly all other states followed in the next two decades (Jenkins 1988: 232). Having established these natural heritage programs, TNC developed close ties with the state agencies in which the data centers were housed, sometimes leading to a revolving-door process in which TNC employees worked for the new state programs and state employees subsequently moved to TNC.

In California, a coalition of environmental groups successfully lobbied the Secretary for Resources to establish the Natural Diversity Data Base in the California Department of Fish and Game (CDFG) in 1979. In 1981, the state legislature incorporated this new program into the Fish and Game Code.[10] The Natural Diversity Data Base, a computerized inventory of the location and condition of biodiversity in California, became widely recognized as the standard inventory of rare and endangered species in the state. It was also closely associated with other programs designed to facilitate the protection of biodiversity within CDFG's Natural Heritage Division. The Significant Natural Areas Program (SNAP), for example, analyzed the Natural Diversity Data Base to identify the specific locations of rare species and natural communities, particularly concentrations thereof. SNAP thus provided a basis for establishing protection priorities. By 1991, the Natural Diversity Data Base included more than 20,000 sites in California, at which approximately 730 rare plant species, 330 rare animal species, 88 rare terrestrial communities, and 40 rare aquatic communities were located. SNAP winnowed this list to establish more than 1700 priority sites.[11]

Marc Hoshovsky, a professional ecologist and member of the Society for Conservation Biology, ran SNAP from 1986 to 1989. His primary task was to develop and maintain the inventory of California's significant natural areas. SNAP's mandate also directed the CDFG to consult and coordinate with other public and private organizations in developing this list and managing significant natural areas.[12] For these purposes, Hoshovsky relied on the Interagency Natural Areas Coordinating Committee (INACC), a group that had been inactive for several years prior to his arrival in 1986. INACC was established in 1983 when the directors of three state agencies and three federal regional offices signed an agreement stating that their organizations would cooperate in preserving California's significant natural areas. The four-page "Master

Agreement Regarding the Preservation of Natural Areas through the California Significant Natural Areas Program" laid out a decision-making process and organizational structure for identifying significant natural areas and recommending management alternatives. The agreement did not modify any of the agencies' existing authorities or funding priorities.

Notably, the only nongovernmental signatory to the Master Agreement was TNC, which had led the movement to establish the Natural Diversity Data Base and similar natural heritage programs throughout the country. Unlike most nonprofit environmental organizations, TNC routinely cooperated with public agencies; it did not challenge them in court or the media. Instead, TNC identified, bought, sold, and managed real estate to protect biodiversity. In doing so, TNC staff worked closely with government agencies—recommending properties for the agencies to acquire, transferring properties to the agencies, and sometimes even managing lands owned by the agencies. Given that INACC sought to establish a list of significant natural areas to protect, and then recommend management alternatives for these areas (such as acquisition, designation as a special management area, or regulation), TNC's participation in this interagency group was certainly appropriate.

Unfortunately, INACC was plagued from its inception by several problems that impeded cooperation. The Master Agreement only committed the agencies to a vaguely defined process, not to tangible products like management plans or land acquisitions. Moreover, having been formally constituted by a nonbinding interagency agreement, INACC's success depended on the personal motivation of its participants to commit to a voluntary, cooperative process. Even though the participants were all professional ecologists, and therefore shared the same goals, they did not initially trust one another. In fact, INACC quickly stalled in its first year, and was essentially moribund between 1983 and 1986, because someone tried to dictate procedures. According to Hoshovsky, who arrived several years later, someone from "Fish and Game came in and said 'thou shalt do it this way,' and everybody said 'bye!' "[13]

In 1986, Hoshovsky resuscitated INACC, with the help of his boss, Chris Unkel, who had briefly run SNAP in the mid-1980s while under contract with TNC. Believing that an interagency group was needed to manage natural areas in California, Unkel visited Tennessee, where a similar interagency natural areas group had been working well together.[14] As Unkel recalled,

> Everybody seemed to feel warm and fuzzy about the group. It tended to create an esprit de corps, a rallying point around which people could focus their energy. A lot of these organizations that participated had similar interests; they had sliced off one part of the natural areas pie or another, or they looked at it from a little different direction, or what have you. And, by having this group, it enabled everybody to sit together and gain a common vision of what it is that they were trying to accomplish.

Unkel's enthusiasm after returning from Tennessee, however, did not spread immediately to other agencies. Hoshovsky recalled INACC's first meeting during its second incarnation:

> It was a really cold, frosty meeting. . . . Everybody was sitting there with their cards held close to their chest, trying to figure out what somebody else was going to do—you know, "What else are they going to make me do at this meeting?"—and it was just not that spirit of cooperation at all. But we persisted with it; and some of the players changed; and after awhile people recognized that we were not going to be telling everybody what to do, that we were actually a more cooperative type of thing.

In sum, even though the ideas behind the Natural Diversity Data Base and SNAP were launched in 1979 and codified in 1981, and even though INACC was formally established in 1983, it was not until the latter part of the decade that a cooperative spirit finally emerged among the participants.

Yet even then, cooperation existed only at the technical level, among professional ecologists. Line managers were virtually absent from INACC meetings. Although agency directors signed the 1983 Master Agreement (and a revised interagency agreement in 1989, which brought two additional agencies into INACC), most of the signatories never attended a single INACC meeting. The original Master Agreement established an Executive Group (similar to the CRMP Executive Council), which was supposed to "meet jointly" with the staff-level technical committee "at least annually" to discuss accomplishments and provide guidance. Yet there is no evidence that the Executive Group ever convened, and the distinction between the executive and technical levels was dropped entirely from the revised agreement in 1989.

The lack of participation by agency directors was not surprising. After all, biodiversity protection was not supported by the traditional constituencies of the INACC agencies, and INACC dealt largely with technical issues related to the selection and management of relatively small parcels of land. Thus, INACC's activities were of relatively minor concern to most agency directors. INACC participants nevertheless wanted the directors to attend the meetings because executive-level attendance "adds clout" and draws in "other not-so-friendly agency heads."[15] Yet, unable to get agency directors to attend, Hoshovsky simply updated them on INACC activities by forwarding copies of meeting minutes.

Only one director attended a single INACC meeting from 1986 to 1988, and his agency was not even a signatory to the original Master Agreement.[16] Ron Stewart, director of the FS Pacific Southwest Forest and Range Experiment Station, delivered opening remarks at INACC's summer meeting in 1988, at which he announced his support for INACC's goals and said he would sign the revised agreement on natural areas coordination. Though he expended little personal effort working with INACC, Stewart's credentials indicate support for INACC's goals. Stewart held a Ph.D. in forest ecology and silviculture from Oregon State University, and served as chair of the Natural Areas Committee of the Society of American Foresters from 1988 to 1990. His support for INACC is also intriguing because he later became the director of the FS regional office in California, in which capacity he signed the MOU on Biodiversity in 1991.[17] While FS experiment stations largely conduct and disseminate research, the regional offices have line authority over land management practices in the national forests. The regional office in California managed 24.3 million acres, or nearly a quarter of the state's nonaquatic surface area.

Hoshovsky recalled that another director was attuned to INACC's activities, supported INACC's goals, and may have even attended one or more INACC meetings in the 1980s. This was Ed Hastey, State Director of the California Office of the BLM, who had line authority over 17 percent of California's land. According to Hoshovsky, Hastey expressed more interest in INACC than any other agency director. "As a matter of fact," Hoshovsky noted, "for a while there I had better access to Ed Hastey than I did to our director, [Peter] Bontadelli, which surprised me!" Evidence of Hastey's support for INACC can also be seen on the signature pages of INACC's founding documents. He was the first director to sign both the 1983 Master Agreement and the 1989 revised agreement on natural areas coordination. Hastey signed the 1983 Master Agreement two weeks before any of the other signatories, and three weeks before the CDFG director. He signed the 1989 revised agreement seven weeks before any of the other signatories, and three and a half months before a CDFG assistant director signed in the director's stead.

Hastey's attention to INACC was indicative of his early leadership on biodiversity issues, which was driven by the constellation of forces impinging on the BLM in California, not by his professional training. Hastey had served twice as State Director of the BLM since 1975. His tenure in that position provided him with a long-term perspective on the effect of environmental lawsuits on BLM operations and how those effects could be mitigated through proactive planning. His professional training certainly did not fit the profile of an ecologically sensitive defender of biodiversity. He was trained as a logging engineer at the University of Washington during the 1950s, a period in which professional forestry schools focused on the efficient harvesting of trees, not habitat protection. Moreover, no one I interviewed identified him as an ecologist, and most ecologists wondered why he put as much effort into INACC and other interagency processes as he did. The short answer is that Hastey saw the potential for a different type of benefit from cooperation than did staff ecologists. The BLM had long been the target of environmental lawsuits, and BLM land provided habitat for many endangered species in California. Thus, BLM line managers had a strong incentive to protect natural areas that provided habitat for these species. Hastey was well aware that protecting these natural areas required interagency cooperation because BLM landholdings were highly fragmented and interspersed with other jurisdictions, not to mention private property.

Under Hastey's leadership, the BLM's multiple-use balancing act in California swung in a new direction. From the perspective of the environmental community, Hastey's intentions were enigmatic because environmentalists had long derided the BLM for giving more attention to resource extraction than to species preservation. While some believed Hastey was earnestly trying to do a better job of protecting natural resources, others distrusted him because of his close ties to local resource users and his vocal opposition to the California Desert Protection Act. Yet Hastey's opposition to the Desert Protection Act was not necessarily incompatible with efforts to protect natural areas. Not only did the Act ultimately transfer several million acres of BLM land in California to the NPS, thereby raising the turf-protection instincts of BLM line managers, it transferred many of the BLM's most pristine natural areas in the Mojave Desert, thereby concentrating ranching, mining, and off-road recreation activities on BLM land.[18]

Moreover, transferring BLM land to the NPS did not necessarily mean that biodiversity would be better protected on these lands without a wilderness declaration. Although required by its organic act to protect natural resources, the NPS had long spent far more money developing roads and tourist attractions within parks than identifying and preserving natural resources. The NPS did not even maintain a national inventory of internal and external threats to park resources, despite repeated recommendations by the U.S. General Accounting Office (1996: 4) to maintain this basic level of information. Moreover, only 3.2 percent of the agency's nationwide staff in 1980 worked in the combined science and resource management fields, a category that also included cultural preservation.[19] In California, the NPS had so few ecologists in the 1980s that an NPS employee at one INACC meeting said the agency could "use help in identifying natural areas" in the national parks because it did not have enough ecologically trained personnel to accomplish the task.[20] In sum, while environmentalists distrusted the BLM and Hastey, there was little evidence that the NPS would be a better steward of these lands.

In light of such staff shortages, state and federal ecologists had much to gain through cooperation, but they had to overcome initial distrust and lack of support from most line managers. INACC's goals were also too ambitious relative to the size of the group. Meeting quarterly, this small group of professional staff working in the Sacramento–San Francisco area was essentially trying to coordinate management practices for more than 1000 significant natural areas on public and private land spread across the entire state.[21] For the most part, INACC participants simply shared information. They discussed specific sites to which they were devoting their personal attention, and shared ideas about acquisition and management options. They also discussed which sites should be on the list of significant natural areas, and to which local, state, and federal agencies the list should be distributed.

Beyond exchanging information, INACC was accomplishing relatively little, which was apparent to Hoshovsky:

> Frankly, after a while, I was getting really frustrated with the Natural Areas Committee because it wasn't doing much, and I figured there's got to be some better way. And what I discovered was that it wasn't doing much because you had ten people trying to deal with the state of California—100 million acres! I talked with some folks over in Tennessee, and said, "We're borrowing this great idea of yours, but it's not working. What's the problem?" And a guy said, "Well, the problem is you've got ten Tennessees in California! Over here in Tennessee, this works great because we all live within an hour of each other, we all know the whole state, we have a commitment to the whole piece of territory we're talking about, and it's easy to get together." Whereas in California, BLM's mostly interested in the desert, the Forest Service is up in the Klamath and Sierra, Fish and Wildlife Service may be down on the coast or in the [Central] Valley; so you have ten people in the same room trying to deal with this whole state, and nobody shares the same common interests. . . . And they all had jobs themselves on the side, and this was just one part of their jobs, and they weren't really able to follow through on a lot of things. So the only way to do that was to really go back to the Tennessee scale, and start with a regional kind of approach.

Accordingly, INACC participants sketched a map that divided the state into bioregions, within which they planned to organize regionally based interagency groups.

INACC's bioregional map, which was crudely sketched in 1990, provided the basis for the official bioregional map of California, which is reproduced in figure 3.1. INACC intended the bioregional boundaries to be fluid because ecologists did not agree on the precise demarcations of California's physiographic provinces, which provided the primary criterion by which INACC defined the bioregions.[22] Because the boundaries did not need to be precise in a scientific sense, INACC participants drew the lines in ways that accommodated existing administrative jurisdictions. It made little sense, for example, to put a small fraction of a national forest in one bioregion and the rest in another, and then expect a FS representative to travel to meetings in two different bioregions. Even the number of bioregions was fluid.[23]

INACC participants presented agency directors with their tentative bioregional map of the state, and suggested starting bioregional INACCs in each of them. Hastey strongly supported this idea, in part because decentralized planning meshed well with BLM's grassroots philosophy.

Figure 3.1
California bioregions. © 1997 California Environmental Resources Evaluation System.

Hastey had long supported CRMP. From his perspective, natural areas were simply another natural resource, and natural areas planning could—and should—follow the CRMP model. In 1989, as he gathered signatures on the revised natural areas agreement, Hastey also sought support for the bioregional idea. Although support for INACC remained weak among other agency directors, Hastey was diplomatic in recalling past events:[24] "We renewed our agreement with the idea that we really needed to put more emphasis into this thing. It was kind of struggling with a lot of folks at the lower level working on it, but not really being very effective. So I got everybody to sign a new agreement on [INACC] and [make] a real commitment to start working on a regional approach." The task of forming and nurturing bioregional INACCs fell largely on Hoshovsky's shoulders. Other than Hastey, line managers provided little or no support.

Regional INACCs subsequently emerged in the Colorado Desert, Mojave, South Coast, and Klamath Bioregions. An additional group formed in the Sacramento Valley Bioregion but quickly lost momentum when participants stopped attending the meetings. Given that field staff viewed reserved lands as a relatively small part of their responsibilities, some regional INACCs expanded their horizons, working toward the long-term goal of developing regional strategies encompassing more than reserved natural areas. In this regard, Hoshovsky wanted regional INACCs to gather the support of line managers.

> The original idea was to have management and technical staff, but managers kept saying "Well, my technical person's going; that's fine; I don't need to be there; he'll tell me what's going on." Well, technical people started coming up with some really creative ideas about what needed to be done, but then they got to a point and said "Wait a minute, why are we wasting our time, because we don't have management support for this?" . . . There was just this real feeling of disempowerment at the technical level. So I figured the only way to do this was to somehow go around to the top because it was the middle management that's been holding things up.

Even regional line managers within Hoshovsky's own agency, the CDFG, did not endorse and support INACC's efforts. At the federal level, the lack of support from line managers was similarly apparent. Hastey's support of regional INACCs was not mirrored by BLM line managers in the field. As Hastey himself noted, "The hard thing is reaching out to people at the lower level of the organization to get them to buy in to this kind of approach, and it takes a lot of work and effort."

By the late 1980s, CRMP and INACC had become well-known acronyms among field staff and professional ecologists in several state and federal agencies. Of the two interagency processes, CRMP was more widely known and productive. Numerous local CRMPs were developing and implementing resource management plans, but they did not focus on natural areas planning, and therefore protected biodiversity haphazardly. INACC, on the other hand, focused solely on significant natural areas, but its narrow objectives hampered the ability of participants to gain the support of line managers, and thereby commit agency resources to interagency efforts. As of 1989, natural areas planning in California was simply not the domain of agency directors or lower-level line managers. Although Hastey gave symbolic and administrative support to CRMP and INACC, his presence was felt less immediately by INACC participants. With the exception of Hastey, it was not clear that line managers provided any support at all, other than by signing the agreements on natural areas planning.

The Timberland Task Force (TTF): Agency Directors Finally Come to the Table

In 1989, the same year Hastey gathered signatures for a new interagency agreement on natural areas coordination and sought the support of other directors for bioregional INACCs, an additional opportunity emerged to rope state and federal agency directors into biodiversity issues. The California State Legislature passed Assembly Bill 1580, which directed the California Resources Agency (roughly the state equivalent of the U.S. Department of the Interior) to develop a better understanding of the relationship between wildlife and forestland habitat.[25] AB1580 also directed the Resources Agency to improve interagency coordination on wildlife management and timber harvest practices, particularly with regard to two agencies housed within its superstructure: the CDFG and the Department of Forestry and Fire Protection (CDF). The formal organizational relationships among these agencies are outlined in figure 3.2. In response to this legislation, the California Resources Agency convened the Timberland Task Force, consisting of top officials representing five state agencies, four federal regional offices, and two interest groups. Most of the TTF agencies were already represented in either CRMP or INACC (see table 3.1), but agency directors routinely attended TTF meetings. Because their attendance was an important step in the evolution of interagency cooperation, it is important to understand why they participated when they could have sent subordinates instead.

By 1989, timberland issues had become highly politicized and polarized throughout the American West. In California, while natural areas planning maintained a relatively low profile during the latter half of the 1980s, political battles over the use of private and public timberlands

Governor

George Deukmejian (Republican), 1983–1990
Pete Wilson (Republican), 1991–1998
Gray Davis (Democrat), 1999–

Secretary for Resources
California Resources Agency
(appointed by Governor)

Gordon Van Vleck, 1983–1990
Douglas Wheeler, 1991–1998
Mary Nichols, 1994–

Director
Department of
Fish and Game
(appointed by Governor)

Peter Bontadelli, 1987–1991
Boyd Gibbons, 1991–1995
Jacqueline Schafer, 1996–1998
Robert Hight, 1999–

Director
Department of Forestry and
Fire Protection
(appointed by Governor)

Hal Walt, 1990–1991
Richard Wilson, 1991–1998
Andrea Tuttle, 1999–

Other agencies housed within the California Resources Agency:

Department of Boating and Waterways
Department of Conservation
Department of Parks and Recreation
Department of Water Resources
California Conservation Corps
California Energy Commission
San Francisco Bay Conservation and Development Commission
Santa Monica Mountains Conservancy
California Tahoe Conservancy
Colorado River Board of California
California Coastal Commission
State Coastal Conservancy
State Lands Commission
State Reclamation Board

Figure 3.2
Organizational chart of the California Resources Agency.
Source: Dates and names compiled by author. The list of agencies is not complete.

resulted in numerous lawsuits against state and federal agencies, along with three statewide ballot initiatives and a series of bills in the state legislature, all of which threatened in various ways to wrest control of timber harvest practices from agency directors and lower-level line managers. This constellation of external pressures largely accounts for executive-level participation in TTF.

Their participation was not due to the state legislation that authorized the TTF. State legislation cannot compel the directors of federal regional offices to participate. AB1580 also explicitly permitted the directors of state agencies to designate subordinates to attend meetings in their stead. Therefore, legislation did not compel their participation. Agency directors were motivated by much broader political and judicial forces impinging on their agencies and their management autonomy.

At the state level, over 40 lawsuits had been filed since the mid-1980s against the CDF and the Board of Forestry (Timberland Task Force, 1993: 18). The Board of Forestry is a small, politically appointed body with policy authority over the CDF. The CDF is both a service agency and a regulatory agency. As a service agency, the CDF manages the largest fire-fighting outfit in the state, providing services for 33 million acres of the state's forest, brush, and rangelands. As a regulator, the agency reviews timber harvest plans, which are required for commercial timber harvests on private land. Because the Resources Agency had certified the rule-making program of the Board of Forestry, as well as CDF's regulatory program for reviewing timber harvest plans, every time the Board of Forestry or CDF was sued the Resources Agency was sued as well.[26] The Board of Forestry and CDF were prime targets for environmental lawsuits because many perceived these agencies to be captured by the timber industry (Nechodom 1994).[27]

In addition to lawsuits, three statewide ballot initiatives in 1990 gave voters the opportunity to approve new administrative procedures for wildlife management and timber harvest planning. All three ballot initiatives failed, as did several bills lumped together in the state legislature as the "Sierra Accord" (1991) and "Grand Accord" (1992). Nevertheless, state forestry officials were concerned that future political actions could strip their management authority over timberland policy.

At the federal level, FS line managers were also increasingly aware of the potential for litigation to wreak havoc on the agency's internal management practices, constraining the ability of line managers to define goals and select alternatives. At the center of this litigation was the northern spotted owl, which depended on old-growth forests for habitat. Because nearly all old-growth forests on private timberlands had been logged in California, federal agencies, particularly the FS, managed most of the owl's remaining habitat. Research indicated the owl was threatened by timber harvests, but high-level FWS officials dragged their feet in listing the owl as a threatened or endangered species under the ESA, while the FS continued timber sales in the owl's old-growth habitat (Yaffee 1994). Environmental activists sued the FWS, which subsequently proposed listing the owl as a threatened species a few months before AB1580 cleared the state legislature. The owl was officially listed in June 1990, four months after the first meeting of the TTF.

Environmental groups did not relent once the owl had been listed. Continuing their confrontational strategy, they sued the FS and BLM for scheduling timber sales without adequately considering the owl's habitat requirements under environmental laws. U.S. District Judge William Dwyer sided with the plaintiffs on May 23, 1991, ruling that the FS had violated several environmental laws, including NFMA, the agency's multiple-use statute. Judge Dwyer placed the blame squarely on high-level agency officials:[28]

> More is involved here than a simple failure by an agency to comply with its governing statute. The most recent violation of NFMA exemplifies a deliberate and systematic refusal by the Forest Service and the FWS to comply with the laws protecting wildlife. This is not the doing of the scientists, foresters, rangers, and others at the working levels of these agencies. It reflects decisions made by higher authorities in the executive branch of government.

Because Judge Dwyer also concluded that the owl was threatened with extinction, he issued an injunction on timber sales in the range of the owl—from Northern California to the Canadian border. In California, the effect was concentrated in the area INACC ecologists called the Klamath Bioregion (see figure 3.1).

As Yaffee (1994) eloquently argued, "all hell" broke loose in the FS following Judge Dwyer's ruling because line managers rapidly lost control over their organizational units. The loss of decision-making authority was particularly acute in Oregon and Washington (within the agency's Pacific Northwest Region), which harbored the greatest concentrations of the owl's habitat. Yet the long-term threat to FS line managers in California was arguably greater because of the potential listing of a similar subspecies—the California spotted owl—found in the Sierra Nevada, the Coast Range south of San Francisco, and the forested mountains of Southern California. FS line managers in California therefore had to deal with the immediate impacts of the northern spotted owl on the agency's national forests in the Klamath Bioregion, as well as the potential impacts of the California spotted owl on most of the agency's other national forests in the state.

On BLM land, litigation over the northern spotted owl severely hampered timber operations in western Oregon. In northwestern California, the BLM managed relatively small parcels of timberland within the owl's range. Nevertheless, the owl symbolized broader concerns regarding listed species on BLM land throughout the state. At that time, BLM staff were completing a study, which found that 78 federally listed species (either threatened or endangered) and 361 candidate species resided on BLM land in California (Bureau of Land Management, 1992). Some of these species, like the owl, had very large ranges.[29]

In light of these impacts on agency operations, state and federal line managers could gain by coordinating their habitat planning and management efforts, particularly for the northern spotted owl.[30] Yet many agency directors seldom—if ever—interacted with one another prior to the TTF. In contrast to the situation with elected officials, who routinely interact in Congress, state legislatures, city councils, and county boards of supervisors, interactions among state and federal agency directors were either idiosyncratic or tied to specific programs. For example, the regional forester—the top FS line manager in California—usually developed ties with the CDF director because the FS had responsibilities with regard to state and private forestry.[31] Yet the regional forester historically had little contact with the directors of other state agencies, in part because the FS Regional Office was in San Francisco, a 90-minute drive from the state capital in Sacramento. While some federal agencies had regional offices in Sacramento (e.g., the BLM, FWS, and BoR), others were in San Francisco (e.g., the FS, NPS, and EPA). For the latter, timely participation in state policymaking processes was problematic.

This physical separation became so potentially costly for the FS in the 1980s that the regional forester assigned two assistants to work permanently in Sacramento as state government liaisons. The immediate issue of concern for the FS was a state bill proposing air-quality legislation that would increase permit fees for prescribed burns on national forests, but the FS liaisons remained in Sacramento to manage other multijurisdiction issues, including biodiversity. Rather than being 90 minutes from Sacramento, the new liaison office was just a few blocks from the Governor's Office, the Resources Agency, and the Capitol building. Jon Kennedy, who ran the liaison office and reported directly to the regional forester in San Francisco, focused his efforts on developing relationships with state and federal agencies. His assistant, Harley Greiman, focused on the state legislature. Kennedy and Greiman were essentially lobbyists. They tracked legislation and agency activities, traded information, attended meetings, testified at hearings, and mailed letters endorsing particular laws, regulations, and other governmental activities. While most federal agencies had individuals who spent at least part of their time monitoring state government activity and attempting to influence state policy, the FS liaison office in Sacramento was unique because it was physically detached from the rest of the agency and emphasized lobby-like activities.

Once established, Kennedy and Greiman turned their attention from air-quality legislation to other issues affecting the FS. Greiman soon noticed a timberlands bill—AB1580—that sought to resolve conflicts between wildlife management and timber harvest practices on private land by bringing state agencies together to work out their differences. Federal lands were not being considered. As Greiman put it, the bill "left out some of the key players [because] forest issues go beyond boundary lines, ownerships, and authorities."[32]

The bill sought to reconcile the conflicting regulations and management practices of the CDF and CDFG, which were often in conflict over wildlife issues. As the trustee for the state's wildlife, the CDFG employed a large number of game wardens who nurtured stocks of game species, and a smaller number of ecologists who sought to preserve biodiversity. The CDF primarily fought wildfires and regulated timber harvests on private lands, activities that have detrimental effects on biodiversity. Timber harvests fragment forest habitat, producing open canopies, which benefit browsing species like deer. But as forests become fragmented into smaller patches of old-growth trees, these patches are increasingly unable to support dependent species, such as the northern spotted owl. Wildfire suppression also threatens species that thrive in the presence of low-intensity fires. While the CDFG had authority to review timber harvest plans, it lacked sufficient staff to review many of these plans.[33] Moreover, because the Board of Forestry established forest practice rules, CDFG staff would have to challenge the Board's rules if they hoped to change the content of timber harvest plans, rather than react to each one submitted.

Yet, even if CDF and CDFG officials worked out their differences in regulating timber harvests on private lands, they might still be in conflict with other agencies, including the FS, BLM, and FWS. Greiman accordingly met with legislative staff to discuss broadening the bill's scope, and persuaded them to amend the bill's language to include federal agencies. The revised bill also called for agency directors (or their designated alternates) to sit down at the table with representatives from the timber industry and the environmental community.[34]

Timber interests heavily debated the bill because it was linked to another bill moving through the legislature that would abolish clearcutting in California. Like all timber bills during this period, the bill was contentious; it barely passed both houses. At the close of the 1989 legislative session, Assembly Speaker Willie Brown attached the Senate version to an appropriations bill distributing cigarette tax money (Assembly Bill 1580), which cleared the legislature on the last day of the legislative year. Kennedy and Greiman then sent a letter to Governor George Deukmejian supporting AB1580.[35] On October 1, 1989, Deukmejian signed the bill, which authorized the TTF and included $400,000 to fund it for two years.

The TTF subsequently met eleven times between February 1990 and January 1993. What was initially conceived as a bilateral process, bringing together two state agencies to work out their differences, emerged as a multilateral process, with high-level line managers and political appointees representing nine state and federal agencies sitting together at a single table. This fact alone was remarkable. As Greiman put it, "Never have we ever had those executives together sitting at the table in one place." While agency staff influenced the character and size of the forum, the directors nevertheless came to the table—and stayed—under their own volition.

Yet they did not have a clear understanding at the outset as to where their common interests might lie. Moreover, some agency directors were troubled by the incongruity between their hierarchical position and the technical nature of the tasks with which the TTF had been charged.[36] According to one staff assistant to the TTF, "What usually happens in a situation like that is the directors send their third or fourth in charge, and it becomes quite a technical project."[37] Litigation over the northern spotted owl, however, had significantly altered their incentive to participate because "there was all of a sudden this growing, tremendous frustration . . . with the Endangered Species Act and the problems that act was causing for their agencies."

None of the directors expressed their frustration with the technical discussions initiated by TTF staff more than BLM State Director Ed Hastey. He believed the energy of top agency officials was being squandered on technical issues better handled separately by staff, rather than being devoted to what he believed to be a much bigger problem:[38]

> The obvious problem in the Klamath . . . was that you had four Forest Service plans; two BLM . . . resource management plans; a habitat conservation plan being developed by Simpson [Timber Company for the northern spotted owl], and one being proposed by the State of California; you had the Governor trying to work out this [Grand] Accord; you had state legislation being proposed; you had the court action; you had the Fish and Wildlife Service doing critical habitat designation; you had a scientific panel working on the spotted owl; and probably another half-a-dozen items that I've forgotten. But the people who were most affected, and most impacted by this, were the people that lived in that Klamath Province, and they didn't have any input or any knowledge as to what the hell was going on. I think the only thing they knew was they were gonna get screwed at the end, but they didn't know how, when, or how bad.

So Hastey prodded the TTF to think more broadly about trying to coordinate planning efforts already underway in the Klamath. He wanted them to develop

> a multi-jurisdictional, multi-habitat, multi-species plan. And because we were concentrating on the Klamath Province, I guess what I envisioned at that time was this huge Klamath Province plan that would deal with private and public land, in terms of timber, so you could actually make trade-offs, maybe setting aside some of the public timber to allow the private timber to be developed. You wouldn't try to do a habitat conservation plan for Simpson, and wouldn't

develop one for some other logging company, but you'd do a large regional approach for that whole province.

Hastey's efforts prompted TTF Chair Gordon Van Vleck, who was then Secretary of the California Resources Agency, to create an Ad Hoc Committee in October 1990 to address these broader concerns.

The Ad Hoc Committee Drafts the MOU on Biodiversity

The new Ad Hoc Committee, chaired by Hastey, immediately began holding its own series of meetings while the TTF continued to work on the technical issues with which it had been charged. In part, the Ad Hoc Committee sought a means by which to continue TTF's work when it officially expired after two years. It was clear to participants that the TTF was unlikely to complete its initial study of old-growth habitat in the Klamath in two years, let alone an additional study of the Sierra Nevada as suggested by the legislation, because the technical tasks alone were immense.[39] The Ad Hoc Committee, however, did not limit its thinking solely to the Klamath and Sierra timberlands, but instead set in motion a coordinated approach for protecting biodiversity throughout California.

The Ad Hoc Committee differed considerably from its parent organization, the TTF, because it was composed of lower-level officials. Some were line managers, like BLM District Manager Al Wright, who was based in the Klamath, and who would soon become Associate State Director under Hastey. Others were professional ecologists, like Marc Hoshovsky, who ran the Significant Natural Areas Program in CDFG and coordinated INACC. Jon Kennedy, the regional forester's liaison in Sacramento, also participated, as did individuals from other public agencies and a few interest groups. Yet, of all the participants in the Ad Hoc Committee, none would have a larger impact on the ultimate shape and character of the MOU on Biodiversity than Robert Ewing.

Ewing headed an analytic unit in CDF known as the Forest and Rangeland Resources Assessment Program (FRRAP).[40] In some respects, FRRAP was relatively powerless within the much larger agency. FRRAP's budget and staff were a small fraction of CDF's. Ewing also did not have line authority over CDF units in the field. In addition, the public constituency for FRRAP's products—which included arcane ecological information, cutting-edge analytic techniques, and technical reports—was small compared with popular support for CDF's fire-fighting activities and the commercial demand for timber. On the other hand, FRRAP staff were highly trained ecologists and policy analysts. Ewing, for example, held a Ph.D. in natural resource policy from the University of California at Berkeley. Thus, FRRAP staff had the analytic tools that the TTF needed to carry out its mandate. FRRAP accordingly assumed a relatively large role in staffing the TTF, providing broad-scale depictions of ecological relationships in the Klamath timberlands.

Some members of the Ad Hoc Committee brought with them a broader perspective than forest issues. Hastey, for example, was much more familiar with rangelands. Hence, when the Ad Hoc Committee first met, Hastey proposed using CRMP as a model for interagency cooperation because the BLM had long relied on CRMP as a tool for managing rangeland conflicts. Hastey strongly believed in the CRMP process because it emphasized participation by local stakeholders, and he suggested developing a new interagency agreement similar to the MOU on CRMP. This did not sit well with staff ecologists like Ewing and Hoshovsky, because CRMP had not proven itself to be a good tool for protecting biodiversity. Individual CRMPs focused on relatively small areas and were not integrated with broader efforts to preserve species, habitats, or ecosystems. In Ewing's words, "Some of us were really dissatisfied with CRMP as a model; not that CRMP doesn't work, but it has much more of a localized focus." Protecting biodiversity required a broader planning process, one that encompassed the entire range of species and the ecosystems of which they are a part.

Once CRMP had been challenged, Hastey stepped aside to allow the Ad Hoc Committee to broaden the discussion. During a couple of brainstorming sessions, the participants generated a list of 17 principles to guide collaboration among state and federal agencies in conserving biological resources. Although some ecologists had initially rejected CRMP as a model, two of these principles echoed the CRMP philosophy, emphasizing cooperation between public and private entities and open communication with all affected interests. Other principles were listed under the heading "Conservation Biology," which included recommendations to minimize habitat fragmentation, mimic natural disturbance regimes, and allow for shifting habitat mosaics. The remaining principles suggested a range of policy tools (such as land acquisitions, mitigation banks, fees, and penalties), and recommended cooperation among agencies at the bioregional scale. As an Ad Hoc Committee status report (Hastey 1990: 2) states,

> In order to achieve the goal of reconciling forest management with wildlife habitat needs in the Klamath Province, the strategy must be defined at the largest appropriate spatial scale, that of the bioregion. The bioregion, characterized by plant and animal assemblages that differ considerably from adjacent regions, includes all the vegetative, land, and wildlife resources that can be affected to achieve the protection goal. The bioregion constitutes the strategy's "theater of operation."

While initially developed for the Klamath Bioregion, the Ad Hoc Committee believed these principles could serve as the basis for a statewide biodiversity strategy.

The 17 principles were simply bullet points, not an integrated strategy. Therefore, when Hastey reported back to the TTF, they received only "lukewarm acceptance," according to Ewing, because "no one quite knew what to do with them." So the Ad Hoc Committee went back to the drawing board during the early months of 1991. Rather than losing momentum, participation actually increased as new individuals joined the group. Around this time, Hastey asked Ewing to draft a new interagency agreement. As Ewing later recalled, Hastey said: "Well, if you don't like the CRMP MOU, why don't you write another one and we'll use that." So Ewing and FRRAP Ecologist Greg Greenwood, along with Hoshovsky from the CDFG, drafted a new agreement. Thus, staff ecologists, not line managers, drafted the MOU on Biodiversity.

Hoshovsky's perspective in drafting the MOU on Biodiversity was shaped by his frustration with regional INACCs in the late 1980s. The regional INACCs had faltered in part due to the lack of support from line managers. Therefore, Hoshovsky welcomed the opportunity to develop a new interagency agreement as a means for exerting top-down pressure on lower-level line managers to support natural areas planning. Hoshovsky believed the MOU on Biodiversity, once signed by agency directors, would provide him with a "door to get down to the middle-management" level of the agencies, which had been impeding interagency cooperation at the local and regional levels.[41] In federal agencies, line managers in the BLM and FS were particularly important because they managed, respectively, 17.1 and 24.3 percent of California. The NPS was less important because it managed less than 5 percent of the state, only slightly more than the Department of Defense. Moreover, individual park units had much more autonomy from the NPS regional office than was the case in the BLM and FS.

Once the document was drafted by staff ecologists, Hastey's forceful leadership again assumed prominence. In this regard, several participants recalled one of the last meetings of the Ad Hoc Committee, during which the draft of the MOU was discussed by an enlarged group of individuals. A CDFG employee, dissatisfied with the MOU's language, wanted the agency's legal staff to review it. As Ewing later recalled, "It was a great meeting because Fish and Game stood up and said: 'We're gonna take this to our legal staff' and 'we're not satisfied with the wording at all.' . . . And Ed said: 'Screw it! I'm just gonna get it signed.' So he personally walked it around Sacramento and got the MOU signed." In gathering the signatures, Hastey was not simply following the lead of a few professional ecologists in CDF and CDFG. While these ecologists hoped the MOU on Biodiversity would provide top-down pressure on federal and state line managers to follow the planning and management principles of conservation biology, the MOU also espoused a bottom-up approach to resource management very much akin to CRMP. Ecologists could point to references in the MOU regarding the preservation of biodiversity, but Hastey could likewise claim that "the most important thing" in the MOU on Biodiversity was CRMP's cooperative, multi-jurisdictional approach to planning.

The MOU on Biodiversity differed from CRMP primarily in focus and scale. CRMP was a generic tool for managing a wide variety of natural resources on relatively small pieces of land. By contrast, the MOU on Biodiversity focused specifically on biological resources, and at a much larger geographic scale than CRMP. Its purview was also broader than INACC, which focused on lands reserved specifically for natural values. Reserve lands were certainly important for protecting biodiversity, but they represented relatively small parcels compared to the planning scale set forth in the MOU.

The MOU emphasized that decisions should be made at several organizational levels: at the state level, through a new executive council; at the bioregional level, through new bioregional councils; and at the watershed or landscape level within each bioregion, through numerous CRMP-like local associations, many of which already existed. The MOU on Biodiversity mirrored the CRMP and INACC agreements in that it established an executive council consisting of the signatories to the agreement, but unlike CRMP and INACC, the new Executive Council on Biological Diversity would meet quarterly rather than annually. Given that few directors ever attended the annual meetings of the CRMP Executive Council, and that the INACC Executive Group never even convened, quarterly meetings would certainly test the motivation of agency directors to participate in cooperative efforts.

While the organizational structure of the MOU on Biodiversity is relatively straightforward, it is less clear what the framers intended by using the scientific term *biological diversity*. Ewing hoped that *biodiversity* would become a measure of environmental quality, which could be exchanged on the open market like pollution permits.

> One of the things I've been real interested in is . . . trying to get some kind of market or voluntary way of dealing with these natural resource issues. And it seemed to me that one thing you need to do is to have a currency of nature, in a sense, if you're going to accomplish that. I mean, if you look at the places where these kinds of solutions tend to work, they are like air or water where you have some kind of measure of quality that you can relate to across sites and maybe across disciplines. And I think sooner or later, if this is really going to work statewide, we need the same kind of measure of vegetation and biological systems. So one of the ideas about biodiversity is that it may actually be this kind of measure that allows us to compare across sites and to . . . get involved in some kind of mitigation trading.

Biodiversity also emphasized that ecological relationships transcended agency jurisdictions and reserve systems. Hoshovsky put it this way:

> Biodiversity looks at genetic diversity, species diversity, and ecosystem diversity. . . . No one agency has all of that as a mandate. . . . So biodiversity broadens it out by saying we're looking at all this different stuff. And we're trying to broaden people's perspectives—it's not just simply another way of saying "habitat management."

Not only did no single agency have a mandate that addressed all aspects of biodiversity, it was not even clear that state and federal laws taken together encompassed biodiversity.

Biodiversity was a relatively new concept, one with which few legislators were familiar when most environmental statutes—including the ESA—became law. Yet now, under the MOU's "Policy and Principles" section, appeared profound new language: "The signatory parties agree to make the maintenance and enhancement of biological diversity a preeminent goal in their protection and management policies." With a single sentence, the drafters sought to change the priorities of a large number of state and federal agencies. This phrase was not slipped in at the last minute. The Ad Hoc Committee, including Hastey, reviewed a draft of the MOU with this sentence before Hastey gathered the signatures of other agency directors. Moreover, it appeared prominently on the first page of the three-page document, and was therefore visible to the signatories.

Some stakeholders were subsequently upset by this "preeminent goal" statement, but it was not the framers' intent for the agencies to abandon or run roughshod over them. Not only were Hastey and the BLM closely aligned with local stakeholders through the CRMP process, the professional staff drafting the MOU also recognized that state and federal agencies could never preserve biodiversity on their own because the agencies lacked the resources to do so. As one state ecologist on the Ad Hoc Committee later put it,[42]

> To be successful, these efforts have to be owned and operated by the folks that are responsible for management on the ground. That's probably *the* big lesson that CRMP has taught us in California—that, if you're going to have conservation on the ground, you have to have ownership in the process from the people that are responsible for managing the resources, whether they're private landowners, whether they're local biologists, or agency people that are down there. Without that, it doesn't make any difference what kind of plan you have in Sacramento or [at the] national level. If you don't have people on the ground that are interested in implementing it, it'll never happen.

Regardless, had staff ecologists drafted a document that ignored local interests, Hastey would not have supported it.

Agency Directors Sign the MOU on Biodiversity

Hastey shepherded the MOU on Biodiversity from draft form to final version with signatures. Yet these signatures did not all arrive with ease. In part, the process was complicated by the election of a new governor, and the subsequent arrival of new political appointees, during the period in which the Ad Hoc Committee developed the MOU on Biodiversity. In the November 1990 general election, Republican Pete Wilson defeated

Dianne Feinstein. Governor Wilson quickly appointed Doug Wheeler as Secretary for Resources, thereby replacing Gordon Van Vleck as TTF Chair. This transition created some uncertainty within the Ad Hoc Committee as to whether the new administration would support its recommendations.

As it turned out, however, the new Secretary for Resources strongly supported the principles embodied in the MOU on Biodiversity. Wheeler previously served as vice president of the World Wildlife Fund and Conservation Foundation, executive director of the Sierra Club, and founder and president of the American Farmland Trust. By contrast, Wheeler's predecessor, Van Vleck, was a rancher and former president of the American National Cattlemen's Association. Van Vleck supported CRMP because private landowners were generally satisfied with the CRMP process as a means for resolving local land-use conflicts, but it was not clear he would have supported the MOU on Biodiversity. Wheeler not only supported it, he subsequently became its most vocal and visible advocate, riding the momentum initiated by other state and federal officials.

All nine directors of the TTF agencies signed the MOU on Biodiversity in 1991 (see table 3.1), and the new Executive Council initially followed TTF's footsteps by focusing on wildlife and timberland issues in the Klamath Bioregion. The Executive Council differed from the TTF by appealing to CRMP's consensus-based decision-making process at the local level, and by adopting INACC's bioregional map (figure 3.1)—and INACC's notion of bioregional planning—as its first order of business.

The Executive Council could have included many more agencies because dozens of state and federal agencies—not to mention local governments—affected biodiversity in California. The drafters of the MOU on Biodiversity recognized this fact, and accordingly included a large number of potential signatories in an early draft.[43] Rather than extending invitations to these agencies, Hastey initially focused only on the TTF agencies. The only deviation from this rule was the California State Lands Commission, whose director asked to participate, and who subsequently became the most vocal critic of the new Executive Council (see chapter 4). In 1992, six additional state and federal agencies were added to the MOU, thereby including all of the CRMP agencies in the Executive Council (see table 3.1).

Hastey intentionally drew the line with the agencies with which he was already familiar through CRMP, INACC, and TTF. He purposely excluded most regulatory agencies because he believed regulators tended to be more dogmatic than land managers, and therefore less coöperative. In his words, "I think some of the regulatory agencies don't look at things in the same light as land management agencies [because] they're so focused on regulations and what's black and white. The only way this thing is gonna work is [if] people . . . give a little bit." Yet some Executive Council agencies had both regulatory and land management functions. The CDFG and FWS managed wildlife refuges and ecological reserves, and each implemented their respective state and federal ESA. Because enforcement of the federal ESA, in particular, had been constraining the decision-making autonomy of Hastey and other line managers, it made sense to include the FWS in a cooperative process to protect biodiversity. Other regulatory agencies, such as the EPA, were less germane. Besides, Hastey bluntly added, "I don't like the EPA."

Not all of the directors wanted to participate. While some were amenable to signing the MOU on Biodiversity, others were lobbied to sign. The director of the University of California Division of Agriculture and Natural Resources (UC/DANR) was a reluctant partner. The UC/DANR would be an important partner because it could facilitate implementation of Executive Council programs. The UC/DANR had an annual budget exceeding $200 million, several research programs addressing California's agricultural and natural resource problems, and a county-based infrastructure spread across the entire state to disseminate research findings to landowners and public officials. Moreover, these county-based Cooperative Extension offices had good working relationships with local stakeholders. Unlike many state and federal agencies, which were distrusted in local communities, the 64 Cooperative Extension offices were perceived locally as providing politically neutral technical assistance, without regulation. UC/DANR's academic research was also more credible than research conducted in other agencies. As Greiman put it, "We really needed the University of California. We needed to have the science and the academic community because without it there would be no credibility." The UC/DANR director eventually signed (or, to be accurate, allowed an assistant to sign the MOU for him), but he did not subsequently participate in the Executive Council.

The state and federal park directors also maintained a low profile. No one from the NPS or the California Department of Parks and Recreation (CDPR) helped draft the MOU on Biodiversity. Both agencies were included in the TTF, but no one from these agencies participated in the Ad Hoc Committee. Moreover, neither agency had been a signatory to CRMP. Historically, the park agencies had been relatively detached from interagency efforts to manage natural resources. For both agencies, participation had been largely at the technical level, through INACC, rather than in the field or through the directors' offices. When Hastey approached the directors of the TTF agencies to sign the MOU on Biodiversity, the state and federal park directors simply added their names; they were neither prime movers nor reluctant partners, a pattern that would continue throughout the 1990s.

Summary

The MOU on Biodiversity emerged from the confluence of three separate interagency processes: CRMP, INACC, and TTF. Each had its own distinct history, agenda, cast of characters, and working relationships. Together, they formed an overlapping set of agencies and individuals working to solve natural resource problems. Other interagency efforts addressed similar issues during the 1980s, but these three provided the basic ideas, forums, and working relationships that gave rise to the MOU on Biodiversity.

CRMP laid out a philosophy for resolving natural resource problems at the local level. CRMP's emphasis on decentralized planning and management, public-private partnerships, and collaborative decision making gained many local supporters. Because participants were pleased with the CRMP process, and because it did not threaten traditional constituencies, agency directors signed the MOU on CRMP to symbolize their support, but they rarely attended the annual meetings of the CRMP Executive Council. Instead, CRMP was largely run by field staff, with assistance from a state-level Technical Advisory Council. CRMP was not designed to address biodiversity issues. Most CRMPs were too small and independent from one another to form a regional or statewide strategy for protecting biodiversity. Nevertheless, CRMP offered a cooperative approach, which satisfied local participants, if not environmental activists. While some professional ecologists were also dissatisfied with CRMP because it failed to address biodiversity at a regional scale, others believed that a CRMP-like process would be a necessary component of any larger effort because most biodiversity occurred on private land and nonreserved public land.

INACC was composed almost entirely of staff ecologists, who brought a statewide and regional planning perspective to biodiversity issues. Focusing initially on reserve lands, INACC participants gradually expanded their focus beyond ecological reserves to begin developing regional conservation plans. Yet INACC accomplished relatively little because INACC ecologists were not supported by line managers within their own agencies. The statewide INACC and regional INACCs also lost momentum in the early 1990s when Hoshovsky, who had done most of the organizational work, began devoting his time and attention to staffing the new Executive Council on Biological Diversity. Because the Executive Council and its professional staff ultimately subsumed much of INACC's agenda, there was less need for an interagency group dealing only with technical matters related to managing natural areas. Moreover, agency directors now seemed motivated to participate in and lend their support to this new, broader effort, which might provide a top-down means for gathering support from lower-level line managers.

The TTF was an important step in the evolution of interagency cooperation because it brought agency directors together at the same table for the first time, and provided the immediate forum within which the MOU on Biodiversity emerged. Although agency directors were not required to attend TTF meetings, several did so because environmental litigation increasingly constrained their management autonomy. It was becoming clear to some agency directors that these new legal obligations might be easier to fulfill through some form of coordinated action. Regardless of the TTF's impact on forest policies (an issue not raised here), it demonstrated to agency directors that they had common interests. The TTF also provided a forum within which they could develop interpersonal working relationships and discuss alternative means for decreasing the burden imposed by newly required tasks.

It was at this point that the interests of staff ecologists and agency directors dovetailed. Neither liked the status quo, and the planning principles of conservation biology offered the possibility of enhancing biodiversity *and* management autonomy. As stated in *The Report of the California Timberland Task Force* (Timberland Task Force, 1993: 31–32),

> When species viability drops to a level where they must be listed, management options for species recovery and opportunities for resource use are decreased. Endangered species acts force an overly narrow concentration on high profile, threatened species. Species recovery plans typically reflect little consideration for maximizing biological diversity. Concerns for significant effects also are generally administered on a site-specific basis and are generally remedial. Finally, reserves are stop-gap emergency measures generally not designed with long-term protection in mind or are designated for other purposes. The task force concluded that it is essential to augment current species and site-level approaches with landscape-level strategies that recognize the linkages within ecosystem processes and the need to maintain biologically diverse forestland.

Notably, the report extolled the MOU as one of TTF's major accomplishments (Timberland Task Force, 1993: 3).

To a large extent, however, the MOU on Biodiversity was really Hastey's accomplishment. He pressured the other directors to look beyond the northern spotted owl to develop a broad, multijurisdiction approach for managing all species. He formed and chaired the Ad Hoc Committee for this purpose, he curtailed debate on the draft MOU when at least one participant threatened to derail the effort by calling in lawyers, and he personally gathered the signatures on the agreement. Line managers in other agencies experienced similar problems with endangered species and believed that interagency coordination provided a means for dealing with these problems, but it was Hastey who provided the leadership. Because of Hastey's leadership on this initiative, some later called him "the godfather" of biodiversity in California.[44] This image is ironic because Hastey was trained as a logging engineer in the 1950s. His background did not suggest he would later promote biodiversity. Yet Hastey, more than any other career line manager or political appointee, saw the potential for joint gains through cooperation on this issue. Without his presence, the MOU on Biodiversity may not have been written, let alone signed by so many state and federal agency directors.

The MOU elevated biodiversity from an obscure academic idea promoted by agency ecologists to a primary management objective—at least on paper. Agency directors signed their name to a document that stated that "the signatory parties agree to make the maintenance and enhancement of biological diversity a preeminent goal in their protection and management policies." This statement did not seem radical for the NPS or its state counterpart, the CDPR, because their dual mandates focus on preservation and recreation. But for the multiple-use agencies, particularly the FS and BLM, this statement appeared to represent a major shift in policy.

If one glanced at a detailed map showing land ownership patterns in California, one might wonder why BLM line managers took the lead in pursuing this coordinated approach to regional planning. Most of the agency's land is concentrated in the arid southeastern portion of the state, which means the BLM primarily manages arid rangelands, not the forest-lands with which the TTF was concerned. In the Klamath Bioregion, BLM landholdings produced relatively little commercial timber compared with FS and private lands. Yet, having served twice as State Director since 1975, Hastey saw a much bigger picture. The MOU on Biodiversity was not simply a means for dealing with the northern spotted owl, timber harvest practices, and uncoordinated planning and management in the Klamath. The owl was only one of 78 listed species—not to mention 361 candidate species—that resided somewhere on BLM property in the state (Bureau of Land Management 1992). It was neither the first, nor would it be the last, listed species with which BLM line managers would have to cope. The owl was simply the first to galvanize high-level interagency activity by fixing the attention of other agency directors.

In sum, it was becoming increasingly difficult for line managers in some state and federal agencies to maintain their autonomy from one another when confronted by scientific knowledge about ecological relationships and litigation over endangered species. The BLM's fragmented parcels of land are difficult to manage, particularly as habitat, because they are intermixed with parcels belonging to other landowners responding to different financial, legal, administrative, and political incentives. BLM staff had long cooperated with individuals in neighboring jurisdictions because of the agency's intermixed ownership patterns. Habitat management was no different in this regard; it had simply become an immediate concern given the presence of numerous listed and candidate species. The NPS, by contrast, managed relatively consolidated lands. Moreover, NPS line managers had not been threatened by court challenges or impending legislation that might constrain their management autonomy. Facing a different set of incentives than BLM line managers, they were much less motivated to participate in interagency groups like CRMP, INACC, and TTF. As we will see in subsequent chapters, these patterns of participation were to continue after the MOU on Biodiversity had been signed.

4

Institutionalizing Cooperation

The MOU on Biodiversity was an important indicator of interagency cooperation among agency directors, but it is only one indicator of cooperation, and it tells us nothing about the degree to which cooperation became institutionalized. This chapter accordingly traces the continuing evolution of cooperation after the directors signed the MOU, focusing specifically on the activities of the Executive Council on Biological Diversity and its professional staff. The discussion begins with the original ten signatory agencies, and then expands to include subsequent signatories. Local governments also enter the picture, first as detractors threatened by the emergence of the MOU and the Executive Council, and then as co-opted participants.

The Formal Role of the Executive Council

The purpose of the Executive Council on Biological Diversity, as stated in the MOU on Biodiversity (appendix B), was "to develop guiding principles and policies, design a statewide strategy to conserve biological diversity, and coordinate implementation of this strategy through regional and local institutions." At the local level, public-private groups would follow the CRMP model, giving specific attention to managing habitats and ecological systems. At the regional level, line managers would mimic the Executive Council by developing regional MOUs and bioregional councils. The Executive Council would provide overall guidance and administrative assistance to both local groups and bioregional councils. This three-tier approach would address biodiversity at several scales, as recommended by conservation biologists, and would allow political input at each level and within each bioregion.

The MOU on Biodiversity did not, however, give the Executive Council or its member agencies additional authorities or budgets. Instead, the MOU encouraged the Executive Council to "seek adequate funding to implement regional strategies," but did not specify whether the directors would be expected to allocate some of their discretionary budgets to Executive Council activities. The agreement also did not specify precisely what the signatories were expected to do. Because the MOU was not legally binding, they were not required to do anything at all to implement its principles. Therefore, the activities discussed in this chapter should be viewed as additional indicators of interagency cooperation, not causal results of the MOU.

The Executive Council met for the first time on December 12, 1991, just three months after the MOU was signed. Since then, it has met quarterly, as specified in the MOU—with the exception of 1992, when there were only three meetings. Participation at these meetings varied greatly. Some directors attended most of the meetings; others attended relatively few. Some spoke frequently at the meetings and contributed agency resources to interagency processes; others were largely bystanders. The following sections analyze attendance and discussion at the quarterly meetings to demonstrate variation in participation among the directors.

Attendance at Executive Council Meetings

As noted in chapter 3, the original ten signatories were not equally committed to the MOU's planning and management principles. This variation is reflected in their subsequent attendance at Executive Council meetings. Tables 4.1 and 4.2 present attendance data from the minutes of the first 13 quarterly meetings. The minutes identify which agency directors attended each meeting, and who sat as designated alternates in the absence of the directors.[1] I ranked the agencies according to an attendance score in the right-hand column of table 4.1, based on the hierarchical status of agency officials attending Executive Council meetings. The third column identifies the number (and percent) of meetings attended by each agency's director. If the director did not attend one or more meetings, the next three columns indicate whether the alternate was the associate director (i.e., the agency's second-in-command),

Table 4.1
Attendance at Executive Council meetings, by highest official (first 13 quarterly meetings, December 1991–March 1995)

Agency	Location of headquarters	Director [× 3]	Associate director [× 2]	Other official [× 1]	No one present [× 0]	Attendance score
California Resources Agency	Sacramento	11 (85%)	2 (15%)			37
U.S. Bureau of Land Management (State Office)	Sacramento	9 (69%)	4 (31%)			35
California Department of Forestry and Fire Protection	Sacramento	10 (77%)	1 (8%)	2 (15%)		34
California Department of Fish and Game	Sacramento	7 (54%)	4 (31%)	2 (15%)		31
U.S. National Park Service (Western Region)	San Francisco	6 (46%)	4 (31%)	2 (15%)	1 (8%)	28
California Department of Parks and Recreation	Sacramento	6 (46%)	2 (15%)	5 (39%)		27
U.S. Forest Service (Pacific Southwest Region)	San Francisco	6 (46%)		7 (54%)		25
U.S. Fish and Wildlife Service (Pacific Region)	Portland, Oregon	3 (23%)		10 (77%)		19
California State Lands Commission	Sacramento	3 (23%)		8 (62%)	2 (15%)	17
University of California (Division of Agriculture and Natural Resources)	Oakland			13 (100%)		13

Source: Compiled by the author from minutes of the quarterly meetings of the Executive Council on Biological Diversity. (Percentages rounded to the nearest whole number.)

Table 4.2
Agency directors attending Executive Council meetings (first 13 quarterly meetings, December 1991–March 1995)

Agency	Location of headquarters	Director	Total meetings attended	Percent attended	Meetings outside Sacramento
California Resources Agency	Sacramento	Doug Wheeler	11 of 13	85	3 of 4
U.S. Bureau of Land Management (State Office)	Sacramento	Ed Hastey	9 of 13	69	3 of 4
California Department of Forestry and Fire Protection	Sacramento	Richard Wilson	10 of 13	77	4 of 4
California Department of Fish and Game	Sacramento	Peter Bontadelli	1 of 1	100	N/A
		Boyd Gibbons	6 of 12	50	1 of 4
U.S. National Park Service (Western Region)	San Francisco	Stanley Albright	6 of 13	46	2 of 4
California Department of Parks and Recreation	Sacramento	Henry Agonia	1 of 1	100	N/A
		Donald Murphy	5 of 12	42	1 of 4
U.S. Forest Service (Pacific Southwest Region)	San Francisco	Ron Stewart	3 of 10	30	0 of 1
		Lynn Sprague	3 of 3	100	3 of 3
U.S. Fish and Wildlife Service (Pacific Region)	Portland, Oregon	Marvin Plenert	3 of 10	30	0 of 1
		Michael Spear	0 of 3	0	0 of 3
California State Lands Commission	Sacramento	Charles Warren	3 of 9	33	N/A
		Robert Hight	0 of 4	0	0 of 4
University of California (Division of Agriculture and Natural Resources)	Oakland	Kenneth Farrell	0 of 13	0	0 of 4

Source: Compiled by the author from minutes of the quarterly meetings of the Executive Council on Biological Diversity. (Percentages rounded to the nearest whole number.)

whether the alternate was some other agency official, or whether no alternate sat at the table. I gave each agency three points for every meeting attended by the director, two points for every meeting at which the associate director was the highest official, and one point for any other official. This method captures the symbolic and administrative importance of sending the associate director as an alternate rather than a lower-level line manager or staff professional.[2] Table 4.2 focuses on agency directors, identifying them by name and providing some indication of their willingness to travel to the four Executive Council meetings held outside Sacramento.[3] The agencies retain their rank from table 4.1.

Two intriguing points emerge from these tables. First, 70 percent of the directors attended at least six of the first thirteen meetings. This is a remarkable statistic given their previous absence in similar forums. Attendance at each quarterly meeting exceeded attendance at the annual meetings of the CRMP Executive Council and the INACC Executive Group. Mere attendance, however, does not indicate whether the directors believed in the value of biodiversity, participated in discussions, or contributed to related interagency activities to protect biodiversity. Moreover, the meetings lasted only a few hours; therefore, the time commitment was not great, except for those directors based in cities other than Sacramento, where most meetings were held.

The variation among agencies raises another intriguing point. The BLM—an agency not traditionally known for protecting biodiversity—emerges near the top. BLM State Director Ed Hastey attended nine of the first thirteen meetings, which is consistent with the formative role he played in bringing the MOU on Biodiversity to fruition. Even more remarkable, Hastey's attendance record might have been perfect, had he not been temporarily reassigned to the BLM Alaska State Office in 1993 and 1994. Due to Hastey's absence from the state, California Resources Secretary Doug Wheeler, who chaired the Executive Council from its inception, had the best attendance record. For both agencies, executive-level presence also loomed large at these meetings because the associate directors regularly accompanied the directors—a fact not captured in table 4.1. If the attendance score counted the attendance of associate directors when directors were present, the Resources Agency and BLM would receive significantly higher scores than the agencies ranked immediately beneath them.

Of the other eight directors, CDF Director Richard Wilson's attendance is also notable. The CDF played an important role in the TTF because the agency was at the center of the timberland-wildlife conflicts from which TTF had been born. Moreover, the agency's ecological analysis unit, the Forest and Rangeland Resources Assessment Program, played a central role in staffing TTF. Therefore, one might expect the CDF director to play a similarly prominent role in the Executive Council on Biological Diversity. Yet Wilson's attendance record was not matched by other forms of high-level participation. The associate director rarely attended these meetings and Wilson was not a discussion leader.

The next four agencies ranked in tables 4.1 and 4.2 had relatively similar attendance records, with their directors attending roughly half the meetings. Yet this is nevertheless remarkable given their previously poor attendance records at executive-level meetings for CRMP and INACC. For three of these four agencies—the NPS, CDFG, and CDPR—the associate directors usually attended when the director was absent. For the FS, the associate director in San Francisco never attended during this period, but the regional forester's liaison in Sacramento, Jon Kennedy, regularly sat as the alternate. While not himself the associate director, Kennedy reported directly to the regional forester.

Participation from the FWS is almost certainly underestimated by the attendance score in table 4.1, which does not factor in travel time. Directors based in Sacramento generally had better attendance records because the first nine meetings were held in the state capital. Given that FWS Regional Director Marvin Plenert flew in from Portland, Oregon, his attendance at three meetings represented a relatively strong showing. Not only did Plenert travel the greatest distance, the FWS Pacific Region covered six states, including Hawaii. By comparison, the regional offices of the BLM and FS basically encompassed a single state. Moreover, Plenert's regular alternate was a high-level line manager in California.[4]

By comparison, the superficially similar attendance record of the California State Lands Commission (CSLC) should be questioned. Given that CSLC Executive Officer Charles Warren was based in Sacramento, his attendance at only the first three meetings indicates relatively weak support for the Executive Council. Notably, Hastey did not ask Warren to sign the MOU on Biodiversity, because Warren had not previously participated in CRMP, INACC, or TTF. Instead, Warren asked to sign the MOU. Warren's departure is intriguing because his demeanor toward the Executive Council quickly became hostile. I will return to this story later in the chapter.

The UC/DANR ranks last because its director, Kenneth Farrell, never attended a single meeting during this period, nor did his associate director. The UC/DANR stands out as the only agency of the original ten not represented on the Executive Council by anyone in the director's office. While I did not interview Farrell, observers generally believed he distanced himself from the Executive Council because the agricultural industry felt threatened by ideas in the MOU on Biodiversity. Regardless of Farrell's actual motives for missing Executive Council meetings, the industry's lobbyists made their position clear. The California Farm Bureau Federation announced its strong opposition to the MOU on Biodiversity and related activities in a 1992 editorial comment:[5]

> Farmers cannot be expected to forfeit their property rights to conserve wildlife habitats and biological communities up and down the state at the behest of a new government entity. We are all for preserving our state's rich natural heritage, but not when it means placing private property in government hands and putting rural Californians in the poor house.

The UC/DANR director could not easily ignore such proclamations. The agricultural industry constituted the agency's primary clientele, and farmers had long been tied to UC/DANR's county-based Cooperative Extension offices in particular.[6] The UC/DANR director accordingly distanced himself from interagency activities his constituents had publicly attacked, and in which he had not been initially interested. As will be seen later in this chapter, the California Forestry Association similarly attacked the MOU and Executive Council, but the state and federal forestry directors participated anyway because environmental lawsuits were foreclosing their management options.

Discussion at Executive Council Meetings

While attendance is an important indicator of participation, it tells us nothing about what participants did or said at the meetings. Did agency directors participate actively in discussions? If so, did they introduce ideas, play the role of critic, or primarily sit back and watch the proceedings? This section analyzes discussions at Executive Council meetings. Again, the minutes serve as the primary source, providing approximately 5 to 10 pages of single-spaced text for each quarterly meeting. Unfortunately, they have several shortcomings, which limit textual analysis. First and foremost, they paraphrase discussions rather than providing direct quotes, which means that otherwise lively debates have been watered down.[7] The minutes are also not inclusive. Some comments are missing; others appear without attribution. Despite these shortcomings, the lengthy text provides ample evidence of speakers, discussion topics, and issue positions.

For example, the minutes provide evidence of the frequency with which directors spoke at the meetings, an additional indicator of participation. Table 4.3 presents statistics on the number of times the minutes attribute a comment to a director, and the average number of attributed comments per meeting attended.[8] The agencies are listed by their attendance rank in table 4.1, which demonstrates that directors who routinely attended meetings also tended to speak more frequently at the meetings they attended.

Wheeler stands out from the others with the largest number, and highest average number, of attributed comments. In part, these statistics can be explained by his formal role as Chair of the Executive Council, in which capacity he called the meetings to order, introduced guest speakers, and provided transitions between agenda items. Therefore, in addition to his personal interest in the discussions, his formal role required him to speak throughout the meetings. Though Wheeler had been in California for less than a year when the Executive Council first met, he quickly stepped into the formal leadership position for several reasons. As Secretary for Resources, he was the formal conduit between the Governor's Office and the numerous departments, boards, and commissions housed within the Resources Agency (see figure 3.2). While Wheeler exerted little line authority over the other political appointees who managed these agencies, his high-profile position in the Wilson administration enhanced the legitimacy and political clout of the Executive Council. Wheeler was also a dynamic public speaker. Ed Hastey was well respected within the BLM, but he lacked Wheeler's exuberant, hortatory style. Therefore, Wheeler was a more likely leader around whom others might rally, particularly within state agencies.

Wheeler and Hastey generally led discussions, while other directors mostly listened. The attendance record of the directors other than Wheeler and Hastey stands out rather than their propensity to

Table 4.3
Number of comments attributed to agency directors at Executive Council meetings (first 13 quarterly meetings, December 1991–March 1995)

Agency	Director	Number of comments (a)	Meetings attended (b)	Average (a/b)
California Resources Agency	Doug Wheeler	60	11	5.45
U.S. Bureau of Land Management (State Office)	Ed Hastey	21	9	2.33
California Department of Forestry and Fire Protection	Richard Wilson	9	10	0.9
California Department of Fish and Game	Peter Bontadelli	2	1	2.0
	Boyd Gibbons	5	6	0.83
U.S. National Park Service (Western Region)	Stanley Albright	3	6	0.5
California Department of Parks and Recreation	Henry Agonia	1	1	1.0
	Donald Murphy	1	5	0.2
U.S. Forest Service (Pacific Southwest Region)	Ron Stewart	4	3	1.33
	Lynn Sprague	0	3	0
U.S. Fish and Wildlife Service (Pacific Region)	Marvin Plenert	3	3	1.0
California State Lands Commission	Charles Warren	14	3	4.67
	Robert Hight	0	0	0
University of California (Division of Agriculture and Natural Resources)	Kenneth Farrell	0	0	0

Source: Compiled by the author from minutes of the quarterly meetings of the Executive Council on Biological Diversity.

participate actively in the proceedings. For example, NPS Regional Director Stanley Albright attended two of the first three meetings in Sacramento, despite being based in San Francisco. Yet the minutes attribute only two comments to him during this period, and only one comment during the next four meetings he attended. Albright's limited participation led one regular participant to comment that Albright "is barely awake when he's at those meetings."[9] Albright's state counterparts in the CDPR were similarly reticent. As discussed in chapters 1 and 3, the state and federal park agencies did not have a historical reputation for developing interagency relationships. Thus, it is not surprising that these directors observed rather than led Executive Council discussions.

The minutes attribute approximately one comment to each director other than Wheeler and Hastey for each meeting they attended. The notable exceptions in this regard were Charles Warren (CSLC), who spoke a great deal at the first three meetings, and Lynn Sprague (FS), who appears from the minutes to have been silent at the meetings he attended. Though silent, Sprague was nevertheless diligent, attending all three Executive Council meetings after replacing Ron Stewart as Regional Forester in 1994. Moreover, the absence of attributions to Sprague in the minutes does not seem substantively important because the Executive Council had become much larger by the time he attended, expanding from 10 to 24 signatories at the fourth meeting, which limited opportunities to speak.

Charles Warren, on the other hand, was loquacious during the Executive Council's first three meetings, then abruptly stopped attending. His intense, albeit brief, participation provides interesting insights into Executive Council dynamics, particularly regarding the very definition of biodiversity, the purpose of the Executive Council, and the role of partisan politics. Because Warren was at the center of some of the only conflicts within the Executive Council during its early years, these conflicts should be examined in depth.

The Outsider Who Raised a Ruckus

Charles Warren was the only original signatory who asked to sign the MOU on Biodiversity. Hastey did not initially ask him to sign it because Warren had not participated in CRMP, INACC, or TTF. Not only was he an outsider, the insiders did not initially believe they needed the CSLC for the Executive Council to be successful. Therefore, it is not surprising that the outsider raised issues that other Executive Council members preferred not to consider, and offered suggestions that were not well received.

Warren did not succeed, for example, in his attempt to expand the Executive Council's purview beyond terrestrial issues, like timber harvest practices, to encompass aquatic concerns, like water quality. At the last meeting he attended, on August 5, 1992, Warren suggested inviting three agencies to join the Executive Council—the EPA, which had responsibilities for water quality under the federal Clean Water Act; the State Water Resources Control Board (SWRCB), which also regulated water quality; and the National Marine Fisheries Service (NMFS), which had responsibilities under the ESA for marine species, including salmon. Hastey, who then chaired a subcommittee considering whether and which additional agencies should be invited to sign the MOU on Biodiversity, did not want to extend membership to these agencies, both because they were regulatory and because the BLM primarily faced terrestrial issues. Accordingly, Hastey countered Warren's suggestion by arguing that membership decisions should be based on past activity in coordinated planning processes (e.g., CRMP and INACC), that the Council's size was a limiting factor in inviting new members, and that water-quality agencies would still be able to participate in bioregional activities without signing the state-level MOU. Warren rebutted that the EPA and SWRCB, in particular, played important roles in achieving biodiversity objectives, and that adding two more agencies would not make a big difference in the size of the Executive Council. Warren's suggestion received some support from other members, but Hastey won the battle, at least in the short run.[10]

Warren's concern for aquatic diversity stemmed from his agency's mission and landholdings. The CSLC served as trustee for more than four million acres of state-owned lands, which were of two types: school lands and sovereign lands. The school lands were granted to the state by the federal government to support public education. Of the original five million acres of school lands in California (sections 16 and 36 in every township), most had been sold or traded. The remaining school lands (almost 600,000 acres) lay scattered across the state, many enclosed within BLM and FS holdings.[11] The sovereign lands consisted of the bottoms of navigable lakes and rivers, and all tidelands and submerged lands from mean high-tide line to three miles offshore. Because the CSLC owned most of California's submerged lands, it also had some responsibility for managing the water above them. This included Lake Tahoe, Mono Lake, San Francisco Bay, and all of the navigable river channels in the state, including the politically controversial delta at the confluence of the Sacramento and San Joaquin Rivers in the Central Valley. Because most of the agency's land was underwater, many of its environmental concerns were aquatic rather than terrestrial.[12]

Given that much of California's biodiversity was aquatic or riparian (i.e., living on the banks of rivers, lakes, or other bodies of water), ignoring aquatic concerns made little sense if Executive Council members were sincere about protecting biodiversity rather than simply terrestrial diversity. Moreover, aquatic habitats were "the most dramatically and completely altered biotic communities in California" (Jensen, Torn, and Harte 1993: 78). Yet the MOU on Biodiversity emerged from the terrestrial concerns of CRMP, INACC, and TTF. Having largely created the Executive Council, Hastey sought to keep the group focused on BLM concerns—that is, managing habitat for endangered species on and around the BLM's highly fragmented lands, most of which represented the driest lands in the state. Because CSLC school lands were intermixed with BLM lands, particularly in the Mojave and Colorado Deserts, it made sense for Hastey to include the agency as one of the original ten signatories when Warren asked to be included. Warren was a wild card, however, because he had not participated in the TTF, CRMP, or INACC. Moreover, Warren was especially risky—if anyone contemplated such dynamics—because the CSLC had been taking an increasingly assertive stance in the 1980s toward the protection of aquatic habitat, an issue that had major implications for the allocation of freshwater, California's most divisive natural resource.

Warren's skirmish with Hastey over aquatic diversity and water-quality agencies was not, however, the only factor prompting his departure. Partisan politics also played a significant role. Warren was a staunch Democrat with a long background in the Democratic Party, where he had been active on environmental issues.[13] Republican Governor Pete Wilson appointed Wheeler and three additional members of the

Executive Council—Richard Wilson (CDF), Boyd Gibbons (CDFG), and Donald Murphy (CDPR). The CSLC, like these other agencies, resided nominally within the Resources Agency, but the director reported to a three-member commission, not to Wheeler, and Democrats represented the majority on this commission. With Governor Wilson's sole appointee on the commission outnumbered, Warren did not report to the Governor, either through the commissioners or through the Secretary for Resources. Unable to control the CSLC through the appointment process, Governor Wilson did so instead by slashing the agency's budget in half.[14] This infuriated Warren, who subsequently became very hostile toward the Wilson administration, including Wheeler. Therefore, regardless of his earlier confrontation with Hastey over aquatic diversity and water-quality agencies, Warren's participation on the Executive Council was soured by his political relationship with the Wilson administration.

In fact, Warren later praised Hastey's intentions regarding the MOU on Biodiversity, despite the aquatic diversity issue, but he came to believe that the Wilson administration was using the Executive Council to "subvert" the ESA by acting as if the agencies were working together to protect species while in fact doing very little:[15]

> It's an announced intention of the Wilson administration to undo the Endangered Species Act. That's a specific policy of the Governor. . . . *Biodiversity* is a word which has portent for desirable results, in my opinion, for wildlife protection, and I think that is a worthwhile approach, but I think the whole effort has been co-opted and nothing is being done with it. You see, it's all right to use biodiversity as a substitute for the Endangered Species Act *if* you intend to do something with biodiversity. But what they've done, they've created the Council, and given it a name, and it has done absolutely nothing, but in the meantime they cite its existence as an alternative to the Endangered Species Act. So, if they're successful in getting the Endangered Species Act either repealed or weakened, then there will be nothing on the books to protect species which are endangered. And I think that's the goal.

Warren summed up his opinion of the Wilson administration's intentions with an off-color simile: "*Biodiversity* in their mouths is like *love* in the mouth of a whore, to use an old phrase. They're not serious . . . and I think it's intentional, so I'm very upset about it. They're giving biodiversity a bad name."

On the one hand, it would be easy to write off Warren's accusations as the rhetorical flourish of a former elected official who, as an agency director, had seen his budget cut in half by the opposing party. On the other hand, the charges are not without merit. Governor Wilson and some of his appointees had indeed proposed major changes to both the state and federal ESAs, and Wilson later made this a major theme of his campaign for the presidency in 1995.[16] Moreover, as will be seen in subsequent chapters, Warren was not the only person who believed that some members of the Wilson administration were trying to subvert the state and federal ESAs. Others also thought members of the administration were pretending to protect biodiversity through the Executive Council while calling for major legislative revisions to weaken the acts and citing the Executive Council's existence and activities as effective substitutes for the ESAs.

I will return to this subversion hypothesis in subsequent chapters, but readers should note an important nuance: Warren argued that the Wilson administration was primarily seeking new legislation, not changes in the enforcement practices of state and federal agencies. Since environmental activists had largely prodded agencies to enforce the federal ESA through lawsuits, they were not about to stand by idly simply because the MOU and Executive Council existed. According to Joan Reiss, who closely followed the Executive Council's activities as Regional Director of the Wilderness Society in California from 1990 to 1993, "You can't subvert the Endangered Species Act. It's the law. And when it's not obeyed the environmentalists will sue again."[17] It is possible, however, to change the laws in ways that decrease the efficacy of lawsuits. Accordingly, rather than trying to convince environmentalists that they no longer needed to sue to prompt enforcement, the Wilson administration—so the story goes—was trying to convince legislators and the general public that certain provisions in both ESAs were no longer needed because biodiversity would be protected through activities initiated or sponsored by the Executive Council.

Warren emphasized his dissatisfaction with the Executive Council by designating a professional staff member to sit at the table as his alternate, rather than sending his associate director or other line manager. In doing so, Warren snubbed the Wilson administration because the directors and associate directors of most other agencies were attending the meetings. Warren's alternate was not, however, new to this interagency milieu. Diana Jacobs, Environmental Specialist for the CSLC, had been a regular participant in INACC.

Staff Participation

Professional staff played an important role in the Executive Council. They assisted the directors, spoke before the Executive Council, sometimes sat as designated alternates, and regularly held their own series of low-profile meetings to plan projects and shape the Executive Council's agenda. Most had academic and professional backgrounds rooted in the ecological sciences. Some were line managers or their liaisons. Unlike staff ecologists, the latter had different reasons for believing their agencies—the BLM and FS, in particular—could no longer operate independently. They hoped to ward off future lawsuits against their agencies by maintaining viable populations of species through coordinated habitat planning and management.

The first staff meetings were basically a continuation of the TTF Ad Hoc Committee. Having developed the MOU on Biodiversity, Ad Hoc Committee members continued working together to bring the MOU's principles to fruition. Now called the Staff Committee, they carried out much of the Executive Council's work behind the scenes, including developing and implementing specific projects. The Staff Committee met frequently, usually holding two or three meetings prior to each Executive Council meeting, for which they reviewed agenda items, speaker lists, and their own presentations. For regular participants, this meant about one interagency meeting every month, in addition to related projects on which they might be working and the daily demands of their jobs.

While most of the original ten agencies were well represented at Staff Committee meetings, two agencies were not: the NPS and California Resources Agency. No one from the NPS attended Staff Committee meetings during the first year. No one from the NPS even appeared on the Staff Committee's contact list until 1994, three years after the MOU on Biodiversity was signed. Although NPS staff had attended INACC meetings, their enthusiasm did not carry over to the Staff Committee. This suggests that NPS staff had some interest in interagency planning and management issues regarding specific parcels of reserved land, but did not view statewide bioregional planning as a priority.[18]

The California Resources Agency was also not well represented at Staff Committee meetings. Yet this was not due to lack of interest. While the agency appears large on an organizational chart (see figure 3.2), Secretary Wheeler had a relatively small staff, and they were preoccupied with other interagency programs. Rather than participating directly, Wheeler and his staff orchestrated the Staff Committee's activities from a distance by placing implicit and explicit bounds on discussion topics.[19] Typically, these bounds were communicated to the Staff Committee through its cochairs, Robert Ewing (CDF) and Susan Cochrane (CDFG).

Ewing and Cochrane managed two analytic units within the Resources Agency superstructure, placing them in similar positions with respect to disseminating ecological information and knowledge to local, state, and federal agencies. Ewing's Forest and Rangeland Resources Assessment Program played a central role in staffing the TTF. Having also drafted the MOU on Biodiversity, Ewing was an obvious candidate to provide staff leadership under the new Executive Council. Cochrane managed CDFG's Natural Heritage Division, which housed Hoshovsky's Significant Natural Areas Program and the Natural Diversity Data Base, among other programs. Although Cochrane played a smaller role in the TTF and Ad Hoc Committee than Ewing, the directors appointed her cochair of the Staff Committee to ward off lingering conflicts between the CDF and CDFG—conflicts that provided the original impetus behind the TTF. Moreover, the mission of CDFG's Natural Heritage Division was more germane to biodiversity protection, thus giving Cochrane cause for concern that Ewing and FRRAP might be encroaching on her division's turf.[20] Ironically, both units were marginalized within their respective agencies, which were oriented toward the production of commodity species (i.e., timber and game) rather than biodiversity. From a strategic standpoint, ecologists in both units stood to gain from working cooperatively on data collection, analysis, and dissemination. Ewing and Cochrane were allies in this professional sense, but nevertheless struggled over bureaucratic turf.

Unlike Executive Council meetings, Staff Committee meetings were not announced to the public. Only attentive observers knew of the Staff Committee's existence and attended its meetings. This effectively limited public participation to a few timber and environmental advocates. The California Forestry Association (CFA) represented the timber industry at both Executive Council and Staff Committee meetings. Unfortunately, CFA Executive Director Bill Dennison, who routinely disseminated scathing critiques of the MOU on Biodiversity and the Executive Council, declined to be interviewed for this project, as did his successor, Gilbert Murray. (In a strange twist, Murray died in 1995 when he opened a letter bomb addressed to Dennison from Unabomber Theodore Kaczynski.)

Environmental advocates on the Staff Committee, by contrast, were willing to be interviewed. Joan Reiss, Regional Director of the Wilderness Society, and John Hopkins, biodiversity coordinator for the Sierra Club, regularly attended Staff Committee meetings. Hopkins believed the staff often decided what the Executive Council should do, with much of the time in staff meetings devoted to planning the next Executive Council meeting. Yet Hopkins also believed staff knew that certain topics were not politically feasible; they would not touch, for example, "anything that sounds top-down."[21] This included developing a statewide strategy to conserve biodiversity, and consistent standards and guidelines, as stated in the MOU on Biodiversity. Reiss believed the Staff Committee's influence was more attenuated because agenda items suggested by staff were routinely rebuffed by the Resources Agency.[22] Reiss believed that certain topics were politically palatable to Wheeler and his staff, but not to the Executive Council as a whole. Contentious issues, in general, were avoided, including the very definition of *biodiversity*. As Hopkins put it, "The political mantra is bottom-up, and let local folks decide what they want to do."

Local Governments and Stakeholders Demand a Greater Role

The MOU on Biodiversity envisioned a decentralized role for local governments and stakeholders. Rather than overlooking local issues and interests, the MOU on Biodiversity stated that "local associations are to be a primary forum for the resolution of local issues and conflicts related to biodiversity concerns." Yet this statement appeared in the penultimate sentence of the three-page document, allowing local governments and stakeholders the opportunity to read into the preceding text a grand strategy by which state and federal agencies intended to usurp local prerogatives by promulgating plans, standards, and guidelines from the top down. Some stakeholders became very upset by this prospect, particularly because the MOU implied that extractive uses of the land would be sacrificed to protect biodiversity. After all, the MOU stated, "The signatory parties agree to make the maintenance and enhancement of biological diversity a preeminent goal in their protection and management policies." Logically, the existence of a preeminent goal implied that other goals would be subordinate. Yet these trade-offs were already occurring in a haphazard way through court enforcement of the ESA and other environmental laws. The MOU did not itself carry the legal authority, administrative will, or political legitimacy to reorient agency missions.

When agency directors and staff held a press conference in September 1991 to announce they had forged a new interagency agreement—the MOU on Biodiversity—that would benefit both species and the economy, they issued a press release, which emphasized that the MOU and the coordinated strategy proposed therein were consistent with Governor Wilson's "preventive government" philosophy and environmental agenda.[23] In the press release, Resources Secretary Doug Wheeler stated,

> Rather than focusing protection efforts on specific species in specific sites, we plan to identify for conservation whole biological and geographical regions. . . . We seek to protect, in a coordinated fashion, all of an area's resources—endangered species, critical habitat, fish and wildlife, and water quality. By doing this we can save more of the natural environment and do so in a manner that is socially and economically viable.

Notably, the press release also added the following statement: "Crucial to this cooperative approach will be the active participation of local government, private industry, and environmental groups."

The press release and press conference generated favorable coverage.[24] Yet, by proudly going public with their efforts to coordinate resource management practices in California, agency officials inadvertently made themselves a conspicuous target in the continuing rural backlash against state and federal control. County supervisors, commodity-based interest groups, and other local stakeholders barraged the agencies with their concerns. Supervisors from rural counties feared further erosion of their limited authority; landowners worried about public control of private property; families fretted about the possibility of mill closures and job losses; and lobbyists for the timber and agricultural industries fanned the flames with speeches and press releases denigrating state and federal bureaucrats for destroying jobs and undermining family values. These stakeholders pressured state legislators to restrain the agencies. Wheeler took the brunt of this criticism as Chair of the Executive Council, even though he had not himself developed the MOU on Biodiversity or gathered the signatures on it. In subsequent speeches recalling the backlash, Wheeler routinely induced knowing laughter from audiences with this quip: "The only thing the public fears more than an uncoordinated bureaucracy is a coordinated bureaucracy."[25]

The reaction of local governments and stakeholders was not directed solely at the MOU and Executive Council. During the same period, Wheeler also orchestrated a high-profile interagency event called the Sierra Summit, which focused on the Sierra Nevada, a mountain range stretching over 400 miles north-south along California's eastern flank. The Sierra Summit was not directly tied to the MOU or the activities of the Executive Council, but it had similar objectives and was viewed by agency officials as a parallel effort. Held near Lake Tahoe on November 21, 1991, just two months after the MOU was signed, the Sierra Summit was initially planned as a one-shot event. Organizers did not intend for it to evolve into a bioregional council, but to many observers, particularly those who did not receive invitations to the event, it appeared that state and federal agencies might be developing a regional plan for the Sierra Nevada. Environmentalists had long dreamt of establishing a new "Range of Light" national park in the Sierra, so local stakeholders were concerned that Sierra Summit participants might discuss land transfers to the NPS or closing national forests to logging, grazing, and other extractive uses.

The ideas behind the Sierra Summit were similar to, and compatible with, the MOU on Biodiversity. Yet different individuals developed the two efforts. The Sierra Summit was not an outgrowth of CRMP, INACC, or TTF. The immediate impetus for the Sierra Summit was a five-day series of articles by Tom Knudson in the *Sacramento Bee*, which appeared in June 1991, three months before the MOU was signed. Titled "Majesty and Tragedy: The Sierra in Peril," Knudson's articles had a powerful two-pronged effect. First, they informed the public that John

Muir's fabled "Range of Light" suffered from several problems, unbeknown to millions of casual viewers who assumed the Sierra's outward appearance indicated ecological health. Second, Knudson characterized the region as administratively fragmented and asserted that no government agency was taking the lead to address these problems. In the last article of his five-part series, Knudson (1991) merged these two themes into a single claim: "The Sierra Nevada, one of the world's great mountain ranges, is slowly dying—and government has done little to help it."[26]

Wheeler used the "Sierra in Peril" series as a springboard, announcing soon after that a meeting would be held to discuss the issues raised in Knudson's articles. At that time, Wheeler also chaired the TTF, to which he announced his intention, and received unanimous approval, to hold the Sierra Summit. This was simply a courtesy, however, because no formal relationship existed between the Sierra Summit and TTF. Nevertheless, participation overlapped between the two processes, and because both efforts were grounded in interagency cooperation and had compatible resource goals, it made sense for the leaders to keep each other informed. Yet participants were later surprised by the degree to which the two processes became politically intertwined in the public's eye.

Because of the backlash to the Sierra Summit by local stakeholders, the Resources Agency held five follow-up workshops in March and April of 1992, during which local stakeholders vented their concerns.[27] About this time, Hastey also began courting county supervisors throughout the state to dispel their concerns about the Executive Council, and to gain their assistance in implementing the MOU. Both follow-up efforts were intended to soften local opposition to interagency activities by co-opting local stakeholders. In both cases, state and federal officials lost a great deal of momentum because, having neglected to issue open invitations to local stakeholders at the outset, they found themselves backpedaling for more than a year to assuage local concerns.

Many local stakeholders were only interested in the Sierra Summit and Executive Council because they did not receive invitations to participate, and wondered whether the existence of interagency cooperation implied a conspiracy at their expense. Far from conspiring to set policies in the absence of public input, Wheeler publicly announced his intention to hold the Sierra Summit. Agency officials were simply caught off guard by the amount of interest in their activities and were unable to accommodate everyone because physical space was limited at the selected facility. The organizers also planned a working meeting rather than an open forum, so they issued invitations to selected individuals. Thus, for example, they mailed announcements to the chair of each county's board of supervisors, not to each supervisor. Although some supervisors were later upset because their colleagues failed to notify them of the event, most were not themselves a major source of the local backlash. As one state official later recalled, "Local governments were suspicious but not hostile, except for a few supervisors who clearly had hostile constituents. . . . In general, they were receptive, curious, and very suspicious."[28] Environmentalists also played a low-key role at the Sierra Summit. Selected environmental representatives received invitations, and a few sat on the steering committee, but the environmental community largely focused on its own parallel conference called Sierra Now.

It was the timber lobbyists—not local government officials or environmentalists—who largely inflamed local passions over the Sierra Summit, the MOU on Biodiversity, and the Executive Council. The California Forestry Association (CFA), in particular, played on the fears of local stakeholders by issuing brash proclamations claiming that state and federal agencies were bent on ruining local communities and families. Forest-product workers were indeed losing their jobs and had cause for concern about their economic future, but this was largely due to environmental lawsuits over timber harvest practices, not to the agencies themselves. The CFA sought to dispel the notion that Sierran timberlands suffered from poor ecological health. One CFA representative called Knudson's articles "an inept, misguided tour through preservationist fantasyland."[29] Yet Knudson's credibility only grew because he won the Pulitzer Prize gold medal for public service in 1992 for his "Sierra in Peril" series. CFA officials also tried to undercut the ecological concepts that academics and agency officials increasingly used. As CFA Executive Director Bill Dennison stated, "We need to work toward change. But, for anyone to impose new regulations because of buzz words such as 'biodiversity' and 'cumulative effects' and 'ancient forests,' just for the sake of that, would be wrong. I think the politics are out ahead of science right now."[30] Later, when the Sierra Summit Steering Committee released its final report in July 1992, Dennison claimed that families in timber communities "see this as just another step towards their degradation."[31] Dennison once even introduced Robert Ewing (CDF) and fellow panelists at a timber conference as "promoters of an agreement that was anti-family, anti-Christian, and was an attempt to overthrow the federal government."[32]

In the state legislature, timber industry allies similarly chiseled away at the credibility of the MOU, Executive Council, and Sierra Summit. Republican State Senator Tim Leslie warned his constituents that "a new layer of bioregional government" might soon be created, one not answerable to the public.[33] He also admonished agency officials at the Executive Council's second meeting not to establish statewide standards in the absence of legislative oversight and public input; data sharing would be a useful contribution, Leslie argued, but not top-down regional planning.[34] Republican Assemblyman David Knowles called the Sierra Summit an "exercise in academic snobbery."[35]

While the Sierra Summit generated much publicity, the Executive Council maintained a lower profile because its goals and activities were more amorphous and diffuse. Yet, because the two processes overlapped conceptually, the Sierra Summit drew critical attention to the MOU and Executive Council. As one Summit organizer summed up the situation,[36]

> We were dogged *unflaggingly, unceasingly* by demagogues from all over. . . . The memories are actually comic because of the way that they acted. Legislators like David Knowles or Tim Leslie, who has apparently moderated his position somewhat since then; Bill Dennison, just about the [long pause], well, I think one of these [pause], well, anyway I'm not going to continue. These types of people decided that this was a power grab; and they, I think more importantly than that, decided that even if it weren't a power grab it made great political hay for them—with their constituents, or their bosses in the case of Bill Dennison—to accuse us of trying to, both with the Summit process and with the Biodiversity MOU, trying to usurp power from local people and to just take over in the name of biodiversity.

Criticism of the Executive Council reached a crescendo at its second meeting on February 28, 1992.[37] The Council subsequently skipped its next meeting, putting a single hole in an otherwise diligent record of quarterly meetings from 1991 to 2001. The backlash from local stakeholders, their state representatives, and the timber industry seriously impeded the Executive Council from accomplishing much during its first years of operation. Moreover, the signatories backpedaled on several central points in the MOU, including the development of consistent statewide goals, standards, and guidelines for the protection of biodiversity.

Because many local stakeholders felt genuinely disenfranchised from the process, it became apparent to some agency staff and directors—particularly Ed Hastey—that the Executive Council would have to bring local government officials to the table before it could effectively begin the coordinating role for which it had been created. The MOU drafters initially envisioned local governments participating at the local and bioregional levels. When the backlash occurred, Hastey came to believe that they should also be represented on the statewide Executive Council itself, and that county supervisors (not county staff) would be the most appropriate representatives. Rural counties were particularly important because they harbored the vast majority of California's undeveloped habitat—not to mention BLM land.

County supervisors were concerned that the MOU on Biodiversity would lead to additional regulation and usurpation of local control by new regional entities. With limited ability to raise revenues, local governments strongly resisted unfunded mandates. Yet Hastey never intended the MOU to be a regulatory mechanism. As a staunch advocate of CRMP, Hastey viewed multijurisdiction planning from the bottom up. He envisioned the Executive Council guiding and encouraging agency staff to participate in CRMP-like efforts to manage large tracts of habitat, rather than acting as a regulatory overlord. For Hastey, the most important language in the MOU had to do with coordination, not biodiversity. Instead of developing new standards and guidelines to protect biodiversity, he sought to make the existing ones consistent.[38] Because Hastey was an ally of local governments and resource users, and was widely acknowledged to be the impetus behind—even the godfather of—the MOU on Biodiversity, it was ironic that local stakeholders initially attacked the MOU and Executive Council.

Hastey personally accepted the blame for not bringing county supervisors into the process sooner: "That was a mistake I think on my part, of not getting the counties cranked in early to it. You know, we just went ahead and set this up with the idea that we'd bring in counties, but never really including them on the draft of the MOU."[39] As head of the

Executive Council's subcommittee on membership, he decided that county supervisors should now be brought on board. Having already gathered the directors' signatures on the MOU, he similarly called county supervisors around the state to sell them on the idea.

County Supervisors Sign a Statement of Intent to Support the MOU on Biodiversity

With 58 counties in California, each containing 5 supervisors, a mechanism was needed to select representatives to sit on the Executive Council. Fortunately, the supervisors had already organized themselves into regional associations within the California State Association of Counties, and these regional associations bore some resemblance to the INACC bioregions adopted by the Executive Council. By selecting one supervisor from each regional association of county supervisors, the Executive Council could avoid the appearance of favoring one region over another, and the size of the Executive Council would not swell to unmanageable proportions.[40]

Yet many supervisors objected to the language in the MOU and wanted to revise it before signing. Hastey rejected this idea because, as he put it, "we went through enough hell getting that thing signed" in the first place. Therefore, rather than signing or revising the MOU, county supervisors signed a much shorter document: "Statement of Intent to Support the Agreement on Biological Diversity" (see appendix C), which was drafted in the BLM State Office. By signing this document, the supervisors appeared to support some of the MOU's principles, and they became official members of the Executive Council.

Once on the Executive Council, county supervisors differed from the signatories of the MOU in another important respect. Agency directors had line authority to pressure subordinates in field offices around the state to support interagency processes at the local and bioregional levels. The supervisors, by comparison, had no authority over the regional associations they represented. They provided an important conduit for information between the local and state levels, but they did not necessarily speak for colleagues within their own counties, let alone other counties within their regional associations. This distinction is important because counties had authority over land-use decisions on private property in unincorporated areas, and rural counties, in particular, contained significant tracts of relatively undisturbed habitat.

While the Executive Council would have little direct effect on county decisions through the supervisors, local government staff increasingly used state and federal databases to make decisions regarding which development projects should be approved, where they should be located, and what mitigation should be required. State and federal officials accordingly sought to centralize this data and make data sets more consistent, to encourage local government staff to use the information. Therefore, it was important that county supervisors endorse the Executive Council's activities by signing the Statement of Intent, if not the MOU on Biodiversity.

Initially, representatives of eight regional associations of county supervisors signed the Statement of Intent in Fall 1992.[41] One year later, the number of local government representatives grew to nine when the Southern California Regional Association of County Supervisors folded and was replaced on the Executive Council by two much larger councils of government (COGs): the Southern California Association of Governments (SCAG) and the San Diego Association of Governments (SANDAG). These COGs were both located in the South Coast Bioregion, where land-use decisions were made primarily by cities rather than counties. The official number of local government representatives on the Executive Council now nearly equaled the ten original agencies, though the effect on the Council's activities was by no means proportional. Participation by most of these representatives was modest. Soon after signing the Statement of Intent, six of eight supervisors attended the Executive Council meeting in December 1992. But attendance immediately plummeted to three supervisors at the next two meetings, and hovered around 50 percent for the next three years. Rather than seeking to participate, most county supervisors simply wanted to know that nothing untoward was happening at the meetings. Accordingly, they attended sporadically, watching the proceedings rather than actively participating in them.

Yet two supervisors attended regularly and strongly supported the Executive Council's activities during these early years. Laurence "Bud" Laurent, a San Luis Obispo County Supervisor representing the South Central Coast Regional Association of County Supervisors, had ecological credentials similar to many participants on the Staff Committee, having previously served for 20 years as a CDFG marine biologist. He was also well respected in the environmental community. The Sierra Club's John Hopkins, for example, called Laurent "one of the rare county supervisors who is not afraid to use the term bioregional planning."[42] Laurent had a perfect attendance record (10 of 10) at Executive Council meetings through March 1995—better than Hastey and Wheeler during the same period, though he had to travel further to accomplish this feat. Laurent also produced a video distributed by the Executive Council, which demonstrated the utility of geographic information systems (GIS) for local government planning.

Art Baggett, a Mariposa County Supervisor representing the Regional Council of Rural County Supervisors, attended seven of nine Executive Council meetings. Baggett had attended the Sierra Summit and hosted one of the Summit's five follow-up workshops. Throughout the Summit process, he pushed the state and federal agencies to involve the counties in bioregional activities, and subsequently acknowledged Hastey's pivotal role in bringing county supervisors into the Executive Council over the objections of some wildlife officials who distrusted local government. Baggett similarly lobbied his fellow supervisors to participate, particularly his more conservative colleagues, with whom he recalled arguing that "we can either be sitting at the table or we can be out in the cold—this stuff's happening, the Endangered Species Act is not gonna go away."[43]

Far from the stereotype of county supervisors who welcome development, logging, and other extractive industries, Baggett and Laurent sought to preserve rural areas for their natural values, including biodiversity. Yet their enthusiasm was not indicative of all 290 supervisors in California, or all supervisors on the Executive Council.[44] Laurent did not believe that he and Baggett represented a growing trend among county supervisors. Moreover, city council members in Southern California did not jump at the opportunity to represent SCAG and SANDAG on the Executive Council. The first two representatives from these COGs volunteered because no one else did; they were neither strong advocates of bioregional planning nor of biodiversity.[45]

Questions therefore lingered as to who had co-opted whom. Some agency staff wondered if the agencies should pressure local governments to do more. In this regard, one staff ecologist even wondered if the MOU had lost whatever teeth it had when the local government associations joined the Executive Council:[46]

> The thing that's really amazing to me is, when the counties were signed on, the hammering we'd get from the legislative offices almost just stopped. At earlier meetings, the legislators, and friends of loggers, and the wives of loggers, or somebody was *always* there and really just being disruptive. Now I don't know if that's a good thing or not, because it may be that they realize there's no threat to their way of life, or at least a perceived problem. It may be in that case it's a dud. If something isn't causing friction politically, then maybe it isn't doing anything.

It is possible that agency directors intended to do more but were beaten back. It is also possible that they intended to do little from the outset, and were simply misunderstood. These interpretations are too general, however, to capture important differences within and among the agencies. As will become apparent throughout the case studies, agency staff generally pushed for more than line managers, but variation also existed among the agencies.

More Agencies Sign the MOU on Biodiversity

Local government associations were not the only new members of the Executive Council during its first year in operation. In August 1992, six additional agency directors, representing three state agencies and three federal regional offices, signed the MOU on Biodiversity. The following month, the director of the California Association of Resource Conservation Districts (CARCD), an association of local governments, also signed the MOU. Unlike the county supervisors, who signed a separate agreement, the new signatories signed the MOU itself, and were already signatories to CRMP. With their addition, the Executive Council included all of the CRMP, INACC, and TTF agencies, with one minor exception (see table 3.1).[47] Hastey had thus maneuvered the Executive Council firmly onto terrestrial terrain, rather than branching out into aquatic diversity as Warren desired. The new agencies nevertheless broadened the Executive Council's horizons because they were tied largely to agriculture rather than timber, which positioned the Executive Council to expand into agricultural issues.

As with the first ten agencies, the new signatories varied in their commitment to the Executive Council and the principles in the MOU on Biodiversity. This can be seen in table 4.4, which uses the same methodology developed in table 4.1 for ranking agency participation based on attendance. The attendance scores in tables 4.4 and 4.1 cannot be directly compared, however, because the new signatories attended a maximum of ten meetings rather than thirteen. Nevertheless, it is clear that the new directors attended far fewer meetings than the original cohort. Four of the six new signatories never attended a single meeting, even though they occupied their offices during the entire period.

The new signatories were all based in Sacramento, where more than half of the meetings were convened; their poor attendance thus indicates they saw little or no utility in participating. Several explanations suggest why this might have been so. First, the Executive Council grew out of the Timberland Task Force. Not only did the Executive Council focus on timberland issues during its early years, agricultural issues ranked low on the list of announced agenda items for upcoming meetings until 1995. Therefore, the new signatories likely did not believe the Executive Council was addressing issues relevant to their concerns. Second, Hastey extended invitations to these agencies because they were already members of CRMP. Yet most of these directors never attended the annual meetings of the CRMP Executive Council, so there was little reason to expect they would start attending quarterly meetings to discuss CRMP-like activities related to nonagricultural issues. The SCS was an exception. SCS field staff worked in both timberland and agricultural counties, and participated routinely in local CRMPs. At the state level, the SCS also alternated leadership with the BLM on the CRMP Executive Council.

Because the other agencies were primarily associated with agriculture, their directors were more susceptible to lobbying from the agricultural industry. As previously noted, the California Farm Bureau opposed the MOU on Biodiversity, bioregional planning, and regulatory attempts to protect biodiversity on private lands. The Farm Bureau's president even issued a press release to this effect in August 1992—the same month these additional directors signed the MOU on Biodiversity.[48] It is no coincidence that the only director from the original ten agencies who did

Table 4.4
Attendance of new signatories at Executive Council meetings, by highest official (10 quarterly meetings, December 1992–March 1995)

Agency	Location of headquarters	Director [× 3]	Associate director [× 2]	Other official [× 1]	No one present [× 0]	Attendance score
U.S. Soil Conservation Service	Sacramento (Davis)	5 (50%)		5 (50%)		20
California Department of Water Resources	Sacramento		5 (50%)	5 (50%)		15
California Department of Conservation	Sacramento	1 (10%)		8 (80%)	1 (10%)	11
California Department of Food and Agriculture	Sacramento			10 (100%)		10
U.S. Bureau of Reclamation	Sacramento			9 (90%)	1 (10%)	9
U.S. Agricultural Stabilization and Conservation Service	Sacramento			4 (40%)	6 (60%)	4

Source: Compiled by the author from minutes of the quarterly meetings of the Executive Council on Biological Diversity. (SCS state headquarters were in Davis, about 15 miles west of Sacramento; in late 1994, the agency was renamed the Natural Resources Conservation Service.)

not attend a single meeting of the Executive Council was also closely aligned with the agricultural industry (see table 4.1).

Poor attendance, however, does not necessarily imply disinterest within these agencies. As table 4.4 indicates, designated alternates routinely attended Executive Council meetings. The California Department of Water Resources (CDWR) is particularly notable in this regard because the associate director attended five of ten meetings. Yet he attended primarily to ensure that CDWR's prerogatives were not being challenged by other agencies. While the agency's attendance score appears relatively high in table 4.4, its reputation for cooperation was very low. Several participants identified CDWR as a relatively independent agency, whose staff and line managers were generally uncooperative, even when legal mandates required coordinated action.[49] CDWR's independent posture was so apparent that even agency directors and associate directors were not reticent in expressing their opinions about the agency in this regard. Doug Wheeler, who as Secretary for Resources was nominally the conduit between the Governor and departments housed within the Resources Agency, including CDWR, noted that it was largely independent of his control because the agency had both a strong constituency and a large budget.[50]

The CDWR managed the State Water Project, which conveyed fresh water to agricultural operations in the Central Valley through more than 700 miles of aqueducts from reservoirs behind 27 dams. Like its federal counterpart, the BoR, which managed the Central Valley Project, the agency had developed strong ties with California's multibillion-dollar agricultural industry and poor track records on environmental issues. Their professional ranks were also dominated by engineers rather than ecologists, which meant they were equipped to build and maintain physical structures, like dams and canals, not to analyze ecological problems that crossed agency jurisdictions. For these reasons, it was unlikely that these agencies would adjust water flows voluntarily to protect aquatic or terrestrial diversity. Their independent posture slowly changed in the 1990s, however, as environmental lawsuits compelled them to protect aquatic diversity by keeping more water in rivers. Yet, until the Executive Council began addressing aquatic diversity and agricultural issues in the late 1990s, there was little reason for line managers in the CDWR and BoR to participate.

The other four agencies differed in that they primarily delivered technical and financial assistance to agricultural interests rather than water. The SCS had long been associated with stemming soil erosion on agricultural lands due to its roots in the Dust Bowl, but SCS conservationists increasingly broadened their concerns to encompass a diverse array of natural resource issues. In the 1990s, SCS line managers and staff began using terms like *Total Resource Management* to characterize their strategic objectives, and, in 1994, the agency received a new name, the Natural Resources Conservation Service, to emphasize its broadening mission.

The Agricultural Stabilization and Conservation Service was also housed in the U.S. Department of Agriculture, with the SCS and FS, where it administered commodity stabilization programs. It also ran programs rewarding farmers for conservation practices, but these programs were relatively small compared to the commodity stabilization programs. Therefore, the agency's line managers and staff demonstrated little interest in the Executive Council's agenda.

The two remaining state agencies similarly lacked the SCS enthusiasm for CRMP, but differed from one another in size, power, and mission. On one end of the spectrum stood the California Department of Food and Agriculture, which regulated the state's gigantic agricultural industry, with annual revenues approaching $20 billion. Its director, Henry Voss, was past president of the California Farm Bureau. Thus, he was an unlikely participant in the Executive Council. At the other end stood the California Department of Conservation, a small agency encompassing several loosely related resource conservation programs. Although similar to the SCS in mission and tasks, the two agencies did not have close ties. Moreover, SCS staff were spread out in field offices across the state, while Department of Conservation staff were largely in Sacramento, which would limit their participation in local and bioregional efforts.

The California Association of Resource Conservation Districts (CARCD) also became a member of the Executive Council at this time. The CARCD is not listed in table 4.4 because it is not a public agency. CARCD is a nonprofit association representing more than 100 resource conservation districts (RCDs) throughout California.[51] RCDs are special districts that administer conservation programs. The CARCD serves as their lobbying arm in Sacramento, and provides support services, such as helping RCDs locate grants for local projects. The most important feature of RCDs is their local character. RCD directors are locally elected volunteers supported on-site by SCS field staff, with whom they often share office space. Like field staff in the SCS and Cooperative Extension (UC/DANR), RCD directors develop strong social ties in the communities within which they live and work, so they tend to be trusted within those communities. According to CDFG ecologist Marc Hoshovsky,

> [RCD] and Cooperative Extension folks have been working with private landowners and private organizations for a long time, as well as with agencies. They've been this perfect liaison between both worlds. And they have a lot of trust from different people. And they've been working on on-the-ground projects. Maybe a lot of them are sort of related to livestock or forestry or something of that sort, but they've got that working relationship which is gonna be real critical to make it happen at the local level. If we had a Forest Service biologist or district ranger come in and say "Okay, we're gonna develop a cooperative watershed project in this area that involves private land and Forest Service land," the private folks will say "We don't trust you. . . . Just stay on your own land."

Therefore, RCD participation would be important for implementing local projects, and, because CARCD represented a geographically dispersed network of offices with close ties to local communities, it was a logical signatory to the MOU on Biodiversity.

Executive Council and Staff Accomplishments

Interagency cooperation involves more than simply signing agreements and attending meetings. Signatures tell us nothing about the intent of the signatories or what parts of the agreement were implemented. Attendance is more significant in these respects, particularly if line managers attended on a regular basis when they had not done so in the past. Attendance is even more significant if documents and participant recollections indicate that line managers followed through by implementing an agreement. Therefore, we should examine whether specific parts of the MOU on Biodiversity were implemented at the local, regional, and state levels. In this section, I briefly summarize some additional indicators of cooperation at the state level, leaving local and regional aspects to subsequent chapters.

While not stated explicitly in the MOU on Biodiversity, one of the major goals of the ecologists who drafted the document was to gain legitimacy and administrative support for their ongoing efforts to preserve biodiversity. Marc Hoshovsky believed INACC had accomplished little because line managers did not support INACC at the regional and state levels. He hoped the MOU and Executive Council would legitimize interagency efforts to protect biodiversity. In this regard, the MOU and Executive Council were remarkably successful, both inside and outside the agencies. In the words of another INACC ecologist,[52]

> Really, the best thing is that [the Executive Council gets] the word "biodiversity" out there; and that biodiversity protection, this thing that the government thinks about, we can point to it and say: "This is the purpose of government; this is a good thing to do." And I don't think that should be undersold. You know, the whole MOU on Biodiversity was a communist plot at the beginning, and now people are more comfortable with it.

More than simply gaining legitimacy, however, staff ecologists also hoped agency directors would apply top-down administrative pressure on lower-level line managers throughout the state to release the necessary resources for implementing interagency projects. As noted in chapter 3, Hoshovsky hoped the MOU would provide an executive-level door to the middle management of these agencies. So what evidence is there that the signatories actually pressured lower-level line managers to support and encourage such activities?

Several directors announced their support of the MOU on Biodiversity by sending memos and attachments to their line managers.[53] Some directors sent these packages out more quickly than others, and used stronger language regarding their expectations. Not surprisingly, Ed Hastey (BLM) was very prompt, sending out his memo (along with copies of the MOU and five other attachments) within a month after the agreement was signed. Ron Stewart (FS) also moved relatively fast, sending his package to all national forest supervisors and staff directors within three months. Boyd Gibbons (CDFG) took much longer. His memo was sent eight months after the MOU on Biodiversity was signed. This slow response recalled earlier experiences with INACC, in which Hastey was the first and the CDFG director was one of the last to sign the 1983 and 1989 natural areas agreements.

The language in these cover memos also suggests the degree to which the directors expected their line managers to follow through in implementing the MOU on Biodiversity. Again at the weak end of the spectrum was the memo from CDFG Director Gibbons (signed for him by an assistant), tepidly stating to the agency's five regional managers: "At the regional level, I suggest that local biodiversity issues and the MOU become an agenda item at upcoming meetings that you have with regional administrators of other agencies or representatives of regional groups, such as the Interagency Natural Areas Coordinating Committee, within your area." The cover memo from Kenneth Farrell (UC/DANR) was similarly weak: "Please review the material and draw it to the attention of your appropriate administrators, faculty, and staff." By comparison, the language in the BLM and FS memos was much stronger. Ed Hastey stated: "I expect BLM to play an active and major role in supporting the efforts of the Executive Council and in the creation and activities of the bioregional councils." He was similarly clear about his expectations for BLM line managers: "I am fully committed to successfully implementing this strategy and need your support in carrying it out." Ron Stewart's memo also contained unequivocal language: "I fully endorse the concepts contained in this MOU, and I believe that we should take immediate steps to begin implementation of them." Stewart's support was all the more remarkable because a FS attorney recommended changing some of the language in the MOU—particularly with regard to the maintenance and enhancement of biodiversity as a "preeminent goal," which the attorney claimed was inconsistent with the agency's multiple-use legislative mandates.[54]

Yet the paper trail appears to end at this point. If line managers in these agencies responded by sending out their own memos—along with copies of the MOU—to their staff, I could not find them. Moreover, interviews with staff indicated that the directors' memos may have had little or no effect in agencies other than the BLM. Several field staff reported they learned about the MOU from individuals outside their agencies, not their line managers. Moreover, if they participated in related interagency activities, they often did so after hours because their line managers did not allocate work time for this purpose.

In some agencies, the MOU did not travel down to staff because the state-level office had little authority over units in the field. This was particularly the case in the NPS. Individual park units were widely recognized to be relatively autonomous from the NPS regional office in San Francisco.[55] Nevertheless, NPS Regional Deputy Director Lew Albert said copies of the MOU were sent to park superintendents:[56]

> We use that document to send to our park superintendents to say, "Park superintendents, we have signed this MOU, and we want you to go out and start talking to local governments, engaging in dialogues on their planning and on your planning and on the actions that you are proposing, and working together to build consensus on somewhat of an ecosystem basis." . . . That was our intention—to build a bottom-up process that way—and that's how we use that MOU.

Yet, more than two years after the MOU had been signed and disseminated, Albert expressed disappointment that it had not spurred park superintendents to engage themselves in local planning processes.

At the other end of the spectrum was the BLM, with Ed Hastey leading a personal crusade to spread the word throughout his far-flung agency. Rather than relying simply on memos, Hastey traveled around the state, speaking with field staff and line managers. According to Associate State Director Al Wright, Hastey "gives a lot of attention to our managers and what they're doing. He does a lot of counseling . . . about what they ought to be looking at, and what they ought to be thinking about—people they ought to be talking to—to try and make sure that we're touching bases with the right people, that we're involving the right people."[57] Others were more pointed about the effect of Hastey's leadership within the BLM. Carl Rountree, who managed the BLM Biological Resources Branch, argued, "If Ed weren't here in Sacramento, and exerting the type of influence, the type of pressure he's bringing to bear on his area managers, I doubt the Bureau would be as actively involved as we are."[58] According to Rountree, top officials in the BLM State Office constantly pushed this agenda: "We keep it before our managers—at *every* management team meeting the subject of ecosystem management comes up." BLM's approach suggests that top-down leadership may be necessary to guide bottom-up decision-making processes. As Rountree put it, "We say ecosystem management is bottom-up; that's the way it works, but in order to start from the bottom up, you gotta go from the top down, . . . getting people to realize that there really is a better way of doing things."

BLM line managers also worked hard to sell these ideas to individuals outside the agency. The most prominent effort in this regard occurred in the San Joaquin Valley Bioregion, where Hastey sponsored an effort to establish a bioregional council by assigning a full-time staff member to coordinate the outreach effort (see chapter 7). Hastey also assigned a full-time staff member to manage the West Mojave Coordinated Management Plan, a CRMP-like cooperative effort in the western portion of the Mojave Bioregion to protect habitat for several species, including the federally listed desert tortoise. The BLM was not the only agency to dedicate staff time to bioregional activities. The FS assigned several individuals to serve as liaisons to local groups in the Klamath Bioregion (see chapter 5). The FWS and CDFG also assigned people in the South Coast Bioregion to work on a program loosely tied to the Executive Council (see chapter 6). Yet these examples tended to be exceptions rather than the rule. Setting staff time aside represented a big commitment on the part of line managers, and only the BLM and FS did so for the express purpose of nurturing the formation of local and bioregional groups as specified in the MOU on Biodiversity.

Agency directors were even more reluctant to place some of their discretionary budgets into a collective fund. Staff raised the idea several times, but the directors did not support it. Though many agencies contributed in-kind support to specific projects, few contributed in cash. The absence of a collective fund, and the paucity of FTE assigned specifically to implement the MOU, indicate the limited extent of multilateral cooperation among agency directors. Contributions to Executive Council activities were largely idiosyncratic and ad hoc. After several years, staff suggested membership dues, either financial or in kind, to provide some stability to Executive Council operations, but again the collective fund was not instituted.[59]

Two fundamental dynamics constrained multilateral cooperation at the state level. First, some directors expressed little or no interest in the activities of the Executive Council, a point demonstrated throughout this chapter by several indicators of participation. Because many directors demonstrated little enthusiasm, we should not expect them to make what several participants called the "hard decisions" to put money or staff aside for group purposes. Second, agency jurisdictions were not spread uniformly across the state; therefore, directors who were serious about cooperative action tended to focus on particular regions of the state, depending on the locations of their jurisdictions and the presence of endangered species therein.

The only project that was germane to the entire state, and that appeared to be of interest to all agencies, was the development of a centralized system for storing and sharing data. Staff were often eager to share data they produced or managed. Yet data sharing was often happenstance because it took time to learn about the kinds of data available for a particular analytic problem. It was also difficult to integrate data sets that had been gathered using different methodologies or stored in different formats. With the advent of GIS, which allows users to layer data sets on a computerized system of geographic grids, such discrepancies became visually obvious.[60] At the same time, it was becoming increasingly difficult for agencies to develop and maintain their own data sets as state and federal budgets tightened in the 1990s. Therefore, it made sense for technical and financial reasons to develop a central clearinghouse, into which data from each agency would be fed, and from which all could draw.

This effort was led by the California Resources Agency, which nominally housed many of these databases within its organizational superstructure. The new clearinghouse was dubbed CERES—the California Environmental Resources Evaluation System.[61] CERES was not an Executive Council project. Although CERES was discussed at Executive Council meetings, it had its own interagency staff and advisory committee. Moreover, CERES was largely a technical effort, with staff supplied almost exclusively from state agencies housed within the Resources Agency. While the Executive Council claimed credit for developing CERES, the effort was administered by the Resources Agency, and, as some participants suggested, it probably would have happened in the absence of the Executive Council because the data problem had become so acute. Yet the Executive Council did guide the development of CERES by emphasizing the importance of making CERES accessible at the local level, and organizing outreach efforts to train local users at agency offices in several bioregions.

The California Resources Agency also led other activities nominally sponsored by the Executive Council. Since 1993, it has published a glossy quarterly newsletter, *California Biodiversity News*, which reports on the activities of the Executive Council, along with local and bioregional efforts.[62] In the Sierra Bioregion, the Resources Agency also moved $100,000 to the CARCD to develop two watershed-based cooperative plans, one each in the northern and southern Sierra.[63] The Resources Agency housed nearly all of the state agencies on the Executive Council, so it was well positioned in a coordinating role. Wheeler also believed strongly in the MOU on Biodiversity. Given his organizational position and personal beliefs, he was the most likely director to undertake these initiatives. His successor, Mary Nichols, has continued his leadership since 1999 under the administration of Governor Gray Davis.

Yet much of the financial and administrative support for Executive Council activities came from Ewing's analytic unit in CDF. Ewing was the lead staff member of the TTF, the primary author of the MOU on Biodiversity, and the cochair of the Staff Committee for the Executive Council. Though Wheeler claimed credit as Executive Council Chair and Resources Secretary, Ewing deserved much of the credit for committing agency resources to interagency processes. Ewing assigned a staff member to arrange logistics for Executive Council meetings and to run the Executive Council help desk, which provided local groups with a single point of contact for questions related to the Council and member agencies. He also funneled money to interagency projects at the bioregional level. FRRAP managed a multiyear contract exceeding $200,000 to form a bioregional council and nurture local groups in the Klamath Bioregion (see chapter 5). By 1995, the Executive Council's annual budget exceeded $500,000, most of which came from FRRAP. Ewing also prodded the directors to fund Executive Council activities, often to no avail.

As these examples suggest, the Executive Council seldom acted as a group in funding or staffing specific projects. It nominally sponsored a number of activities, but most were undertaken by a handful of agencies, if not a single agency. The group itself served primarily as a forum for discussion. Because the Executive Council tended to endorse activities rather than jointly pursue them, it is difficult to attribute specific activities to the existence of the Council itself. The individual projects nevertheless provide an additional indicator of cooperative effort emanating from each agency.

Yet some state-level activities almost certainly occurred because of the Executive Council and the MOU on Biodiversity. Notable in this regard were a series of interagency workshops for line managers designed to spur them to develop cooperative relationships at the regional and local levels. The first of these meetings, held in Sacramento on January 25, 1993, was attended by more than 100 managers from around the state, representing most of the signatory agencies. In addition to listening to lectures about the importance of biodiversity, the participants also watched a series of agency directors give short speeches announcing their support of the MOU on Biodiversity and indicating their expectations regarding its implementation.[64] One observer, struck by the sheer novelty of this meeting, thought it even more remarkable than the Executive Council meetings:[65]

> It's *staggering* to think about all those [directors] from different appointing authorities and responsible for different bureaucracies and different constituencies to be in the same room together [at Executive Council meetings], let alone trying to coordinate policy in any proactive way—recognizing that they all march to different enabling requirements and other legal mandates and budget constraints and everything else. Even more staggering, I think, was a conference . . . in which all the [line managers] came together to meet and to share that same experience.

This daylong meeting was indeed unprecedented, and more would follow.

Agency staff and directors subsequently developed a series of regional workshops to encourage active involvement from regional line managers and staff. Yet regional line managers showed little enthusiasm for this idea, and, as will be seen in the bioregional chapters, their resistance to the MOU on Biodiversity constituted one of several factors plaguing the formation of bioregional councils. The state-level organizers therefore directed their attention to the local level, where they convened a series of workshops for field staff and local government officials.[66] They also formed an interagency liaison team in 1994 to coordinate Executive Council assistance at the local level. They hoped the new liaison team would "sunset with the formation of regional managers' committees," but it continued to operate for several years because the bioregional layer of line managers did not coalesce.[67]

Summary

Executive Council meetings were primarily a forum for discussion that legitimated biodiversity preservation as a goal and encouraged coordination as a means to achieve that goal. The Executive Council met diligently after its inception in 1991 but accomplished few of the grand objectives laid out in the MOU on Biodiversity. It did not "develop guiding principles" or "design a statewide strategy" to conserve biodiversity as stated in the MOU, nor did it "recommend consistent statewide goals for the protection of biological diversity" or "recommend consistent statewide standards and guidelines" to meet those goals. It did not even attempt to define *biodiversity* until 1996.[68] While these shortcomings disappointed staff ecologists and environmental activists who hoped for more, they relieved many local government officials and private stakeholders who feared that state and federal agencies were preparing to usurp local control and regulate private property.

The Executive Council turned out to be a discussion forum, pure and simple, but it was an important forum because it elevated the protection of biodiversity on agency agendas, while emphasizing the importance of cooperation at the local, state, and bioregional levels. By meeting quarterly, agency directors raised the profile of an otherwise obscure academic concept, which had not previously been linked with agency missions. It also inspired some directors to pressure lower-level line managers—through memos, workshops, and persuasive leadership—to involve themselves in bioregional and local initiatives. Yet line managers proved to be a stumbling block in most agencies, particularly at the bioregional level. Therefore, the Executive Council appealed directly to existing local groups, even nurturing the formation of new groups in targeted bioregions like the Klamath and Sierra.

Agency directors also listened to presentations by local groups at Executive Council meetings. This local emphasis was accelerated in 1994, when the Executive Council began holding its quarterly meetings outside Sacramento, in communities such as El Portal, Morro Bay, and Napa. At this time, it also initiated Local Group Forums—daylong events held the day before Executive Council meetings. These forums facilitated networking among local groups and informed agency directors and staff about their status and needs. The Executive Council picked these locations because they were invited and because local cooperative efforts had been relatively successful. In doing so, the Executive Council sought to put its imprimatur on success rather than stir activity within recalcitrant communities. Although local groups were screened in advance to ensure their activities fit the profile council members sought to endorse, their presentations nevertheless gave a state-level voice to local government officials, private citizens, and agency field staff. In listening to these presentations, and endorsing local groups, the directors provided symbolic support to ongoing interagency efforts and public-private partnerships, even if the Executive Council did not play a direct role in these efforts. As of 2001, the Executive Council was still holding its meetings in local communities, and still convening Local Group Forums.

Multilateral cooperation among agency directors was limited largely to attendance and participation at meetings. The directors did not pool their budgets or staff. They did not develop uniform goals, consistent standards and guidelines, or a statewide strategy to protect biodiversity. This does not mean that the Executive Council was a sham. Several agency directors earnestly sought means to protect biodiversity to forestall court enforcement of environmental laws, which threatened to wrest control of their management autonomy. Nor does this mean they sought to subvert the ESA, as Warren charged, by merely acting as if they were protecting species to use the Executive Council's existence as a pretense for lobbying legislators to weaken the ESA. While Governor Wilson clearly sought to weaken or overturn both the state and federal ESAs, I found no evidence that any of his political appointees on the Executive Council attempted to use this forum for that purpose. This does not foreclose the possibility that other political appointees, either in the Governor's Office or signatory agencies, used the council's existence as a rhetorical lobbying ploy while limiting the ability of agency staff to protect biodiversity at the local, regional, or state levels.[69]

The dearth of concrete accomplishments at the state level should instead be interpreted from three perspectives. First, powerful interests in the timber and agricultural industries seriously impeded the Executive Council during its first two years. Combined with the rural backlash from local governments, it is surprising the Executive Council even survived, let alone thrived. Second, habitat is distributed locally and regionally in California, not statewide. Because line managers were primarily seeking a low-cost means to comply with environmental laws, we should expect to find concrete examples of cooperation for managing interagency dependencies primarily at the local and regional levels. Third, the MOU on Biodiversity, with all of its language about statewide goals, standards, and guidelines, was drafted by staff ecologists who sought to bring the planning and management principles of conservation biology into decision-making processes. Agency directors never debated the MOU's language; they simply signed the document when Hastey asked them to do so. Arguably, many of the signatories likely expected little, if anything, to be accomplished in the wake of the MOU on Biodiversity because most interagency agreements are soon forgotten, even by the signatories themselves.

Yet the MOU was not soon forgotten. Instead, the Executive Council grew in size and stature. Initially formed with ten state agencies and federal regional offices in 1991, the Council added six more agencies in 1992, along with several associations of local governments. In 1994, nine more state agencies and federal regional offices joined, thereby expanding the Executive Council to thirty-five members. Table 4.5 lists member organizations chronologically through 1994, using the official names of the agencies when they joined. By this point, the composition was largely solidified, and the membership has been relatively stable since then, with a few agencies joining and a few low-participation agencies leaving. As of 2002, there were thirty-nine members.[70]

With the 1994 additions, the character of the Executive Council changed significantly. None of the new agencies had participated in the TTF or were signatories to the CRMP or INACC agreements. Therefore, these agencies were unfamiliar partners because they did not share the same history. The EPA, in particular, was a notable addition. Ed Hastey had previously rejected Charles Warren's motion to add the EPA, which regulates water quality under the Clean Water Act. Yet the EPA's role in the Executive Council was not regulatory. Several participants had long sought to bring the EPA into the Executive Council because the agency had large sums of grant money to bestow on local resource conservation efforts led by RCDs and other organizations. Nevertheless, adding the EPA suggested the Executive Council was expanding from terrestrial diversity to aquatic diversity. With the additions of the California Coastal Commission and Coastal Conservancy, the Executive Council was also

Table 4.5
Members of the Executive Council, 1991–1994

U.S. Bureau of Land Management, State Office (1991)
U.S. Fish and Wildlife Service, Pacific Region (1991)
U.S. National Park Service, Western Region (1991)
U.S. Forest Service, Pacific Southwest Region (1991)
California Resources Agency (1991)
California Department of Fish and Game (1991)
California Department of Forestry and Fire Protection (1991)
California Department of Parks and Recreation (1991)
California State Lands Commission (1991)
University of California, Division of Agriculture and Natural Resources (1991)
U.S. Soil Conservation Service (1992)
U.S. Agricultural Stabilization and Conservation Service, State Office (1992)
U.S. Bureau of Reclamation, Mid-Pacific Region (1992)
California Department of Conservation (1992)
California Department of Water Resources (1992)
California Department of Food and Agriculture (1992)
California Association of Resource Conservation Districts (1992)
North Coastal Counties Supervisors Association (1992)
Sacramento–Mother Lode Regional Association of County Supervisors (1992)
Northern California Counties Supervisors Association (1992)
San Joaquin Valley Regional Association of County Supervisors (1992)
South Central Coast Regional Association of County Supervisors (1992)
Central Coast Regional Association of County Supervisors (1992)
Regional Council of Rural County Supervisors (1992)
San Diego Association of Governments (1993)
Southern California Association of Governments (1993)
U.S. Environmental Protection Agency, Region IX (1994)
U.S. National Biological Service, Western Region (1994)
U.S. Geological Survey, Regional Office (1994)
U.S. Bureau of Mines, Western Field Operation Center (1994)
California Environmental Protection Agency (1994)
California Coastal Conservancy (1994)
California Coastal Commission (1994)
California Department of Transportation (1994)
California Energy Commission (1994)

positioned to move into coastal issues. Indeed, the Council's formal agenda expanded significantly in 1995 from the original focus on timberlands to include agricultural, coastal, and aquatic issues.

The Executive Council also changed its name at this time from the "California Executive Council on Biological Diversity" to the "California Biodiversity Council" (the name it retained into the twenty-first century). This name change was symbolic in at least two respects. First, whether intended or not, dropping the word *Executive* connoted the fact that many agencies were represented at Council meetings by staff rather than directors. This did not mean that overall attendance was trailing off. In fact, attendance from the original agencies remained strong. Second, the core participants formed a new organization within the Council called the "Executive Committee." In doing so, they acknowledged that Council meetings had become primarily a forum for presentations rather than conducting business, and created a new group for the latter. The new Executive Committee met eight times a year rather than quarterly and was empowered to conduct Council business, recommend policy, and perform other duties assigned by the Council—including developing a budget and implementation strategy. Not surprisingly, this nine-member Executive Committee was initially dominated by the most active original agencies, with five individuals representing the California Resources Agency, CDFG, CDF, BLM, and FS. These five representatives held three-year terms, while the other four members held two-year terms, with two positions reserved for local government associations and two assigned at-large from the remaining agencies and associations. Though not all members of the Executive Committee were agency directors, they were all high-level officials "with the authority to speak directly for their organizations."[71] This state-level structure remained unchanged as of 2002, with the new Executive Committee, the renamed California Biodiversity Council, and the Staff Committee still meeting actively on a regular basis.

In the mid-1990s, the real action turned to the bioregions, within which state-level officials sought to encourage the formation of new bioregional councils. Rather than supplanting existing cooperative efforts at the local level, they hoped to link these efforts together through new bioregional councils that would develop regional biodiversity strategies. They hoped these bioregional councils would provide ecological information and administrative support to local groups while creating a forum for planning and management on a regional scale. Staff ecologists also hoped that top-down pressure within the agencies would be forthcoming, encouraging line managers to support cooperative efforts at both the local and regional levels. As specified in the MOU on Biodiversity, line managers in field offices throughout the state were expected to develop regional memoranda of understanding and establish bioregional councils. These bioregional councils would then look to the Executive Council for guidance and "actively encourage the development of watershed or landscape associations to assist in implementing regional strategies." On paper, this regional strategy appeared to rely on hierarchical control, but the underlying intent was to encourage local decision making while linking local cooperative efforts into mutually beneficial efforts at the regional level.

The bioregional councils, however, failed to materialize. This was not due to lack of effort on the part of agency staff or directors in Sacramento. Instead, three statewide problems seriously impeded the formation of bioregional councils. First, local government officials in California generally opposed the idea of regional government, particularly when raised by state and federal agencies. Second, line managers within the bioregions showed little enthusiasm for implementing the MOU on Biodiversity. Third, INACC ecologists developed the bioregional map (figure 3.1) based on California's physiographic features. Although the bioregions included minor border adjustments to conform to existing jurisdictions, the bioregional boundaries did not conform to the existing scale of social planning processes. The bioregions made sense to ecologists, but not to the inhabitants of local communities.

While the nascent bioregional councils sputtered and died, local associations thrived. Organized around watersheds and communities, local groups emerged in great numbers throughout the state during the 1980s and 1990s. In part, the groups were encouraged by the MOU on Biodiversity and Executive Council, which brought legitimacy and some concrete assistance to their efforts. They were also driven by some of the same factors that motivated agency officials in Sacramento, including court enforcement of environmental laws, particularly the federal ESA. Yet, unlike their state-level counterparts, members of local groups were driven by strong concerns about the communities in which they lived.

At the local level, the scale of cooperation was much different. Living closer to the land and in smaller communities, these individuals were willing to volunteer their own time, and not simply the resources of their particular agencies.

It was here, at the local and regional levels, that Ed Hastey expected county supervisors to play a lead role. The BLM had long been closely aligned with local governments, particularly rural counties. Some staff ecologists even recognized that local governments and private landowners were a necessary component of any strategy to protect biodiversity since public agencies did not manage enough habitat to protect ecosystems. Because the MOU on Biodiversity arose out of the activities of the TTF, which focused on the Klamath, and because the Executive Council sponsored its first regional initiative in the Klamath Bioregion, the Klamath is the most appropriate place to begin.

5

The Klamath Bioregion: Local Cooperation and the Demise of the Bioregional Ideal

Conflicts over timber harvest practices in the Klamath Bioregion provided the immediate impetus for agency directors to sign the MOU on Biodiversity in 1991 (chapter 3). The Executive Council accordingly chose the Klamath as the venue for its initial effort to organize a bioregional council, as outlined in the MOU. Their strategy was to hire independent contractors to organize bioregional meetings, which they envisioned would evolve into a formal bioregional council. This bioregional council would then coordinate the planning and management activities of state and federal agencies, in concert with the numerous local groups of public and private actors that were already organized throughout the mountainous Klamath Bioregion (see figure 3.1).

Yet it was unlikely that a bioregional council would emerge in the Klamath in the 1990s. After all, why would line managers, local government officials, and private stakeholders participate in a new regional organization covering the entire Klamath for the purpose of preserving biodiversity? Regional line managers, like their state-level counterparts, valued their management autonomy and had long shown little interest in biodiversity. Local government officials also showed little inclination to share decision-making authority with a new regional organization. Moreover, the geographic scale of the Klamath did not mesh well with existing social planning processes at the local level. Nevertheless, this case provides insight into the willingness of line managers in a bioregion to cooperate across agency jurisdictions, and it provides additional evidence regarding the willingness of state-level line managers to support cooperative efforts at the local and bioregional levels. Notably, line managers in agencies facing litigation under the ESA and other environmental laws—like the BLM and FS—tended to cooperate and support field staff more than others. But most cooperation occurred at the local level, within local communities, where field staff played a central role.

Natural and Social History of the Klamath Bioregion

The Klamath Bioregion—like all of California's bioregions—is a physiographic province. In a physiographic province, climate, altitude, and soil combine to favor particular plant species, which in turn provide habitat for distinct assemblages of birds, insects, and mammals. The Klamath is not homogeneous in these respects, but it does have distinguishing characteristics. Its rugged mountain ranges capture large amounts of moisture moving inland off the Pacific Ocean, giving rise to lush evergreen forests of towering conifers. This single characteristic largely distinguishes the Klamath Bioregion from the sagebrush steppes of the Modoc Bioregion to the east, the low-lying grasslands of the Sacramento Valley Bioregion to the southeast, and the oak savannas rimming the San Francisco Bay to the south.

Because physiographic provinces tend to merge into one another, without abrupt boundaries, INACC ecologists massaged the bioregional boundaries at the margins so they would be more congruous with prominent administrative and political jurisdictions. This avoided, for example, putting 90 percent of a national forest in one bioregion and 10 percent in another. The California border provided the only firm bioregional boundary. Despite these acknowledgments of political and administrative realities, the INACC bioregions were based primarily on biogeography, an ecological discipline. They were not related to the counterculture bioregional movement, which had been gathering momentum in Northern California since the 1970s. Bioregionalism was a social movement emphasizing the importance of individuals living in a place, understanding that place, and caring for it in a personal way (Andruss et al. 1990; McGinnis 1999). Because *bioregion* sounded like *bioregionalism*, some observers wondered if the agencies were associating themselves with the back-to-the-land ideology of rural environmentalists. To the contrary, agency staff derived the idea of using bioregions as an organizational tool from ecological knowledge, not environmental philosophy.

The Klamath Province has long fascinated ecologists and nature writers because it is biologically one of the richest areas in North America, containing a unique mix of species found nowhere else on earth (Norse 1990; Wallace 1983). The mountainous climate is relatively mild, allowing floral species from the north and south to intermix. Rainforest trees of the Pacific Northwest, like Alaska cedar, reach their southernmost extent on north-facing slopes in the Klamath, while drought-tolerant trees thriving to the south survive on southern slopes. Seventeen conifer species have been counted in a single square mile in the Klamath, a possible world's record (Wallace 1983: 77). Coast redwood, which once grew over much of the Northern Hemisphere when the climate was milder and more humid, survives as a relict species in the Klamath, where it has retreated to the fog belt along the Northern California coast. Coast redwood is the world's tallest tree, and it grows remarkably straight, giving redwood forests a cathedral-like atmosphere. Further inland grow Douglas fir, the world's third tallest tree, and one of the most valuable on the world timber market.

Despite the Klamath's great biodiversity, few species generated economic wealth for the modern human communities that settled there, and these species were rapidly depleted. Sea otters were hunted to extinction on the California coast north of San Francisco during the nineteenth century, their thick pelts prized for garments. Annual salmon runs once filled rivers throughout the Pacific Northwest, sparking stories suggesting one could walk across rivers on their backs, but these tall tales evaporated with the salmon. In 1994, only a few thousand coho salmon returned from the ocean to spawn in the Klamath Bioregion. Overfishing was not the only problem. Logging, cattle grazing, and dams degraded spawning habitat. Local communities depended on these and other species for their economic livelihood, but global markets and absentee landowners largely determined the rate of resource consumption. This was particularly true for the forests.

In the nineteenth century, forests throughout the American West appeared limitless to settlers, but they were rapidly depleted in the twentieth century. Timber companies acquired large tracts of land granted to railroad syndicates by the federal government. Conifer forests were cut down at staggering rates on private land before World War II, as the timber industry pressured the FS to keep national forest timber off the market to bolster prices. After World War II, the situation reversed; the FS rapidly opened the national forests to logging, providing lumber for the postwar housing boom and other markets as the supply of timber from private lands dwindled (Andrews 1999: 193–194). Coast redwood forests, originally covering two million acres, primarily in California, were particularly decimated. By 1988, fewer than 20,000 acres of old-growth redwood forest were left on private land (Jensen, Torn, and Harte 1993: 131). On public and private land, less than 4 percent of the original extent of virgin redwood forests remained by the 1990s, with only 2.5 percent protected (Ricketts et al. 1999: 242).

As species, these trees were not endangered because they regenerated, but the ecological structure of the forests changed significantly after harvesting. The forests became less complex, with fewer plant species and even-age canopies, which provided a narrower range of habitat for wildlife, particularly species that depended on the structural characteristics of old-growth forests (Norse 1990). Old-growth forests have a multilayered canopy, including downed logs and standing snags interlaced by a complex network of fungi that nurtures the entire forest. Because it takes about 200 years for a coniferous forest to acquire old-growth characteristics, and because large-scale timber harvesting began in the Klamath in the late nineteenth century, the only old-growth forests remaining in the late twentieth century were uncut "virgin" forests.

Some rural communities depended on old-growth forests for their economic livelihood because their lumber mills were designed to process large-diameter logs. Old-growth logging also required new roads, thereby creating additional jobs in road construction and maintenance. Yet old-growth logging was based on a finite resource. By the 1990s, old-growth forests were nearly gone on private lands in the Klamath, and most timber companies had long since turned to the national forests. Public lands provided an economic bonanza for many rural communities, in part because the FS subsidized the cost of new logging roads, making it much cheaper for timber companies to haul the logs out. County governments encouraged this logging because they received 25 percent of the revenues from timber sales on national forests, which provided a significant source of revenue for many rural counties.[1]

Social and economic restructuring in the Klamath began before the depletion of old-growth forests. Urban emigrants hoping to live closer to the land flooded into the Klamath during the 1960s and 1970s. To them, industrial logging practices—like clearcuts, slash burning, and herbicide spraying—were aesthetically displeasing and unhealthy, setting the stage for political conflict between new immigrants and longtime residents over timber harvest practices. With the economic boom in the 1980s, the rate of harvesting increased throughout the West, alarming national environmental organizations and local activists, who became more confrontational.

Environmental activists chained themselves to trees, sat on logging roads, and pulled up survey stakes. Some pursued more violent strategies, the most notorious of which involved hammering metal spikes into trees as booby traps to disable chainsaws, shatter milling blades, and terrorize loggers and mill workers with flying metal. Protests in the Klamath reached a crescendo in 1990 with Redwood Summer, when environmentalists converged on coastal communities in a highly organized series of rallies, blockades, and other actions to preserve the last unprotected stands of coast redwood. That fall, a statewide initiative, dubbed Forests Forever, also appeared on the ballot. Among other things, it called for a moratorium on old-growth logging in California. Though Forests Forever failed at the polls, these and other events received statewide and national media attention, leading to several bills in the state legislature, including the TTF bill (see chapter 3).[2]

Litigation and the Northern Spotted Owl

Protests, legislation, and administrative appeals brought incremental change in logging practices, but profound change came through the courts. Environmentalists pursued several legal avenues to slow, if not halt, timber harvests—including NEPA, NFMA, and the ESA. To use the ESA, they first needed evidence that one or more wildlife species depended on old-growth forests as habitat. The FWS maintained a nationwide list of candidate species, but these species could not be listed without supporting evidence. As was the case with many species across the country, academics stepped forward with the evidence.

In 1968, Eric Forsman, a wildlife biology undergraduate at Oregon State University, heard an owl while working at a FS guard station in Oregon. He mimicked the call, and watched the owl fly into a nearby yard. This rare sighting of a northern spotted owl inspired Forsman to learn more about this subspecies.[3] In 1972, he began rigorous studies of the owl as a graduate student at Oregon State University, and subsequently peppered federal officials with memos and letters pointing out that the first 37 nest sights he found were in old-growth forests, most in proposed timber sales (Yaffee 1994: 14–19). Forsman continued gathering data but made little progress in convincing federal officials to protect the owl's habitat. Meanwhile, in 1973—five years after his first encounter with the owl—the ESA became law.

Like most wildlife-habitat relationships, the owl's dependence on old-growth forests was not completely understood. Old-growth canopies appeared to provide nesting sites, shelter from predators, and reduced competition from other species (Norse 1990). Yet the owl could also be found in second-growth forests, a fact that provided ammunition for logging proponents (Bonnett and Zimmerman 1991: 112). But the owl did not thrive in most second-growth forests, in part due to predation by the great horned owl and competitive displacement by the barred owl. Ecologists use the terms *source* and *sink* to describe this phenomenon. A source is an area in which a species tends to reproduce and from which it migrates, while a sink is an area in which immigration exceeds emigration. A species' population declines as sources shrink and the landscape becomes dominated by sinks. For the northern spotted owl, old-growth forests appeared to be a source; everything else appeared to be a sink. Researchers did not know, however, how many acres of old-growth forest were necessary to maintain viable populations of the owl, how large each habitat patch should be to support nesting pairs, or whether populations isolated in dispersed habitat patches would intermix and thereby remain genetically viable. The future of the owl and the timber industry hinged on these technical questions.

In 1981, the FWS declared the owl "vulnerable" to extinction, but did not list it as threatened or endangered. Meanwhile, studies of the owl expanded in both the private and public sectors. The FS started its own research program, in part because Forsman's studies constituted the only owl research on national forests, and Forsman had been proclaiming that FS timber sales threatened the owl. Unfortunately for the agency, FS line managers did not follow recommendations emanating from research staff. The late 1980s witnessed the flowering of conservation biology, and FS scientists—including Jack Ward Thomas and Jerry Franklin—were part of this new wave of research linking ecological knowledge with resource planning and management. FS line managers consistently resisted their recommendations for habitat management because they wanted to maintain control of the timber harvest program (Yaffee 1994). Rather than preserving the owl's habitat, timber harvesting accelerated during the 1980s, further transforming national forests into sinks.

So environmentalists went to court, with empirical data behind them. The number of cases involving the northern spotted owl was large, covering many aspects of the ESA, but some of the highlights are important.[4] In 1986, an obscure environmental group named Green World, which apparently operated out of a Boston phone booth, petitioned the FWS to list the owl as an endangered species. National environmental groups had been reluctant to submit a petition because they did not believe public opinion was sufficient to counterbalance the political ramifications of listing the owl. But the poorly prepared Green World petition forced their hand, so the Sierra Club Legal Defense Fund filed a petition on behalf of 29 environmental organizations (Yaffee 1994: 108–109). In 1987, the FWS decided against listing the owl, primarily for political and administrative reasons, not biological reasons, and even altered data to fit the decision (U.S. General Accounting Office, 1989). The Sierra Club Legal Defense Fund accordingly sued the FWS, because the ESA requires listing decisions to be justified "solely on the basis of the best scientific and commercial data available."[5] In 1988, U.S. District Judge Thomas Zilly voided the agency's decision, which he held to be arbitrary and capricious, because the FWS disregarded expert opinion. He remanded the case to the agency, ordering it to produce evidence supporting its decision.

Meanwhile, another coalition of environmental groups, led by the Audubon Society, pursued legal action under NEPA and NFMA, seeking injunctive relief against FS timber sales in owl habitat. In March 1989, U.S. District Judge William Dwyer issued a preliminary injunction on 139 planned timber sales; in May he extended the injunction indefinitely (Yaffee 1994: 114). One year later, the FWS listed the owl as threatened, but it did not designate critical habitat, as required by the ESA, so environmentalists sued again. In 1991, Judge Zilly ruled in favor of the plaintiffs, and the FWS subsequently proposed 11.6 million acres as critical owl habitat, including 6.5 million acres of FS land. The big hammer fell in May 1991, when Judge Dwyer issued a permanent injunction on FS timber sales throughout the owl's range until the FS developed an acceptable habitat management plan and environmental impact statement. Thus, in the end, it was not the ESA that brought the FS to its knees, but rather NEPA's procedural guidelines for assessing the environmental impacts of federal actions and the vague biodiversity mandate in NFMA.[6]

The ultimate monkey wrench had been thrown into the machinery of the timber industry. Timber harvests on federal land in the owl's range all but ceased for three years.[7] This included 23 million acres on national forests in all three states and 2.4 million acres on BLM land, primarily in Oregon. The effect on timber-dependent communities in the Klamath Bioregion was devastating, particularly in rural counties dominated by national forests, like Trinity County. The FS managed 71 percent of Trinity County, with approximately half the county lying within a single administrative unit: the Shasta-Trinity National Forest. Trinity County is larger than Rhode Island but had fewer than 14,000 year-round residents—and no stoplights. With the local economy driven by timber harvests, summer tourism, and illicit marijuana plantations, the logging injunction had a tremendous impact. Timber harvests on national forests in Trinity County peaked at 215 million board feet (mmbf) in 1988, declined to 56 mmbf in 1992, and were projected in 1994 at no more than 25 mmbf for the foreseeable future.[8] As unemployment crested above 20 percent, the county lost millions of dollars in revenue-sharing payments from timber sales on federal land, constraining the county's ability to provide social services. Trinity County's receipts dropped from $7.5 to $5.3 million between 1990 and 1993, and would have plunged more precipitously had not the method for calculating payments been restructured after the injunction to allow for gradual declines.

The FS was not the only agency constrained by litigation over the owl's habitat requirements, though it certainly was hit hardest by the logging injunction. The BLM was also hammered in Oregon, where it managed large tracts of timberland. State agencies could be sued as well. The CDF, for example, could be sued under the ESA for permitting logging that harmed the owl on private land. It was therefore no coincidence that agency directors signed the MOU on Biodiversity just four months after

Judge Dwyer's injunction. Nor was it surprising that the Executive Council focused initially on timberland issues in the Klamath, where other forest-dependent species—like the marbled murrelet—might become "the next spotted owl."

The North Coast INACC: Cooperation among Staff Ecologists

As these high-profile events transpired, a little-known group of agency ecologists coalesced in 1990 along the coastal portion of the Klamath. Spawned by the statewide INACC, the North Coast INACC was also a technical group of staff ecologists. Its members primarily shared information about the condition of species and ecosystems, and worked toward the long-term goal of designing regional reserve systems. Yet the group did not attempt to cover the entire Klamath Bioregion. The North Coast INACC was based in the logging port of Eureka, where most of the participants lived and worked. At 17 million acres, the Klamath Bioregion was roughly the size of Maine, and was divided by a mountain range that slowed east-west travel. Thus, even for ecologists, the geographic scale of the Klamath posed organizational challenges.

The North Coast INACC's goals were clearly stated at the top of the group's newsletter. The first goal, to improve communication and coordination between local, state, and federal agencies and conservation groups, was relatively modest, though not easy to achieve. The second goal was politically ambitious: "To develop a regional natural diversity conservation plan." For this, local government support was crucial, because any proposed natural reserve system would likely include public acquisitions of private land and restricted use of public lands, both of which would erode local tax bases. Yet few local government officials attended these meetings. Representatives from conservation groups and the timber industry participated, but local government staff and elected officials were largely absent.

Line managers were also missing from North Coast INACC meetings, and they showed little support for the group's activities. This was true even for Redwood National Park, one of the few national parks in the country established primarily for ecological restoration rather than tourism. One NPS ecologist, who laid out the benefits of the North Coast INACC in a memo to his superiors, was subsequently disappointed by the lack of support.[9] Without the support of line managers, like park superintendents, the North Coast INACC made little headway in changing agency practices and establishing interagency preserves. The only line managers who regularly attended North Coast INACC meetings came from the BLM.

When the state-level Executive Council held its first meeting in December 1991, the North Coast INACC was the closest thing to a bioregional council in the Klamath, but it operated primarily on the coast rather than in the interior, and line managers, local government officials, and other stakeholders were not well represented. It was a technical group of staff ecologists in field offices, not a political or administrative body. Because it did not approximate the model of a bioregional council laid out in the MOU on Biodiversity, state-level staff and line managers decided to start from scratch, and attempted to build a bioregional council from the ground up.

The Klamath Bioregion Project: The Executive Council's First Bioregional Strategy in California

Local groups, organized primarily within watersheds, already flourished in the Klamath Bioregion. However, they focused on local issues, and their members generally knew little about the activities of groups in neighboring watersheds—let alone groups working in distant portions of the Klamath. Numerous agency planning processes were also underway, but with the exception of the BLM, they were largely insular. State-level officials sought to link these local groups and agency planning processes together by creating a larger, coordinated effort to manage habitat on a regional scale, and they wanted to begin with a clean slate, uncontaminated by local perceptions of past agency activities. Therefore, as a prelude to establishing a new bioregional council, they hired independent contractors to convene two bioregional meetings in the Klamath to announce the idea and see how it would be received.

Contracting out was sensible for two reasons. First, local governments in California routinely worried that state and federal agencies sought to usurp their authority, and landowners were concerned about increased regulation of private land. These concerns were particularly apparent in rural areas like the Klamath, where home rule was a popular issue and distrust of state and federal agencies was rampant. In this context, independent contractors were more likely to be trusted by local stakeholders than agency officials were. Second, agency officials were not trained as professional organizers or dispute mediators. Therefore, by relying on independent contractors, state-level officials could draw on the expertise of professional organizers, while giving the project a neutral appearance. Yet they were unable to monitor the contractors on a daily basis, and depended on third-party reports to hold them accountable.

Thus began the Klamath Bioregion Project, which to those involved became a long-running soap opera, rife with rumors, innuendo, character assassinations, and even threats of lawsuits—the very thing agency officials hoped to prevent through coordinated action. Ironically, local stakeholders came to believe that the contractors were manipulating the process in pursuit of their own agenda, an agenda separate from that of the Executive Council, and contradictory to what state-level line managers and staff intended.

The Initial Bioregional Meetings

Before agency directors signed the MOU on Biodiversity in September 1991, the TTF sought to build local support for its efforts to reconcile agency practices. This was not simply an issue of gaining legitimacy, though public support was certainly on their minds and would soon prove crucial after the debilitating backlash generated by the Sierra Summit (chapter 4). Of more fundamental concern was the very reason agency directors had become involved in the first place: to ward off lawsuits by maintaining viable populations of species through coordinated habitat planning and management. This required support and participation from those already involved—including public-private partnerships, like CRMPs; watershed organizations, like the Mattole Restoration Council; interagency technical groups, like the North Coast INACC; and intra-agency planning efforts focused specifically on the northern spotted owl.

Therefore, the agencies began an outreach effort to involve local officials and other stakeholders in the formation of a bioregional council, which would set goals and a strategy for the bioregion.[10] The initial contract for coordinating this outreach effort went to Ted Trzyna, who ran an independent organization in Sacramento called the California

Institute of Public Affairs. At that time, the California Resources Agency pledged $20,000, and the BLM and FS together pledged $20,000, for the first five months.[11] Trzyna accordingly planned two bioregional meetings in the Klamath, but he lost the contract over allegations of misappropriated funds before he did much more than organize the meetings. Rumors also circulated that Trzyna received the contract because of a personal relationship with Robert Ewing, the lead TTF staff member. The veracity of these allegations was less important than what their very existence indicated about the political cross fire within which the Klamath Bioregion Project operated. From the beginning, the project was highly politicized because of the perceived threats it posed to the timber industry, local stakeholders, and line managers in the bioregion.

Before being forced out, Trzyna asked two individuals to assist him with the project. Tim Wallace and Jerry Moles thereafter became synonymous with the Klamath Bioregion Project, nurturing it from a shoestring budget to a substantial enterprise. Wallace was a resource economist with Cooperative Extension on the University of California, Berkeley, campus. On paper, he reported to UC/DANR director Kenneth Farrell, who was an official member of the Executive Council, though Wallace had no direct contact with him. Moles, who did most of the legwork in the Klamath, was an anthropologist who had once worked at the UC Davis campus near Sacramento. Thus, unlike Trzyna, Wallace and Moles were not organizationally independent of the Executive Council, though they acted independently by pursuing their own agenda in the Klamath.

At the first bioregional meeting, held in Redding in October 1991, they tested the waters to see whether participants were interested in the MOU on Biodiversity and the idea of a bioregional council. With about 80 people in attendance, they elicited a laundry list of more than 100 concerns related to resource management in the Klamath, and asked the participants what functions a bioregional council might serve in resolving these concerns.[12] At the second bioregional meeting, held in Eureka in January 1992, Wallace did a surprising thing in front of a similar audience. He tossed a copy of the MOU on Biodiversity into a wastebasket, and told the audience that the MOU need not guide bioregional meetings. This was surprising because the MOU legitimated agency participation. It also laid out the relationship between cooperative efforts at the local, regional, and statewide levels. Moreover, state-level officials were promoting the MOU as the umbrella framework for bioregional activities sponsored by the Executive Council throughout California, including the Klamath Bioregion Project.

Yet, as Wallace later explained, he and Moles did not want the MOU on Biodiversity to play a central role in the Klamath Bioregion Project:[13]

> The only reference we made to the MOU was: "You ought to put it in the waste basket." This got a lot of agency people upset. But our interpretation was that this was an *interagency* memo to encourage cooperation and coordination amongst the agencies, to help when asked by local people, on problems that were technically in the bailiwick of whichever agency was asked.

In other words, the contractors envisioned the Klamath Bioregion Project to be a grassroots effort. Rather than encouraging interagency coordination *with* public involvement, they sought to place agency officials in the role of attentive bystanders responding to local concerns. This was an extreme interpretation of the organizational design implicit in the MOU on Biodiversity, if not an outright rejection of it.

Agency staff attending the Eureka meeting were shocked by this incident. To members of the North Coast INACC, in particular, the MOU was an empowering document that legitimized their efforts to preserve biodiversity, and they believed it signaled top-down support within their agencies. In Redwood National Park, Environmental Scientist Lee Purkerson believed the MOU "literally authorized all kinds of joint, cooperative efforts"—including exchanging agency funds and personnel, which Purkerson did with BLM field staff and line managers, without seeking authorization from his park superintendent.[14] Because field staff broadly interpreted the MOU, word traveled quickly to Sacramento reporting the wastebasket incident to the Executive Council and Staff Committee, and Wallace was rebuked. Having been chastened by state-level officials, Wallace redefined his role, stating that he was assuming "leadership" for the University of California's effort to "carry out" the MOU in the Klamath.[15]

The Eureka meeting produced another turn of events, one with much greater portent for the future of a bioregional council in the Klamath. Those attending the meeting thought the bioregion was too big to serve as the framework for regular meetings. So they divided themselves into four groups representing four subregions of the Klamath Bioregion, and met henceforth as separate groups. The idea of a bioregional council built from the bottom up began to unravel at this second bioregional meeting, with the precipitating cause being the seemingly innocuous issue of travel time. Driving between the major population centers in the Klamath required several hours, with up to a day required to drive from one end of the Klamath to the other. Thus, bioregional meetings would be an expensive and time-consuming commitment, particularly for people volunteering their services on a regular basis. Moreover, because most participants were primarily interested in issues related to their local communities, driving time was inversely related to the issues of greatest concern.

Ironically, even ecologists who supported bioregional planning and management in theory found the scale of the Klamath Bioregion difficult in practice because regionwide meetings required extensive travel. The North Coast INACC met on the coast; it did not cover the entire Klamath Bioregion. No matter where regionwide meetings might be held in the Klamath, most participants would have to spend the better part of a day simply traveling to and from meetings. Travel requirements quickly dissuaded even the most willing individuals from participating on a bioregional council.

Subregional Groups Replace the Stillborn Bioregional Council

The new subregional groups met regularly in the Klamath Bioregion's population centers, which were also the county seats. County supervisors also provided local leadership. Therefore, the subregional groups gradually identified themselves with counties—the original form of regional government in California. Hence, there was the Siskiyou-Shasta group, based primarily in the Siskiyou County seat of Yreka near the Oregon border, but which also included participants from Shasta County to the south. There was the Trinity Bio Region Group, which met alternately in Weaverville, the Trinity County seat, and the small logging community of Hayfork. There was the Northern Klamath Bioregional Council on the North Coast, which met in Eureka, the Humboldt County seat, but also included a few participants from Del Norte County to the north. And there was the Sonoma-Garberville group, which differed in that it covered three counties (Sonoma, Lake, and Mendocino) and was organized around a freeway corridor running north south between San Francisco and Eureka.

Two more groups also emerged the following year, one in Del Norte County and the other in Shasta County, both of which spun off from existing subregional groups.[16] This left Sonoma-Garberville as the only subregional group in the Klamath covering more than one county. Yet the Sonoma-Garberville group did not coalesce and quickly lost momentum without spinning off county-based groups.[17] With three counties involved, participants could not reach consensus on the important problems, let alone solutions to those problems. Moreover, the northern spotted owl did not provide a lightning rod for the group's attention because most of this subregion lay south of the owl's range and was in private ownership; hence, group participants were not greatly affected by the logging injunction on federal land. The collapse of the Sonoma-Garberville group also suggested that a freeway corridor does not necessarily provide a sense of community, even if it reduces travel time for participants.

The only surviving subregional group that nominally covered more than one county was the new group in Shasta County, which dubbed itself the Shasta-Tehama Bioregional Council after spinning off from the Siskiyou-Shasta group. Though the Shasta County group nominally included Tehama County on its southern flank, it routinely met in the Shasta County seat of Redding and was composed overwhelmingly of participants from Shasta County. Tehama County was formally included because it was in Congressman Vic Fazio's district, and Fazio was then, according to one participant, "a powerhouse in Congress."[18] The Shasta-Tehama Bioregional Council was quite different from the other subregional groups because it was not a direct outgrowth of the Klamath Bioregion Project. Instead, the group emerged when the Clinton administration introduced a new program to ease the timber conflicts through economic revitalization grants. Shasta County accordingly wanted a stronger voice in Congress than that provided by its own member, Wally Herger.

The more a subregional group identified itself with a particular jurisdiction—such as a county or national forest—the less it could serve as a means for blurring the boundaries between all jurisdictions. The subregional groups that persisted the longest and cohered the strongest identified themselves with a county, incorporated the county's name into the name of the group, and were led by a county supervisor. BLM State Director Ed Hastey indirectly encouraged this link by promoting county supervisors as local leaders within all of California's bioregions. Hastey strongly supported the idea of bioregional councils, yet he also supported county leadership. This undermined the formation of a bioregional council in the Klamath because county supervisors did not want to relinquish local authority to a regional body, particularly one that they did not create and that promoted the seemingly radical idea of biodiversity preservation.

Some subregional groups also focused their substantive attention on national forests within their counties. Group cohesion and longevity were strongly correlated with the percentage of county land in national forests. National parks, BLM land, and state-owned land were distributed in relatively small parcels in the Klamath, generated little direct income for counties, and were little affected by the logging injunction. The Sonoma-Garberville group and the Northern Klamath Bioregional Council operated in counties that were mostly under private ownership, so they were not directly affected by the logging injunction on federal land. Lacking this focus, these subregional groups were overwhelmed by a broad agenda, including the highly contentious issue of private property rights. Property owners feared that the bioregional process represented a new effort to regulate their property. In Humboldt County, for example, the Northern Klamath Bioregional Council operated in fits and starts because private landowners created a ruckus at meetings.

In sum, the subregional groups conformed more closely with existing jurisdictions than with biogeographic criteria, though local participants were chary in discussing this point. They were also run by individuals already participating in local groups, such as the North Coast INACC, watershed organizations, and CRMPs. For the most part, these participants were pleased that state-level officials were paying closer attention to their efforts, and they hoped that administrative and financial support would be forthcoming, but they were not prepared to cede local control over resource management issues. In this regard, Wallace and Moles repeatedly emphasized grassroots control and passivity on the part of state and federal agencies.

The Klamath Bioregion Project Receives New Funding and Pursues a New Strategy

Following the Eureka meeting and the formation of the subregional groups, the immediate problem Wallace and Moles faced was financial support. Agency funding for the Klamath Bioregion Project dried up after Trzyna was forced out over allegations about misappropriated funds, with the agencies investing only half the money originally pledged. Though Wallace's position within the UC/DANR was secure, Moles had to volunteer his services in the absence of funding. With four subregional groups then existing, Wallace and Moles turned to their colleagues in Cooperative Extension for staff support. Moles, in particular, needed the assistance of county-based extension advisors to ensure that a broad range of local stakeholders attended subregional meetings. Without their assistance, he would have to spend all of his time simply getting these groups off the ground, which meant developing contact lists, selling the bioregional idea to local communities, organizing subregional meetings, and then driving throughout the Klamath to facilitate the meetings.

Extension advisors disseminate university research to farmers, ranchers, foresters, and other resource users. Some live and work in a single community for their entire career, developing strong personal relationships within their community. They report to regional UC/DANR line managers but are relatively independent in how they define problems and carry out day-to-day tasks. This organizational system has its strengths and weaknesses. On the one hand, they are trusted members of local communities because they understand local issues. On the other hand, they tend to be agents of the community rather than agents of the state.

Wallace had no line authority over extension advisors in field offices. UC/DANR line managers were also indifferent to the bioregional effort; they neither distributed the MOU on Biodiversity to field staff nor provided administrative resources or direction. Therefore, extension advisors participated in the subregional groups for personal and professional reasons, not because they were told to do so by their superiors. In this regard, several extension advisors latched on to the Klamath Bioregion Project, with two assuming responsibility for organizing the subregional groups in their counties. So Moles turned his attention to the other groups. Of these, the Siskiyou-Shasta group was dissatisfied with his facilitation methods, and soon replaced him with a volunteer facilitator. That left the group in Trinity County, in the heart of the Klamath Bioregion. Moles accordingly moved in October 1993 from Humboldt County to Hayfork, a timber-dependent community in Trinity County, where logging families and environmentalists were enacting their own version of the Hatfields versus the McCoys.

By this time, the Klamath Bioregion Project had secured stable funding. Moles had considered foundation support in early 1992, which would have ended the link with the Executive Council, but that summer the CDF issued a contract for $89,000, which was renewed in 1993 for $158,000.[19] The CDF coordinator for both contracts was Robert Ewing, who staffed the TTF, coauthored the MOU on Biodiversity, and co-chaired the Staff Committee supporting the Executive Council.

With the new contract, the idea of a Klamath Bioregional Council was officially dead. The contract did not even mention it, stating instead that the "goal of the project is to continue the initiation, coordination, and support of the four Klamath regional subgroups" so the groups could work with the Executive Council in implementing the MOU on Biodiversity. Moreover, in moving to Trinity County, Moles all but abandoned his organizational efforts in other subregions, and the Klamath Bioregion Project was essentially narrowed to a single county. Communication among the subregional groups declined, with members of those groups taking it upon themselves to keep channels open.[20] Moles occasionally attended meetings in other subregions, but he did so with less frequency, in part because his presence was not welcomed.

Within a year, Moles would also be kicked out of the Trinity County group. The basic issue was local control. Local stakeholders and county supervisors resisted meddling by outsiders, whether from public agencies, interest groups, or corporations. As long as the Klamath Bioregion Project was funneling money and administrative support into the region, Moles was welcomed. As soon as he attempted to control the agenda, rather than facilitate the flow of information and the discussion of ideas, he was attacked. Moles became a lightning rod in Trinity County because he did not maintain his neutrality. In a community where economic depression fueled political conflict, local stakeholders quickly pounced on him once he took sides.

The Trinity Bio Region Group

The Trinity County group was slow to blossom but became one of the largest and most active subregional groups in the Klamath. It was initially composed almost entirely of agency officials from the BLM and FS, along with county supervisors from both Trinity and Humboldt Counties. Numbering about a dozen members during its first year, this group looked broadly at the entire Trinity River watershed, which extended west into Humboldt County, where the Trinity River passes through the Hoopa Valley Indian Reservation and joins the Klamath River. The group acted primarily as an advisory council, developing strategies to involve a broad array of local stakeholders as outlined in the MOU on Biodiversity. This was not a simple matter given local divisiveness following the jarring dislocations of the logging injunction, so their plan for a larger meeting of local stakeholders was pushed back to the end of 1992. Meanwhile, participation from Humboldt County waned and the group's geographic focus shifted east to Trinity County.

At that time, the Klamath Bioregion Project played a small role in Trinity County. County supervisors and agency staff ran the meetings and agreed the supervisors should assume a lead role in implementing the MOU.[21] This was precisely what Hastey desired. At that time, he was diluting the local backlash against the MOU on Biodiversity by inviting regional associations of county supervisors to join the Executive Council, while promoting county leadership in all of California's bioregions. In Trinity County, Supervisor Arnold Whitridge initially played the lead role. Whitridge, a Yale dropout, was relatively new to the area, having settled two decades earlier during the back-to-the-land movement. He differed from traditional county supervisors in the Klamath, most of whom were associated with resource extraction. According to Whitridge, "I wanted to get as far away as I could from civilization" and "get a place of my own that I could take care of."[22]

It was not until the second year that Moles began to devote all of his attention to Trinity County, stepping in to facilitate the meetings, which now included several dozen local stakeholders, who formally dubbed themselves the Trinity Bio Region Group (TBRG). With local stakeholders now in the driver's seat, TBRG met several hours every three weeks, rotating between Hayfork and the county seat of Weaverville. While this may not seem like a full-time challenge, it is important to put

Moles's organizational efforts in context. Trinity County was riven by disputes among environmentalists and timber workers. With 74 percent of the county federally owned, the rural economy was hit especially hard by the logging injunction. Unemployment in Trinity County peaked at 24 percent following the injunction—extremely high even for a county whose unemployment rate had not dipped below 11 percent in ten years.[23] Intense emotions were vented routinely at TBRG meetings, where personal attacks were the norm and individuals sometimes cried or stormed out. Therefore, Moles had to exert substantial energy outside the meetings, mending hurt feelings and assuaging concerns about the state-sponsored Klamath Bioregion Project. He also resided in Hayfork, a lumber town of less than 2500 people. Because Hayfork was not on Trinity County's main highway, it could not count on tourism for new jobs, so it depended on resource extraction.

Despite the prevailing animosity, TBRG was not the first forum to bring environmentalists and resource users together. TBRG's profile was high, occurring within the context of the northern spotted owl, but other cooperative efforts already existed in Trinity County, which provided social infrastructure for TBRG itself. Of these, the Trinity River Restoration Project (TRRP), a multimillion-dollar program funded by Congress, was the most significant.[24] Salmon and steelhead runs in the Trinity River had been decimated in the 1970s by the Central Valley Project, a multiriver project constructed and operated by the BoR, which included a diversion system that moved water east to the Sacramento Valley rather than allowing it to flow west to the Pacific Ocean. Because the Trinity River was dammed upstream, with about 90 percent of the water in the upper river system diverted, historic water flow was not flushing silt in the mainstem out to sea. Therefore, several CRMPs were formed to control the terrestrial erosion that was choking spawning beds. The owl's image spilled over into CRMP meetings, where some individuals couched their participation in terms of preserving salmon and steelhead habitat to avoid similar legal stalemates.[25] The TRRP project leader also believed the CRMP process was necessary for building public acceptance of restoration alternatives, and he viewed TBRG in the same light.[26]

With several cooperative efforts already underway in Trinity County, Moles was not starting from scratch. Indeed, state and federal officials wanted the MOU on Biodiversity and the Klamath Bioregion Project to integrate local partnerships with agency planning efforts, and TBRG included many crossover participants. Yet the big issue in Trinity County in the early 1990s was not river restoration; it was unemployment and the political controversy surrounding job creation after the logging injunction. Local environmentalists were adamant that lost jobs should be replaced through economic diversification and retraining rather than renewed timber harvests. Timber workers simply wanted their old jobs back; few were pleased by the prospect of gathering alternative forest products, like edible mushrooms and decorative pinecones—even if some pinecones sold for more than $10 apiece in Japan. Thus, TBRG became a forum for debating the private use of public lands, with individuals searching for consensus-based strategies to expand resource extraction in national forests.

Participants credited Moles and Wallace for bringing a broad array of local stakeholders to the meetings, and allowing participants to discover for themselves that they shared broad goals regarding the socioeconomic health of their communities and the ecological health of the surrounding landscape. The group process began to break down, however, when TBRG tried to develop consensus-based plans and proposals. Consensus was difficult to achieve in this highly contentious atmosphere, so different factions produced their own proposals, some of which were presented to the agencies as TBRG proposals, though the group never reached consensus on them.

It was at this point that the Klamath Bioregion Project began to lose credibility because participants believed that Moles favored some proposals over others. As recalled by one agency official, a long-time TBRG participant who gave Moles the benefit of the doubt,[27]

> Anybody from the outside really has to try and appear neutral and not favoring or leaning any way or the other, that they're here to try and help the local people as a group come to a common solution or find that common ground. Now I think where things start to fall apart is when you start to be perceived by any one of these groups as being on the other side. Whether it's a matter of where you live or whatever—maybe just some of your own sympathy, if somehow you start sympathizing more with one side than the other—I think that really gets in the way and can cause problems, and I think that's probably what happened to Jerry. He was seen by some as not being neutral or equally on everyone's side. And maybe it's impossible to do that; maybe no one could come in here and keep everybody happy. It's just like a baseball pitcher: you're good for four or five

innings and then you gotta leave and somebody else has to come in and you take it as a rotation.

The situation was aggravated when federal and state money started flowing to fund local projects, because the funds were doled out on a competitive basis and hoarded by those who received them.

Money and Cooperation, Part 1: The President's Forest Plan

When President Clinton assumed office in 1993, he promised to defuse the Western timber conflicts and quickly convened a conference on April 2, 1993, in Portland, Oregon. Local stakeholders from Northern California, including Trinity County, received invitations. At the close of the Portland Forest Conference, participants were instructed to return to their communities and develop consensus-based projects for managing public and private lands in the owl's range. One TBRG participant took this charge seriously. Nadine Bailey, a logger's wife from Hayfork and ardent timber industry supporter, received a great deal of attention at the Portland Forest Conference from the White House and the news media. Bailey annoyed her timber allies by accepting Clinton's challenge and extending an olive branch to local environmentalists. Some accused her of selling out, but she believed cooperation was the only option: "It became obvious to me that unless federal environmental laws are changed—and it looks as though they won't be—we'll have to work within them."[28]

Two months after the Portland Forest Conference, the Clinton administration released its draft Forest Plan, which sought to protect old-growth-dependent species while allowing limited logging to resume on public lands. The Forest Plan also provided financial assistance to affected communities. White House officials planned to submit a revised version of this plan to Judge Dwyer, who they hoped would accept it as a habitat management plan for the owl and lift the logging injunction he had imposed two years earlier. Therefore, the Clinton administration sought political support for the plan, which represented a compromise that neither environmentalists nor the timber industry liked. Vice President Al Gore called Bailey after the Forest Conference to applaud TBRG's efforts and urge support for the President's Forest Plan. TBRG was now in the national limelight. Bailey had direct access to Vice

President Gore and to Doug Wheeler, the California State Secretary for Resources.

Yet TBRG's attention was now turning to Portland, where a new federal office was established to implement the President's Forest Plan. The Sacramento-based Executive Council had sponsored the Klamath Bioregion Project to integrate planning and management processes in the Klamath, but the Clinton administration was now dangling much bigger carrots in front of the subregional groups. From their perspective, it was not clear what the Executive Council even wanted, other than vague calls for more coordination, nor was it clear whether the agencies would ultimately listen to their recommendations. The Siskiyou-Shasta group, for example, asked the Executive Council for a written agreement that the signatory agencies would follow agreements reached by the subregional groups, but they did not get it. The President's Forest Plan, on the other hand, offered the groups significant financial incentives, along with opportunities for public participation in national forest planning. But it created a competitive funding system, which compounded the very balkanization that state-level officials in California had sought to overcome in the first place.

Clinton's Forest Plan had two components, one ecological and one economic.[29] Each operated under a MOU laying out a framework for interagency cooperation, and each was coordinated by the U.S. Office of Forestry and Economic Development in Portland, a new agency established in late 1993 that reported directly to the White House. The ecological component was known as "Option 9" because it was based on the ninth of ten alternatives for managing federal lands and multiple species in the owl's range. It covered more than 24 million acres of federal land in three states, about one-third of which were already protected from timber harvests because they were inside national parks and wilderness areas. The remaining land was divided into three types: (1) old-growth reserves and riparian areas, in which logging would be restricted to protect the owl and other species; (2) matrix lands, in which traditional logging practices would continue under revised standards and guidelines; and (3) adaptive management areas (AMAs), within which experimental harvesting techniques would be practiced and monitored, with extensive public input.

The AMAs were of great interest to local communities because public participation was encouraged in developing AMA plans. Trinity County received the largest AMA. At 488,000 acres, the Hayfork AMA contained 32.5 percent of the total land area in the ten-unit AMA system. TBRG subsequently devoted much of its time to developing plans for the AMA, convincing national forest supervisors to adopt the plans, and lobbying for federal money to fund planned activities. Proposed actions included shaded fuel breaks along existing roads to keep fires from spreading, and salvage logging in burned areas to provide timber for local mills. Though local communities competed for AMA funding, this competition did not divide communities internally because, as in Trinity County, there was only one AMA in the vicinity.

The economic part of the President's Forest Plan caused much greater havoc within TBRG. To assuage opposition to Option 9, President Clinton created the Northwest Economic Adjustment Initiative, which funneled $1.2 billion into Oregon, Washington, and California for economic revitalization, diversification, and infrastructure development in communities affected by the logging injunction. Because this money was doled out on a competitive basis over several years, it fueled distrust within communities as individuals, factions, and coalitions scrambled for it. The situation was aggravated in Trinity County because few people understood how funding decisions were made, which raised suspicions about why some proposals were funded while others were not.

These proposals were reviewed at the local and state levels by interagency groups, known as Community Economic Revitalization Teams (CERTs). Each county appointed one supervisor to sit on the California State CERT, and each county developed its own strategy to forward proposals to the State CERT. In Trinity County, the board of supervisors organized the Trinity County CERT. While some TBRG participants wondered about undue influence within the Trinity County CERT, most of the filtering actually occurred at the state level, where counties were placed in the awkward position of competing against each other for California's share of the money.[30]

In Humboldt County, the Northern Klamath Bioregional Council consciously avoided conflicts of interest within the group by staying out of the CERT funding process. The Shasta-Tehama Bioregional Council, on the other hand, embraced the CERT process. It was a relatively late entrant on the scene, spinning off from the Siskiyou-Shasta group, following the emergence of the President's Forest Plan, to use the CERT funding process as an engine for economic growth and revitalization. The Shasta County Board of Supervisors authorized the group to run the CERT process in Shasta County. Led by Shasta County Supervisor Francie Sullivan, who sat on the California State CERT, the group reached consensus quickly on proposals and was not divided by factions submitting their own proposals.

The CERT money was intended to fund projects that would assist dislocated timber workers and their families, develop small businesses and infrastructure in affected communities, or promote ecosystem-friendly jobs in the woods. Yet political influence sometimes overshadowed the funding criteria. The city of Los Molinos, for example, received a $1.4 million grant for a water and sewer project, one of the first projects funded. Yet Los Molinos is in the Sacramento Valley—well outside the owl's range. It lies on the eastern side of Tehama County, surrounded by the agricultural lands of the Sacramento Valley. Los Molinos hardly fit the formal funding criteria. Somebody was clearly pulling strings for the city and Tehama County because the State CERT received more than 800 proposals that first year, giving decision makers plenty of proposals from which to choose.

In Trinity County, five proposals were funded, including $1.2 million for a sewer project in Weaverville. None of TBRG's proposals were funded, but suspicions arose when some TBRG participants were funded. The primary targets of suspicion were Lynn and Jim Jungwirth. The Jungwirths championed a new organization in Hayfork, the Watershed Research and Training Center (Watershed Center), which opened in 1993, to create local jobs through the development of new forestry techniques, economic diversification, and retraining. The Jungwirths were regular TBRG participants and their idea for the Watershed Center had previously been included in a larger TBRG proposal submitted directly to the White House after the Portland Forest Conference.[31] Some TBRG members subsequently believed that TBRG should have a controlling interest in the Watershed Center because the original idea had been pitched in a TBRG proposal.

Yet this pie-in-the-sky proposal for $190 million was not funded, so TBRG members subsequently developed competing plans as they jockeyed for position before the Executive Council in Sacramento and the U.S. Office of Forestry and Economic Development in Portland. The Trinity County Board of Supervisors expressed concern at this point that some TBRG participants were passing themselves off as representing Trinity County in Sacramento, without the blessings of county supervisors.[32] When the CERT funding apparatus was implemented, and competition for money from the Northwest Economic Adjustment Initiative accelerated during the latter half of 1993, the rift between TBRG and the Watershed Center grew. The Jungwirths pitched their own proposal and received nearly $100,000 for the Watershed Center. By November, the split was complete, with Jim Jungwirth proclaiming that the Watershed Center no longer had any connection with TBRG.[33] The unified vision of the Klamath Bioregion Project was now collapsing within the county itself. Competition over money was tearing at the weak fabric of cooperation in Trinity County, but the situation was about to get much worse.

Money and Cooperation, Part 2: The Klamath Bioregion Project

The Klamath Bioregion Project also brought money into Trinity County, much of it raised by Moles. Ironically, this contributed to his downfall in TBRG, rather than raising his stature, because he did not allow TBRG to control how the money was spent. Most notably, Moles raised seed money for the Klamath Province GIS Project, which brought computers, software, and a GIS consultant to Trinity County. Yet he did not turn the project over to TBRG, or to any other subregional group in the Klamath, even though these groups were—in the absence of a bioregional council—the *raison d'être* of the Klamath Bioregion Project and his CDF contract. Instead, he housed the new GIS Project in the Watershed Center, which was becoming increasingly estranged from TBRG.

To get the Klamath Province GIS Project off the ground in 1993, Moles and Wallace raised seed money from the FS ($25,000), BoR ($25,000), and FWS ($10,000).[34] Moles's second-year CDF contract for the Klamath Bioregion Project also included roughly $70,000 for the Klamath Province GIS Project.[35] This represented significant funding and intellectual capital for a sparsely populated county, and TBRG participants were pleased to have the new Klamath Province GIS Project located in Trinity County. Yet, by housing the computer equipment and GIS consultant in the Watershed Center, and by volunteering his services as an advisor to the Watershed Center, Moles closely associated himself and the Klamath Bioregion Project with the Watershed Center. In other words, he was putting all of his eggs not simply in one county, but in one organization, and this organization was not TBRG, the subregional group that his CDF contract was intended to support and that represented the broadest array of local stakeholders.

As conflicts between TBRG and the Watershed Center intensified, the Trinity River Restoration Project remained neutral, funneling federal money to both organizations. In TBRG, this money went to Bailey, Truman & Associates, a new consulting firm representing the environmental-timber alliance, to cover coordination expenses for TBRG.[36] As Bailey, Truman & Associates increasingly assumed administrative responsibilities for TBRG, Moles's role diminished. Meanwhile, his role in the Watershed Center grew. As conflicts intensified, an outside mediation specialist was called in to help resolve the dispute. Rather than backing away from the conflict, Moles dug in his heels and stacked the deck prior to mediation by writing a letter to the mediator that favored the Jungwirths and the Watershed Center.[37]

The ostensible purpose of this letter was to provide background on the dispute in Trinity County as a prelude to mediation. But the text of the letter was one-sided, lauding the efforts of the Jungwirths while claiming that others were intentionally stirring up trouble to undermine the Watershed Center. Moles thereby undercut some members of the community behind the scenes, while proclaiming at public meetings that he was creating a "big tent" and "level playing field" for all local stakeholders. Unfortunately for Moles, the mediator—Betsy Watson—did not treat his letter with strict confidentiality. At that time, Watson was working with another subregional group, the Northern Klamath Bioregional Council in Humboldt County, and she shared her file with a leader of that group to gain some perspective on the events occurring in Trinity County. This individual was Kim Rodrigues, the Cooperative Extension advisor in Humboldt County, who had volunteered to assist the Klamath Bioregion Project when Moles and Wallace asked extension advisors for help in 1992.

Moles and Wallace had initially organized some of the subregional meetings in Humboldt County, but participants there did not believe their techniques were effective. Therefore, as in Siskiyou County, they were pushed aside as facilitators. Yet local participants still expected Moles and Wallace to play a coordinating role between the subregional groups and with the Executive Council, as called for in their contract. Instead, Moles and Wallace largely disappeared from Humboldt County. By 1993, the Northern Klamath Bioregional Council was concerned that Moles and Wallace had distanced themselves from Humboldt County and were making exaggerated claims about their accomplishments to the Executive Council. Moles and Wallace also refused requests by this subregional group for documents that would verify their responsibilities and claims. The group's steering committee even wrote a letter to Resources Secretary Doug Wheeler expressing these concerns.[38]

Having already become concerned about the relationship between the Klamath Bioregion Project and the subregional groups, Rodrigues decided that TBRG should know about the content of Moles's letter to the mediator because it indicated that TBRG's facilitator had taken sides in a local dispute. So she copied the letter and sent it to TBRG. This was an extraordinary move. Not only did Rodrigues knowingly compromise confidentiality, she did so to expose the activities of colleagues in her own agency. As a county-based advisor in Cooperative Extension, her loyalties were very much with local communities, including the neighboring community in Trinity County. Therefore, she willingly risked her standing within her own agency (UC/DANR) to protect community interests.

Moles tried to claim the moral high ground, and retain his credibility within TBRG, by arguing that Rodrigues stole the letter. This line of reasoning, however, was rebuked by the mediator and by several TBRG participants, in part because she could not steal something to which she had been given access.[39] Instead, Rodrigues was supported by most TBRG participants, including field staff in other agencies, because they already suspected that Moles was operating behind their backs, manipulating outcomes in the name of TBRG while acting on behalf of the Watershed Center.[40] Because Moles was a new resident in the community and received his consulting fee through CDF headquarters in Sacramento, his actions were additionally suspect. Having already been replaced as TBRG coordinator, TBRG participants decided that Moles should absent himself from all meetings for six months. Some participants even placed calls to Sacramento, threatening to sue the agencies if he was not removed. Moles's contract with CDF expired before his six-month expulsion ended, and he left Trinity County.

Cooperative Extension Takes Over the Klamath Bioregion Project
In the summer of 1994, three years after conceiving the Klamath Bioregion Project, state officials pursued a new tack. Rather than hiring a new project coordinator, they funneled money directly to local and subregional groups through Cooperative Extension. TBRG and the Northern Klamath Bioregional Council supported this new strategy as the problems with Moles and Wallace intensified. Their support indicated that extension advisors were still trusted in local communities within the bioregion, even though UC/DANR's reputation had been tarnished by Wallace and Moles.

Yet the idea of a bioregional council was no longer on the radar screen. The money was simply dispersed to fund specific projects at the local and subregional levels. Moreover, three of the first fourteen projects funded in 1994 were located in other bioregions—one each in the Sierra, Sacramento Valley, and Modoc Bioregions. Without denying the merits of these projects (most of which focused on educational outreach), they were small compared to projects funded through the federal CERT process. The Klamath Bioregion Project dispensed roughly $120,000 each year—less than 1 percent of the funding coming to Northern California from the President's Northwest Economic Adjustment Initiative. This provided little leverage for the Executive Council to shape the behavior of subregional groups. The state funds likely served symbolic purposes, but they were not an effective policy tool for promoting regional integration, particularly in the absence of a coordinating body, such as a bioregional council.

Why Did the Agencies Not Play a Greater Role?

One interpretation of this case is straightforward: state-level efforts to encourage cooperation were stymied by travel time at the bioregional level, and were broadsided at the subregional level by competitive incentives for funding under the President's Forest Plan. It did not help matters that the contractors also pursued their own agenda, which undermined the MOU on Biodiversity and subregional cooperation. While this interpretation is certainly plausible, it does not explain why the agencies did not play a greater role, particularly on habitat planning and management, which was the fundamental purpose of the MOU on Biodiversity, Executive Council, and Klamath Bioregion Project. Thus, two additional explanations seem necessary. One revisits the subversion hypothesis from chapter 4. The other focuses on line-manager intransigence within the bioregion.

The Subversion Hypothesis Revisited

Governor Wilson's political appointees were of two minds on biodiversity issues, sometimes pursuing incompatible policy initiatives at the state level and within the bioregions.[41] Some were serious about the principles in the MOU on Biodiversity. This faction included Resources Secretary Wheeler, who chaired the Executive Council. Having come from the World Wildlife Fund and Conservation Foundation, his enthusiasm for the MOU, Executive Council, and bioregional activities seems undeniable. Other political appointees perceived the Klamath Bioregion Project in a different light. They viewed the subregional groups as a means for funneling federal money into California under the Northwest Economic Adjustment Initiative, but did not support regional planning efforts to protect biodiversity.

The latter included Terry Barlin Gorton, who sought to weaken environmental regulations. Governor Wilson initially appointed Gorton to the State Board of Forestry, which she chaired from 1990 to 1993. During her tenure, Gorton worked to dismantle forest practice rules to ease the regulatory burden on private landowners and timber companies. In 1993, Wilson created a new position for her in the California Resources Agency: Assistant Secretary for Forestry and Rural Economic Development. In that capacity, she chaired the California State CERT, which made her the lead individual in California for dispersing federal money from the President's Northwest Economic Adjustment Initiative. Ostensibly an assistant to Secretary Wheeler, she routinely worked at cross-purposes to him. Wheeler's presence in the Resources Agency allowed Governor Wilson to claim environmental credentials, but Gorton was the political insider—she was married to Wilson's longtime campaign manager, George Gorton. She had also served as Wilson's lawyer during his 1990 gubernatorial campaign. Therefore, she had much better access to the Governor's Office than did Wheeler. Environmentalists accordingly believed that Wilson placed Gorton in the Resources Agency to moderate Wheeler's initiatives, some of which ran counter to the governor's beliefs and the desires of his campaign donors in the timber, agriculture, and building industries. During the 1994 reelection campaign, for example, one environmentalist argued that Gorton served as a watchdog in the Resources Agency: "She's there to keep an eye on the boys so that they don't get too greedy. Wilson's campaign manager is running the show."[42]

Governor Wilson made similar appointments in other Executive Council agencies, including the CDFG, giving the top job to individuals with environmental credentials while appointing others with weak environmental records as assistant or deputy director.[43] It was at this level, in some state agencies, that the subversion hypothesis best applied, and many pointed to Gorton's behavior on timber issues as evidence. Because Gorton did not write or sign the MOU on Biodiversity, did not participate on the Executive Council, and did not create the Klamath Bioregion Project, the strong version of the subversion hypothesis does not apply. Rather, the evidence points to a weak version of the hypothesis, in which she used these initiatives for other purposes.

In this regard, several individuals believed that Gorton used the subregional groups for rhetorical purposes, claiming to federal officials that the groups were an effective substitute for implementing Option 9 in California. As chair of the State CERT, Gorton fought for California's share of the $1.2 billion in federal aid under the President's Forest Plan, but she also used her frequent visits to the U.S. Office of Forestry and Economic Development in Portland as an opportunity to argue that California should be held to different regulatory standards than Oregon and Washington because the subregional groups provided effective environmental protections. On hearing about Gorton's lobbying strategy, Kim Rodrigues, the Cooperative Extension advisor working with the Northern Klamath Bioregional Council in Humboldt County, said, "I've heard that argument, and that's just crazy because we're not there yet; and to call it successful before we are is going to doom it."[44] Nevertheless, Gorton's argument carried weight since federal officials in Portland seldom traveled to California, and thus had little firsthand knowledge of the slow pace at which subregional groups were developing plans and projects, let alone getting FS officials to take them seriously. Gorton's lobbying efforts appeared to pay off because the ecological component of the President's Forest Plan was changed to allow 22 percent more logging in California, while placing further restrictions on logging in Oregon and Washington.[45]

There is no direct evidence that Gorton sought to use the existence of the subregional groups to subvert Option 9 implementation in California. Nevertheless, the indirect evidence is compelling. Gorton was already known for weakening environmental regulations on private timberland, and she advocated weakening the state and federal ESAs. Gorton also lobbied federal officials for reduced Option 9 standards, based in part on the existence of the subregional groups. If Gorton and other political appointees indeed viewed the subregional groups as a rhetorical ploy, the weak version of the subversion hypothesis would explain why state agencies provided little assistance to the subregional groups. Other than the CDF contracts for the Klamath Bioregion Project, which originated in CDF's ecological analysis unit (FRRAP), state agencies provided little support to the subregional groups.

Line Managers in Most Agencies Saw Few Benefits in Cooperation

Easily overlooked among the other problems in the Klamath was a fundamental tension in the bioregional strategy throughout California: many line managers in most agencies resisted implementing the MOU on Biodiversity. In the Klamath, line managers found the very idea of a bioregional council threatening and did not support staff participation in subregional groups. This was not the case in all agencies, but line managers in most agencies saw little utility in the bioregional strategy. Some did not even announce the existence of the MOU to their staff. They also resisted state-level efforts to convene an interagency meeting of regional line managers until 1995—four years after agency directors signed the MOU on Biodiversity and initiated the Klamath Bioregion Project.[46] In light of such resistance and the departure of the Klamath Bioregion Project coordinators, state-level officials established their own liaison team in 1994 to communicate directly with local and subregional groups.

Among state agencies, resistance and indifference to the bioregional effort were ubiquitous in the CDFG and CDF, the two agencies whose conflicting management practices provided the initial impetus behind the TTF. Field staff in both agencies received little or no support from their line managers. In the CDF, one line manager made his position explicit in a memo to Moles, stating that "technical assistance will be limited" due to previous programmatic commitments.[47] Far from a show of support, this line manager essentially stated that he would not encourage his staff to participate. Indeed, CDF foresters in the field bluntly reported that their line managers neither allowed them official time to participate in the subregional groups nor offered the groups administrative assistance. Attendance records also indicate that CDF staff routinely attended subregional meetings, while line managers did not.

In the CDFG, line managers were no more forthcoming. The CDFG director did not even distribute the MOU on Biodiversity to his regional managers until eight months after the MOU was signed, under a cover memo indicating tepid support for the agreement (chapter 4). Not surprisingly, his limited enthusiasm failed to inspire line managers to support the regional initiative. CDFG line managers in the Klamath did not announce the existence of the MOU to their staff. Instead, CDFG field staff typically heard about the MOU from sources outside their agency. One field staff member, who regularly participated in the Northern Klamath Bioregional Council, reported that his hierarchical superiors did not support his participation in the group, and, in a candid letter, he expressed his frustrations about the absence of management support in several agencies.[48] Management support for the bioregional effort within the CDFG was so weak that individuals outside the agency even commented on it. Rodrigues recalled attending a large meeting of CDFG staff around 1993, in which "more than half the room had never even heard of the MOU, let alone could they say they could support it because they didn't have any inkling it had been signed or what it meant in terms of changes to them."[49]

This does not mean that all field staff wanted to participate and were frustrated in doing so by line managers, but they participated to a much greater degree than did line managers. Some field staff would not have participated regardless of management support. Even among ecologists, enthusiasm was not uniform. Though many ecologists initially saw the subregional groups as a means for promoting interagency planning and management, some became frustrated by the slow pace of consensus-based decision making. This included several members of the North Coast INACC, who were core members of the Northern Klamath Bioregional Council. On the eastern side of the Klamath Bioregion, one CDFG ecologist initially attended subregional meetings on a regular basis, but stopped participating because the time commitment did not yield proportionate results in species and habitat protection. In his words,[50]

> It's a matter of priorities. To be honest with you, what motivates me is having success in doing something for resources on the ground that I can count, and that is either in terms of acres, numbers of individuals, [or] dollars to purchase habitat. And I couldn't count the benefits in that fashion there. As a matter of fact, I felt that although I was making contacts with the community it really wasn't productive in establishing resource conservation programs. Ideally it is, theoretically it is, but on the ground I never saw a project that I thought was very good for conservation of endangered species, of watersheds, of water quality, of habitat, I mean you name it.

Though this ecologist did not disassociate himself from the group, he thought his time was better spent implementing existing laws and regulations.

While ecologists generally expressed ambivalence, with optimism gradually overcome by frustration, other field staff maintained enthusiasm in the subregional groups because they were concerned about the socioeconomic health of the communities in which they lived and worked. Rural field staff, in particular, saw the groups as vehicles for uniting local communities and rejuvenating local economies, regardless of their ecological views. The logging injunction had a great impact on some rural communities, rippling through local economies, aggravating social problems like alcoholism and domestic abuse, and highlighting ideological divisions. According to one CDF forester in the Klamath,[51]

> Personally, this is my twelfth year here, so I feel a real strong affinity or bond to this area, and I feel a responsibility as a private citizen to be part of it. So I think I probably have increased my involvement due to my long-term residence here. . . . It's not uncommon [in CDF] to have people in forester positions to be in places ten years or more.

Rural field staff tended to think of their role in cooperative efforts as members of a local community who were bringing their agency-based and professional expertise to a group representing that community, rather than as agency officials who spoke for their agency with the authority of their agency behind them.

The FS long discouraged such affinities by rotating forest rangers from one location to another (Kaufman 1960). Yet even the relatively impermeable FS has responded to changing demands and constraints. The FS was one of the few agencies providing management support in the Klamath, largely due to top-down pressure by Regional Forester Ron Stewart, an ecologist whose agency suffered the brunt of the logging injunction in California. Nevertheless, FS line managers were used to controlling decision-making processes, so FS participation was not consistent. By comparison, BLM line managers and field staff participated in all of the subregional groups. Participation came naturally to BLM officials because CRMP was ingrained in the agency's culture. Hastey also pressured his line managers to participate in bioregional efforts. In the Klamath, BLM line managers even staked their careers on ecological preservation. The Redding Resource Area management plan for 1993 to 2008, for example, envisioned selling BLM's heavily grazed and fragmented parcels in the foothills, while purchasing and consolidating holdings around ecologically significant riparian zones and wetlands.

Associate State Director Al Wright also spread the MOU on Biodiversity far and wide in the BLM. Wright previously served as BLM District Manager in the Klamath, and like Hastey, he was frustrated by the legal battles emanating from the timber conflicts, and by the fragmented ownership and planning processes that exacerbated the problem. When he became Associate State Director in 1991, Wright joined Hastey's effort to promote the MOU throughout the state. In most other agencies, line managers were either antagonistic or indifferent to the MOU. In the BLM, they were socialized to it. As Wright said, "My guess is that you'll have a hard time finding a field person who works for BLM who doesn't know it's important to Ed and I."[52] For BLM line managers and staff, the MOU on Biodiversity became gospel under Hastey's leadership, complementing the agency's historical role in CRMP.

In the FS, line managers were much less used to cooperating across jurisdictions. Moreover, the FS was an organizational supertanker—it was large, seemingly impermeable to external influences, and unable to change directions quickly. When the logging injunction left the agency rudderless and aground in the Pacific Northwest, Regional Forester Stewart used the opportunity to change agency behavior in the Klamath. As an ecologist, Stewart supported the principles in the MOU on

Biodiversity, and he announced his support to all forest supervisors in a strongly worded memo, seemingly ignoring the counsel of a FS attorney who argued that the MOU was inconsistent with the agency's multiple-use mandates (chapter 4). Stewart even sent a follow-up memo to forest supervisors, requesting status reports on their outreach efforts.[53] In the Klamath, he specifically instructed forest supervisors to assign liaisons to the subregional groups. These liaisons did not have line authority, but they routinely attended subregional meetings and provided a direct link between the subregional groups and forest supervisors. For an agency known for its insularity (and at that time downsizing under the Clinton administration), this was not a trivial gesture. National forest supervisors also provided administrative support to the groups, covering fees for meeting rooms and out-of-town speakers, along with duplication and mailing costs. The BLM provided fewer administrative resources because the agency operated on a much smaller budget, but BLM line managers attended subregional meetings with much greater frequency than FS line managers.

In sum, line managers in most agencies resisted cooperation in the Klamath because they wanted to preserve their autonomy. In this regard, they were similar to the directors of state agencies and federal regional offices. Line managers also did not have preexisting forums within which to interact, like the North Coast INACC or CRMPs, making it more difficult for them to develop social capital. Therefore, they needed leadership from agency directors. In the BLM, Ed Hastey had been touting the benefits of cooperation long before the MOU on Biodiversity, and continued to do so thereafter. In the FS, Ron Stewart also played a pivotal role during his stint as Regional Forester from 1990 to 1994. Without Stewart, a research ecologist, it is not clear that forest supervisors and district rangers would have provided any support to the subregional groups.

Summary

Cooperative planning and management occurred in the Klamath, but not at the bioregional level, and not without regard for existing jurisdictions. The size of the bioregion made traveling to meetings difficult, which led participants to meet as subregional groups. These groups subsequently identified themselves with counties, and focused their policy initiatives primarily on national forests within those counties. While jurisdictional concerns did not disappear in the wake of the MOU on Biodiversity and the Klamath Bioregion Project, some public officials did look more broadly beyond their jurisdictions, developing a new appreciation of the bureaucratic landscape around them. In this respect, interagency and public-private cooperation varied both within and among agencies. Within agencies, field staff cooperated much more than did line managers, while the enthusiasm of ecologists typically waned as they became frustrated by the slow progress of the subregional groups in developing projects to preserve biodiversity. Enthusiasm remained strong among field staff who saw the groups as a forum for improving the socioeconomic stability of the communities in which they lived.

Line managers resisted cooperation at the regional, subregional, and local levels for the same reasons as their state-level counterparts. They sought to preserve their autonomy to maintain predictability and certainty for their organizational units. They were accordingly wary of the subregional groups because these groups demanded increased participation in agency planning processes. The Klamath Bioregion Project contractors aggravated this situation by repeatedly stating that agency officials should play a passive role, and that local stakeholders should increasingly control agency decision-making processes. This was precisely what most line managers feared. They supported cooperation if it could enhance or maintain their autonomy, not threaten it. They had to believe that cooperation would buffer their organizational units from uncertainty, not subject them to additional, unpredictable forces.

Among the agencies, the BLM stood out. Line managers in other agencies generally resisted cooperation unless prodded by agency directors. In the state agencies, leadership was virtually absent. Resources Secretary Wheeler supported the bioregional effort, but his authority over the departments housed in the Resources Agency was limited. In part, Wheeler was constrained by other political appointees in the Wilson administration, many of whom did not support efforts to preserve biodiversity. Hence, they largely ignored the MOU and did not encourage line managers in the Klamath to support the bioregional effort. This meant that field staff in pivotal state agencies such as the CDF and CDFG typically learned about the MOU and the subregional groups from outside sources rather than through formal chains of command, and were not allocated time and resources to participate. Funding for the Klamath Bioregion Project came from CDF's analytic unit (FRRAP), but Ewing did not have line authority over CDF units in the field.

Interagency cooperation in the Klamath Bioregion was remarkably similar to cooperative dynamics at the state level in Sacramento, with one important exception. Field staff were not involved with the Executive Council or its Staff Committee. Agency directors and staff in Sacramento were far removed from rural communities. This is not to say they were unconcerned about the socioeconomic health of these communities, but rather that they participated in interagency efforts for different reasons than field staff, many of whom willingly volunteered their time for interagency and public-private efforts because of their personal ties to the communities in which they lived and the natural resources on which these communities depended. Living in a place creates different motivations than professional training or managerial duties, thereby blurring the boundaries between agency jurisdictions and between the public and private sectors.

All three sets of actors, however, were fundamentally driven by a common concern. Something caused them enough grief to make lengthy meetings, travel, and associated efforts worthwhile. In this regard, court enforcement of federal statutes undergirded cooperative efforts in the Klamath. Judge Dwyer's injunction virtually shut down logging operations on public lands in the range of the northern spotted owl for several years. The injunction, and President Clinton's subsequent Forest Plan, gave the subregional groups a focus. As one BLM staff member in the Klamath put it,[54]

> The [northern spotted owl] resulted in economic trouble to those people, and it's the trouble with economics that got their attention. They weren't sitting back with a lemonade in their hand on their veranda in the afternoon saying "That endangered species is really in trouble, let's go fix it." No, it was "That endangered species has got our job! What are we going to do to pay the bills next month?" And so they got their heads together and started to think of things.

In counties containing relatively little federal timberland, the subregional groups did not have a clear issue around which to cohere. Meanwhile, line managers in most agencies simply hoped the whole thing would blow over. Long-term cooperative planning and management threatened their autonomy, and from their organizationally parochial and myopic perspective, it was not clear that investing energy in regional or subregional efforts would increase their autonomy in the long run.

In the next chapter, we move from Northern to Southern California. In many respects, cooperative dynamics were similar in the South Coast Bioregion, but state-level officials pursued a different strategy. Rather than hiring contractors to nurture cooperative groups within the bioregion, they hitched the bioregional strategy directly to federal regulations under the ESA. This made the benefits from cooperative planning and management much clearer to local stakeholders and line managers. State and federal political appointees also removed some line managers from decision-making processes, thereby allowing field staff to communicate directly with agency directors and political appointees in Sacramento and Washington. By clarifying the benefits of cooperation, and by removing levels of hierarchy within some agencies, regional and subregional cooperation progressed much further in the South Coast Bioregion than it did in the Klamath.

6

The South Coast Bioregion: Making Regional Cooperation Work through Regulation

In the South Coast Bioregion (see figure 3.1), state-level officials did not try to create a bioregional council. In part, their enthusiasm for bioregional councils as intermediary organizations between local groups and the Executive Council had been tempered by the unsuccessful effort to create a bioregional council in the Klamath. Yet that failure does not adequately explain why they pursued a different strategy in the South Coast, for at least two reasons. First, state-level officials did attempt to create a bioregional council in the San Joaquin Valley Bioregion after the Klamath initiative had failed (chapter 7), so the idea of bioregional councils was not yet dead. Second, events in the South Coast evolved separately from the timber conflicts in the Klamath, which had given rise to the MOU on Biodiversity and Executive Council. Thus, it is likely that regional cooperation in the South Coast would have proceeded on its own course without the MOU, and without direction or support from the Executive Council. To understand why regional cooperation proceeded on a different course in the South Coast during the 1990s, we need to step back to the 1980s.

Before providing this history, it is important to note that cooperative planning and management in the South Coast did not cover the entire bioregion. State-level officials focused on a 6000-square-mile area within the bioregion that contained habitat for several wildlife species the FWS might soon list under the ESA. In doing so, they created a new state program, called Natural Communities Conservation Planning (NCCP), which became the largest cooperative effort to preserve biodiversity in the South Coast Bioregion. NCCP was the counterpart to the Klamath Bioregion Project, but differed in several respects. NCCP covered a smaller area than the Klamath Bioregion Project, it was much larger in administrative overhead, and it was ultimately a much more successful effort.

Yet NCCP got off to a slow start because it lacked a clear incentive for participation. As initially designed, it was intended to be a proactive and voluntary program. Landowners, developers, local governments, and public agencies were encouraged to enroll some of their lands to preserve habitat, on the assumption that enough habitat would be enrolled to maintain species' populations prior to listings, so that the FWS would not need to list these species. Yet without listed species, there was no immediate regulatory threat to motivate private or public actors to make the necessary sacrifices associated with enrolling their lands. It was not until the FWS actually listed a species that significant participation began. Thus, a major theme of this case is the role of an imminent regulatory hammer to provide the fundamental incentive for public and private actors to do something they would not otherwise do. Ecological knowledge and the planning principles of conservation biology were widely accepted by potential participants, but relatively few would make the necessary sacrifices to set aside or acquire lands to preserve biodiversity without the looming shadow of lawsuits under the ESA.

Natural and Social History of the South Coast Bioregion

The South Coast Bioregion is arid and bathed in sun most of the year. Unlike the Klamath Bioregion, where abundant precipitation nourishes tall coniferous forests, the South Coast Bioregion has a relatively short rainy season. The average annual precipitation in Los Angeles is just 12.4 inches, though precipitation increases several fold on the inland mountains (Bakker 1984: 350). Native plants are therefore drought resistant. During the hot, dry summer months, grasses and flowering plants turn brown. Oak trees and drought-tolerant conifers retain their color but grow in smaller stands than their counterparts in Northern California.

These plant species are not distributed randomly over the landscape. They tend to mingle in distinct assemblages, or what ecologists call natural communities. The South Coast Bioregion contains more than a dozen natural communities, including oak woodland, pine forest, chaparral, grassland, and sage scrub. Each natural community provides habitat for a different set of wildlife species, and each is shaped by differences in soil, altitude, slope face, microclimate, and disturbance regimes. These natural communities tend to drift over the landscape through time, following natural and human disturbances. They also weave together to form complex habitat mosaics.

One natural community—coastal sage scrub (CSS)—became the focus of regional conservation efforts in the South Coast Bioregion. CSS is not popularly recognized like some natural communities, such as oak woodland or pine forest, because it is not particularly attractive to humans. From a distance, it looks like a dense thicket of cacti and woody shrubs. Few would venture into CSS for an afternoon walk without a well-cleared path because it lies on the landscape like an impenetrable thicket of brambles. It also dries out and turns brown during the summer, so it is less visually appealing than some natural communities. Moreover, CSS is not easily recognized because its name does not accurately depict its distribution or composition. "Coastal sage scrub" is not confined to the coast; it grows inland—up to 3000 feet in elevation—as well as on the coastal plain. It also differs in composition depending on location. In the South Coast Bioregion, ecologists consider CSS to be a different subtype than similar CSS communities to the north, because it includes cacti such as prickly pear and coast cholla. Yet it also contains typical CSS species, such as California sage, black sage, white sage, California buckwheat, evergreen shrubs, and numerous wildflowers.

Humans have been much more interested in the ground on which CSS grows than in the plants of which it is composed. CSS occurs on some of the most highly priced real estate in the world and has disappeared rapidly in the wake of human development. As the 1980s drew to a close, CSS was becoming the counterpart to old-growth forests in the Klamath Bioregion. Yet, unlike timber harvests, which changed the structural characteristics of forests but left the soil relatively intact for native plant species to regenerate, bulldozers scraped CSS clean from the earth. Nor were CSS plants harvested for commercial use; instead, the entire natural community, and all wildlife species residing in it, was simply replaced by asphalt, houses, and exotic lawn and garden species. Ecologists did not even ponder whether second-growth CSS provided suitable habitat for CSS-dependent species because little CSS survived or regenerated where bulldozers operated. Once the landscape was transformed into subdivisions, shopping malls, and freeways, CSS was as good as gone forever.

Tiny patches remained in and around housing developments, but they were too small and fragmented to provide suitable habitat for CSS-dependent species.

During the 1980s, California's population increased by approximately 670,000 people per year, with more than half that growth occurring in Southern California (McCaull 1994: 283). As development spread out from the urban cores, natural communities gradually disappeared from the landscape. By the end of the decade, up to 90 percent of presettlement CSS was gone, with the remaining patches highly fragmented.[1] Many associated plant species grew in other natural communities and in developed areas, but some wildlife species had evolved in close concert with CSS, and depended on the structural characteristics of CSS for their survival. None of these wildlife species were federally listed, but several sat on the candidate list. If listed, they could wreak havoc on the development industry. The ESA's prohibition on take (covered in chapter 1) does not cover plant species per se, but bulldozing CSS could be construed as habitat modification that harmed listed wildlife species, in the same way that harvesting old-growth forests harmed the northern spotted owl.

While no CSS-dependent species were federally listed during the 1980s, other wildlife species in the South Coast Bioregion were listed because their habitat was severely diminished and fragmented. Conservation efforts for two of these species—the least Bell's vireo and the Stephens' kangaroo rat—provided formative experiences for local, state, and federal officials. These earlier planning efforts produced very different outcomes, but both influenced the subsequent regional planning effort for CSS. Therefore, to understand the evolution of regional cooperation in the 1990s, we must start with habitat conservation efforts at a smaller scale in the 1980s. To understand these efforts, we need to examine the 1982 amendments to the ESA, which established the legal basis for habitat conservation planning on nonfederal land.

Habitat Conservation Planning under the Endangered Species Act

In the South Coast Bioregion, development occurred primarily on private land, and land-use authority primarily rested with local governments. This meant that landowners, developers, and local governments operated under the shadow of the Section 9 prohibition on take. Prior to 1982, the ESA was unyielding with regard to taking endangered fish and wildlife species for economic purposes. Only scientific research and conservation activities constituted permissible take.

This near-absolute ban on take posed economic, political, and ecological problems. Economically, if one knew about the presence of an endangered species of fish or wildlife on private property, the ESA essentially implied an order to cease activities that might cause take. Although the FWS lacked staff to monitor such activities, environmentalists stood in the wings waiting to sue landowners and developers for such infringements, and to sue local and state agencies for permitting them to occur. Politically, the prohibition on take was a time bomb because the ESA lacked a release mechanism to allow limited economic activity to occur. For this reason, economic interests lobbied hard to keep species off the list, which necessarily politicized the listing process. Environmentalists also picked their fights carefully. They did not petition to list every species for which data supported a listing; instead, they typically focused on charismatic species, which limited the ability of property rights advocates to frame endangered species issues as pitting "rats against people" or "bugs against jobs." Ecologically, the absolute prohibition on take was also not entirely sensible. Endangered species suffered from the cumulative impacts of many activities, not simply the few activities someone happened to notice. Therefore, many ecologists argued that it would be more effective to preserve habitat over the long run by acquiring property and adopting formal land-use restrictions than by blocking bulldozers at each site or punishing individuals after habitat was altered, perhaps irreparably. In other words, it made more sense to develop and implement a plan to preserve habitat than to track individual activities eating away at the habitat on a site-by-site, project-by-project basis.

As the 1970s came to a close, economic, political, and ecological interests dovetailed when a novel idea emerged to preserve butterfly habitat near San Francisco. Development creeping up the slope of San Bruno Mountain had been a political issue for years, but it was framed in terms of open space and growth control, not species protection. The San Bruno conflict assumed a dramatically new form in 1975 when the FWS listed the mission blue butterfly as an endangered species and a local environmental group threatened legal action to stop residential and commercial development in the butterfly's habitat. In 1978, the FWS proposed listing an additional species, the Callippe silverspot butterfly. Backed into a corner, the primary landowner and developer struck a deal with environmentalists, agreeing to set aside approximately 2000 of its 3500 acres on San Bruno Mountain as butterfly habitat and open space in return for being allowed to develop the remaining acres. The logic was simple. The developer would be allowed to take butterflies by building on part of the mountain because ecologists endorsed the plan as a means for protecting sufficient habitat to maintain viable populations of both species. In other words, economic development would be allowed to destroy some of the habitat because credible ecologists believed the plan would preserve sufficient habitat to guarantee the species' long-term survival.[2]

This agreement led to the first habitat conservation plan (HCP), but it could not be implemented until Congress amended the ESA to authorize the FWS to issue a new kind of permit that would allow take. When Congress reauthorized the ESA in 1982, new language in Section 10 authorized the FWS to issue permits to nonfederal actors who submit a satisfactory HCP.[3] Taking endangered animal species for economic purposes was no longer prohibited absolutely. Take was now permitted if it was "incidental to, and not the purpose of, the carrying out of an otherwise lawful activity."[4] Hence, the coveted permit to implement an HCP is known as an "incidental take permit." The 1982 ESA amendments established common ground between economic and environmental interests by allowing incidental take during the course of economic activities, while creating an incentive for private actors and local and state agencies to preserve habitat for the long-term survival of endangered species. In other words, Section 10 reframed endangered species debates from "species *versus* jobs" to "species *and* jobs," thereby providing a legal mechanism to avoid political impasses.

To receive a permit, applicants must submit an HCP that specifies: (1) the likely impacts resulting from incidental take; (2) the steps applicants will take to minimize and mitigate such impacts, and the funding that will be available to implement such steps; (3) the alternative actions applicants considered and the reasons such alternatives are not being used; and (4) any other measures that the FWS may require as being necessary or appropriate for purposes of the plan.[5] How applicants meet these conditions is left largely to them. For example, applicants define the planning area, choose the number of species covered, decide who will participate, and select the policy tools for habitat protection. Thus, they can write an HCP covering one acre or a million acres; they can focus on one species or many species; they can submit an HCP individually or with multiple partners; they can request extensive public input or largely ignore it; and they can select from numerous policy tools to implement the plan. Large HCPs typically establish core preserve areas, within which few human uses are allowed, surrounded by buffer zones of less restricted use, but there are numerous ways to acquire, regulate, restore, monitor, enforce, or otherwise manage these areas. To a large extent, this is determined by the applicants, subject to approval by the FWS. This discretion empowers applicants to be creative, and to tailor solutions to local problems.

The 1982 amendments also increased the incentive for interagency and public-private cooperation under the ESA by expanding the focus from site-specific and project-specific impacts to cumulative impacts. While applicants do not have to plan for an entire habitat, HCPs must be based on credible ecological studies of large portions of the habitat because the FWS needs a scientific basis on which to determine that an incidental take permit will not drive the species to extinction.[6] If adequate studies do not exist, applicants must pay for new studies while preparing their HCP. These studies sometimes uncover other activities degrading the habitat, which provides an additional basis for collective action. The FWS can also issue a single permit to a formal coalition of applicants, such as a joint-power authority, a regional council of governments, or a public-private partnership. Thus, HCPs can bring numerous public agencies and private actors to the table to develop long-term plans covering multiple jurisdictions.[7]

Section 10(a) was by no means the last word on species and habitat preservation. Federal agencies, for example, do not have a clear incentive to participate in HCPs because Section 10(a) provides a waiver from the strict prohibition on take under Section 9, not the consultation requirement for federal agencies under Section 7. HCPs are also time consuming and expensive to prepare, and these costs might be to no avail if the FWS listed additional species in the planning area or if new information arose indicating that additional protection was needed. In 1994, the FWS began to provide assurances that additional regulatory burdens would not be made, but the early HCPs were not covered by these assurances, so the potential benefits to applicants were less clear.[8] Moreover, HCPs did not necessarily plan for an entire habitat. In this regard, "habitat conservation plan" has always been a misnomer because applicants select a planning area, not a habitat per se.

HCPs opened a new avenue for preserving biodiversity on nonfederal land, but they did not immediately proliferate. The FWS issued only 14 incidental take permits in the first decade following the 1982 amendments (1983–1992)—one each in Texas and Florida and 12 in California. This slow start is not surprising given that the FWS did not publish regulations for implementing the Section 10 permit program until 1985, did not distribute draft HCP guidelines until 1990, and did not provide legal assurances until 1994. The HCP program then expanded rapidly during the 1990s. As of August 1996, 179 incidental take permits had been issued and approximately 200 HCPs were being developed. Some of these newer HCPs also covered much larger planning areas than their predecessors. Whereas most of the early HCPs covered less than 1000 acres, 25 of the latter HCPs exceeded 100,000 acres (U.S. Fish and Wildlife Service and National Marine Fisheries Service, 1996: i). With the new regulations, guidelines, and strong support from the Clinton administration, HCPs spread rapidly. By April 2002, the FWS had approved 379 HCPs, covering 30 million acres and protecting more than two hundred listed species.[9]

The growth in number and size of HCPs indicates that permit applicants increasingly perceived benefits from large-scale, multispecies cooperative planning, and events in the South Coast Bioregion suggest why. In the 1980s, Section 10(a) was a bold experiment, and California provided the primary laboratory. A few years after the first incidental take permit was issued for San Bruno Mountain in Northern California, several single-species HCPs emerged in Southern California. Two of these cooperative efforts provided important lessons for those who would later develop the regional conservation effort for CSS. From the least Bell's vireo HCP, participants learned that planning at two geographic scales is both possible and desirable, if not without potential snags. From the Stephens' kangaroo rat HCP, they learned why single-species planning could be a costly mistake. Together, these lessons shaped the emergence of regionwide cooperative planning to preserve biodiversity in the South Coast Bioregion.

The Least Bell's Vireo HCP: A Positive Example of Two-Tiered Planning

The least Bell's vireo is a migratory songbird that winters in Baja California, Mexico, and summers in Southern California. It is a small gray bird, not one of the charismatic megafauna that normally attract widespread public attention. It lives in dense streamside vegetation, where willow and wild rose provide typical nesting sites. In western San Diego County, its habitat appeared on maps as thin riparian corridors running from the mountains to the ocean. This habitat had been greatly altered by highways, dams, and flood-control projects. These human activities forced the vireo into marginal nesting areas, where it was more vulnerable to parasitism by the brown-headed cowbird, which lays its eggs in active vireo nests, thereby displacing vireo chicks (BioSystems Analysis, Inc., 1994: 216, 224–225). During the 1980s, the U.S. population of the vireo dropped to roughly 350 pairs.[10] In 1986, the FWS listed it as an endangered species. At that time, San Diego County contained important remnants of the vireo's habitat, and the western half of the county was undergoing rapid growth.

When the FWS listed the vireo, the San Diego Association of Governments (SANDAG) initiated a multijurisdiction HCP covering numerous development activities in the county. SANDAG's unique two-tiered approach combined regional and local planning components. At the regional level, a 30-member task force oversaw the entire planning process, including data collection throughout the vireo's range. The task force also established a recovery goal of 5000 breeding pairs, the minimum viable population suggested in a FWS recovery plan. At the local level, individual HCP advisory committees coordinated stakeholders who developed detailed plans in four river basins. (Two additional river basins were not included in the HCP because they were primarily on federal land, and thus fell under Section 7 of the ESA.) The local HCP advisory committees selected specific planning mechanisms—such as land acquisition, regulatory mechanisms, and management practices—to meet the population targets for each river basin established by the regional HCP task force. With this two-tiered approach, the HCP took a broader view than other HCP processes then underway by planning for large blocks of habitat within the county, rather than narrowly focusing on specific sites or projects. It was a novel approach for HCPs at the time, and it followed logically from the regional perspective of the lead agency, SANDAG, a county-based council of local governments.

State and federal agency staff were enthusiastic about this innovative approach. Kent Smith, an endangered species specialist for CDFG, served on the regional task force overseeing the HCP. Smith participated in other HCPs as well, so he appreciated the unique two-tiered approach. He also believed this HCP represented a general process that could be applied elsewhere and at broader scales.[11] This was a learning experience for Smith, who began to think seriously about developing an institutionalized process to plan for and manage entire habitats, natural communities, and ecosystems. For him, it was the multitiered planning process that mattered, not the details in the HCP itself.

At the same time, a new method for targeting conservation efforts was being developed by Mike Scott, a FWS ecologist in California. Scott's method came to be known as *gap analysis* because it searches for gaps in biodiversity protection across entire landscapes. Using GIS programs, planners could layer data on vegetation types, species distribution, and jurisdictional boundaries. This approach allowed them to analyze which natural areas were already protected, how they were protected, and where additional protection was needed. Smith subsequently helped Scott develop a national gap analysis program, which was implemented in California by Frank Davis at the University of California, Santa Barbara.

With the advent of GIS programs and gap analysis, a technical means was now available for analyzing entire habitats, natural communities, and ecosystems throughout a bioregion. Moreover, the least Bell's vireo planning process suggested an institutional means for developing multitiered, multispecies conservation plans at these scales. Hence, there was no longer a technical excuse for practicing "emergency room" planning and management—that is, waiting until species like the California condor were on their deathbeds, at which point recovery is expensive and difficult. Ecologists now believed they could preserve biodiversity within a given area through a single planning exercise, thereby keeping species off the threatened and endangered lists and obviating the need for costly recovery efforts. As Smith later described the overall planning objective, "What we need to do is try and get on the front edge of the extinction curve; and try, through cooperative planning and implementation programs, to paint a picture of where we want to be in a hundred years or two-hundred years from now, and develop a strategy for moving in that direction."[12]

Despite the enthusiasm of state and federal ecologists, the least Bell's vireo HCP was not without snags. One of the four river-basin plans was abandoned due to strong landowner opposition. Some participants also groused that the planning process itself was unnecessary because projects in riparian areas generally required a separate permit under Section 404 of the Clean Water Act, which might provide the functional equivalent of an incidental take permit, because the federal agency issuing the 404 permit must consult with the FWS under Section 7 of the ESA (Beatley 1994: 123).[13] Indeed, some participants pulled out because they pursued a Section 7 nexus in lieu of an incidental take permit. In the end, there was no single HCP covering the vireo's entire habitat, but rather multiple HCPs, each covering several species in a specific planning area. Because participants had an incentive to pull out of the larger planning process if their legal requirements could be met more quickly under Section 7 or by preparing their own HCP, incentives would have to be developed to encourage long-term participation in an integrated planning process at multiple scales.

Despite these problems, the two-tiered planning process was unique. Moreover, it was initiated within the public sector by a local council of governments, rather than by private developers or landowners, which indicated an important role for public agencies in coordinating complex HCPs. It also demonstrated the possibility of a broader, comprehensive look at habitat conservation, rather than simply mitigating the effects of individual projects on a site-by-site basis. If successful, this comprehensive planning approach could avoid protracted battles over each project, reduce uncertainty among applicants about mitigation requirements, expedite permit issuance, and result in a more effective reserve system. This was the vision, at any rate, that emerged from the least Bell's vireo planning process in western San Diego County.

The Stephens' Kangaroo Rat HCP: A Negative Example of Single-Species Planning

Meanwhile, in western Riverside County, events unfolded differently for another listed species, the Stephens' kangaroo rat. As the spotted owl crisis loomed for the timber industry in the Klamath, the image of a "rat" suggested a similar crisis for the development industry in Southern California. Rather than providing a positive example, participants and observers learned why single-species planning could be a big mistake.

Kangaroo rats are more closely related to squirrels than to rats or mice. The Stephens' kangaroo rat looks like a big gerbil, with a long tail and large hind legs, which it uses to bound across the land like a kangaroo. It is endemic to the South Coast Bioregion and nearly endemic to Riverside County. Its relatively small range coincides roughly with the western fifth of Riverside County, spilling over into San Bernardino County to the north and San Diego County to the south. It occurs primarily in grassland but can also be found in sparsely vegetated CSS. Its preferred grassland habitat is mostly on private land, which had either been converted to agriculture in the past or was experiencing rapid suburban sprawl in the 1970s and 1980s. Because it is nocturnal, spending the day in underground burrows, most people in western Riverside County had never seen one. Thus, it was very easy for proponents of development to frame the issue as pitting "rats" against people.[14]

By the 1980s, about two-thirds of its habitat was gone, and the remaining habitat was highly fragmented. With its population declining, the FWS listed the Stephens' kangaroo rat as an endangered species in 1988—on Halloween. It had been on the state list since 1971, but state laws proved inadequate to protect the species in the face of development, so the FWS listed it (BioSystems Analysis, Inc., 1994: 75). As with the least Bell's vireo, the state listing was largely symbolic because developers went on with their business. By contrast, local government officials began planning for the species even as the FWS was still considering the listing. By the time the FWS officially listed the Stephens' kangaroo rat, Riverside County had already enacted an emergency mitigation fee ordinance, charging $1950 an acre for development within the species' historic range, to pay for biological and land-use studies. It had also formed a technical advisory committee, consisting of state and federal staff, local planning staff, and representatives from the environmental and development communities. In 1989, a formal HCP steering committee was formed, chaired by an attorney who had already developed a reputation for successfully coordinating the Coachella Valley fringe-toed lizard HCP in Riverside County near Palm Springs.[15]

With all this experience and social infrastructure in place, what went wrong? Rather than providing a complete account, I will simply draw out the most important lesson for conservation planning at broader scales, as perceived by participants and observers in the South Coast Bioregion. This lesson might be called "the single-species trap." While many species were at risk in western Riverside County and could be listed in the near future, the planning process bound the participants to a single species.

In 1990, Riverside County and six cities in the county created a joint-power authority, the Riverside County Habitat Conservation Agency (RCHCA). As the lead agency for the HCP, this new agency would hold the incidental take permit, which would cover all development activities in the HCP within the member jurisdictions. The RCHCA thus played a similar role to SANDAG as a coordinating body, though it was a single-purpose joint-power authority rather than a multipurpose council of governments. Among other things, the RCHCA collected mitigation fees from developers, and used these fees to pay for biological studies of, and acquire habitat for, the Stephens' kangaroo rat. Yet the RCHCA charter was locked into single-species planning, so the agency could not use these fees to study or buy habitat for other species. In another twist of fate, the RCHCA adopted a proportional voting system, weighted by the percent of habitat in each jurisdiction. This voting system gave Riverside County veto power over all the cities, which meant the county essentially ran the RCHCA. One might presume that a county would be more inclined than cities to take a regional perspective, but Riverside County was strongly influenced by the concerns of developers and landowners. As other species loomed on the horizon, local chapters of the Building Industry Association and the Farm Bureau pressured Riverside County to complete the single-species plan before starting plans for other species.

Meanwhile, the RCHCA struck a short-term deal with the FWS in 1990. Developers had been under a full moratorium since the FWS listed the Stephens' kangaroo rat in 1988, but the HCP was still far from complete. To keep participants at the table, the FWS issued a two-year interim take permit to the RCHCA in 1990 to release some land for development. This interim permit allowed for incidental take on up to 4400 acres in the planning area. Though seemingly insignificant compared to the final HCP, which covered 540,000 acres, this interim permit riled environmentalists because it authorized development before scientific studies were completed. Moreover, researchers were denied access to some private property, which impeded the collection of data on the population and distribution of the Stephens' kangaroo rat. The interim permit and limited access to private property created technical uncertainty, as well as political animosity, over whether crucial habitat was being destroyed during the interim period. By 1994, the RCHCA had been sued five times—three times by environmentalists and twice by developers (Brooks 1994: 10). Nevertheless, the RCHCA collected approximately $36 million in mitigation fees by 1994, and purchased nearly 6000 acres of habitat to be included in the eventual reserve system (Brooks 1994: 11). Thus, habitat was being protected, but without complete data, it was not clear if it was the best habitat.

In the midst of this animosity, the RCHCA remained locked into single-species planning and was unable to address the CSS issue as it arose. Unlike the least Bell's vireo HCP, in which local stakeholders developed specific plans in each river basin, the RCHCA was responsible for the entire planning area, and it could only collect mitigation fees for and spend money on habitat for the Stephens' kangaroo rat. The agency's charter would have to be rewritten and reauthorized for it to collect mitigation fees from developers who destroyed CSS—or other natural communities—or to use fees already collected to study or acquire such lands. The longer the Stephens' kangaroo rat HCP was delayed, the longer it would be until the RCHCA received a permanent incidental take permit, and the longer everyone would have to wait for additional land to be released for development. With the single-species HCP well underway, the development industry pressured Riverside County and the RCHCA to push ahead rather than broaden the planning effort to include other species and habitats.

While logical from the short-term perspective of most developers, this sequential approach to planning was inappropriate for long-term preservation efforts in western Riverside County and throughout the South Coast Bioregion because the habitat of the Stephens' kangaroo rat was interlaced with CSS. In other words, CSS was in the HCP planning area. Therefore, should any CSS-dependent species be listed, participants would have to start a new planning process for those species. Not only would this be administratively inefficient and politically annoying, it might have perverse effects on biodiversity. Efforts to enhance the population of one species might harm other species.[16] Though Riverside County and RCHCA staff saw some utility in moving toward multiple-species planning, they were under pressure to finish the single-species HCP first.

In sum, events in western Riverside County symbolized the basic problem with single-species planning, which can trap participants into a specific course of action. Even if a single-species HCP is carried out on a multijurisdictional basis, and even if it accounts for most of the habitat, it does not obviate the need to plan for other species that might be listed. Single-species planning is akin to fighting brush fires one at a time, allowing some fires to become worse while others are fought. With many species in Southern California on the candidate list and headed for the threatened or endangered list, it was sensible to plan for many species simultaneously, perhaps even maintaining healthy populations of candidate species so they would not be listed in the first place. With development rapidly eating away at remaining habitats, additional listings and lawsuits were inevitable so long as the ESA remained intact.

A Few Listed Species—and Many More Knocking at the Door

For local governments and the development industry in the South Coast Bioregion, the ESA loomed as a convincing threat to continued development. The least Bell's vireo and Stephens' kangaroo rat were first listed under the California Endangered Species Act (CESA), but the subsequent federal listings prompted concerted action to preserve both species. CESA was relatively ineffectual in promoting species protection and cooperation, for several reasons. First, as a state law, CESA did not cover federal agencies. Second, listing decisions under CESA were determined by a politically appointed body, the California Fish and Game Commission, which held public hearings that allowed easy access for nonbiological criteria (such as economic considerations) to enter the listing process. Third, environmentalists did not use CESA as a significant legal tool to slow or halt development in court, which meant that local, state, and private actors did not fear legal enforcement under CESA as they do under the ESA.

By the late 1980s, the federal ESA had become a substantive sledgehammer for endangered wildlife species, while CESA was perceived largely as a procedural paper tiger. In the South Coast Bioregion, relatively few wildlife species were then federally listed, but mounting data indicated that numerous species might soon move from the candidate list to the threatened or endangered list. The writing was on the wall for all who cared to see. An oncoming wave of federally listed species was on the horizon, and the Stephens' kangaroo rat provided a regionwide symbol of the potential problems that might arise if public officials and private actors waited until these species were listed, and then dealt with them sequentially. In the Pacific Northwest, the northern spotted owl was also being litigated on several fronts at that time, including cases that compelled the FWS to list the owl in 1990 and designate 11.6 million acres as critical habitat (chapter 5). Thus, it was becoming clear that large areas could fall under the ESA's protective regime, depending on a species' range. Nobody wanted additional species to be listed—except for the most avidly litigious environmentalists, who sought to use the ESA as a growth-control tool to stop development in Southern California. For state and federal ecologists, listings implied failure to protect habitat before it was depleted, at which point recovery is difficult. For the development industry, listings brought uncertainty to business operations. For local governments, listings threatened tax revenues and were an administrative burden for planning departments.

This was the context in which the development industry tried to pull itself together, along with local, state, and federal agencies, to preserve habitat proactively so the FWS would not need to list any wildlife species. Given the likelihood of future listings and generally accepted ecological knowledge, farsighted individuals in the South Coast Bioregion believed it was better to plan proactively than to react as species were listed. In this regard, the prominent actors differed significantly from those in the Klamath Bioregion, where the FS proceeded recklessly with timber sales despite repeated warnings from its own research staff and others that timber harvest practices were driving the northern spotted owl to extinction. In the South Coast Bioregion, some developers, landowners, and local governments began to heed the advice of state and federal wildlife ecologists that proactive steps should be taken to preserve habitat.

For state and federal ecologists, proactive habitat planning and management would be a dream come true. FWS ecologists knew the ESA kicked into gear too late, after most habitat had been destroyed. The FWS also lacked enforcement capacity to protect all of the remaining habitat on public and private land.[17] Some environmental groups, however, continued to push for listings because they were dubious that landowners, developers, and local governments would protect habitat proactively in the absence of federal listings. Others, like TNC, believed that proactive planning was possible. TNC did not sue public agencies or private actors. Instead, it relied on cooperative strategies to preserve biodiversity. TNC staff worked closely with developers, landowners, and local governments, providing them with ecological information and knowledge, while disseminating the planning principles of conservation biology. This advice might have fallen on deaf ears were it not for horror stories surrounding the Stephens' kangaroo rat, northern spotted owl, and other listed species.

Meanwhile, research ecologists provided data indicating that certain species were in trouble. In this respect, the data lurking in the minds of many belonged to Jonathan Atwood, an ornithologist who studied small, insect-eating birds known as gnatcatchers. In the 1980s, Atwood wrote his dissertation at UCLA on gnatcatcher taxonomy. In 1988, he published the results, arguing that gnatcatchers in Southern California and Baja California should be considered a separate species from those in the Sonoran and Chihuahuan Deserts. The American Ornithologists Union agreed with Atwood's conclusion and recognized the California gnatcatcher as distinct from the widespread black-tailed gnatcatcher of mainland Mexico and the southwestern U.S. Atwood's research also indicated that the California gnatcatcher was closely associated with CSS and that it was highly susceptible to habitat fragmentation. Its kittenlike mewing call was heard more often than the bird was seen because it seldom flies above—let alone out of—CSS. By Atwood's count, roughly 2500 breeding pairs of gnatcatchers remained in Southern California, which did not bode well in light of development plans.

Natural Communities Conservation Planning

The gnatcatcher was one of several wildlife species thought to depend on CSS. None were federally listed, but if the FWS listed any of them the development industry would be hit hard because nearly 80 percent of CSS occurred on private land. An endangered listing would likely lead to a moratorium on development, as occurred in western Riverside County, until HCPs were completed and the FWS issued incidental take permits. Rather than wait for listings to occur, some developers asked state officials to create a new program to preserve CSS before any wildlife species were listed. This state-sponsored program became known as Natural Communities Conservation Planning (NCCP). Though designed as a generic program that could be applied in other areas and habitats, the pilot project focused on CSS in the South Coast because developers specifically asked Governor Wilson to do so. NCCP subsequently became the largest cooperative effort to preserve biodiversity in the South Coast, and was the counterpart to the Klamath Bioregion Project. Yet the idea and impetus for NCCP came from the local level, not the state level.

Orange County: The Birthplace of Regional Planning in the South Coast Bioregion

Orange County is the smallest of six counties in the South Coast Bioregion, covering less land than the historic range of the Stephens' kangaroo rat in western Riverside County. While Orange County shares its eastern border with western Riverside County, the Santa Ana Mountains form a natural barrier, which effectively isolated Orange County residents from the Stephens' kangaroo rat and associated litigation. Developers and landowners in Orange County nevertheless followed events in neighboring Riverside County. Orange County was also unique in another important respect: it was dominated by two large landowners. Hence, countywide cooperation would likely be easier in Orange County, providing a potential springboard for regional cooperation.

The largest landowner was the Irvine Company, which owned roughly 63,000 acres, or one-sixth of the entire county. The other large landowner was the Santa Margarita Company. Prior to the U.S. conquest of California in 1846, Mexico and Spain granted large ranchos to settlers (Robinson 1948). Some of these estates, including the Irvine Ranch, remained in family ownership well into the twentieth century. During World War II, the Santa Margarita Ranch was split in half by the federal government, which purchased the southern half of the estate through eminent domain to create Camp Pendleton Marine Base. When the cities of Los Angeles and San Diego boomed after the war, the Irvine and Santa Margarita families turned their attention from ranching and agriculture to development. Orange County soon became rich, Republican, and dotted with large houses connected by wide boulevards, on which expensive cars moved from hillside homes to valley offices. By the 1990s, the Irvine business district was the second largest business area in California—larger even than downtown San Francisco and San Diego.

As a development firm, the Irvine Company was extraordinarily wealthy and technically sophisticated. Unlike smaller development firms, it managed a huge tract of land, half of which it planned to set aside as open space even before endangered species issues arose. The Irvine Company could also afford to hire consultants and a large in-house staff to plan development projects. Thus, it was well aware that it owned large patches of CSS, but it was in no hurry to develop the land quickly, prior to listings, because it did not want to saturate the housing market. The Irvine Company also had a peculiar corporate culture. Though no longer family owned and operated, Irvine Company executives nurtured a corporate culture emphasizing land stewardship and project excellence. They believed their development projects were the best anywhere, and they wanted to retain natural values as part of these projects. Therefore, unlike some developers in Southern California, who received bad press for willfully destroying habitat while flaunting state and federal environmental laws, Irvine Company executives wanted to confront endangered species problems with integrity and technical sophistication. They also wanted government assistance.

Vice President Monica Florian oversaw environmental and land-use issues for the firm. She was well aware of events in Riverside County to the east, where a single listed species halted development on private land in 1988, and where an interim take permit opened only 4400 acres to development in 1990. Further to the east, in the Mojave Desert, Florian also watched developers struggle with the desert tortoise, a federally listed species impeding development in the high desert interior. As she recalled,[18]

> The first thing that captured my attention was the Desert Tortoise, which had no impact on the Irvine Company or our lands, but that was a mess! I mean, that just seemed to be a mess for everybody. And then the one even more disturbing was to see what happened in the kangaroo rat situation with Riverside County. . . . It looked like a mess for everybody. Nobody was getting anything that they needed or wanted on any side of the issue, and it looked like it was chaos. And it looked like it was costing a lot of money in the public and private sectors and nothing was coming of it.

To Florian, it became obvious that development on Irvine Company land would be severely constrained if any of several CSS-dependent wildlife species were listed. Nearly one-third of its 63,000 acres contained CSS, much of which was on the coastal plains and foothills favored for development projects.

As conservation efforts for the desert tortoise and Stephens' kangaroo rat took shape east of Orange County during the late 1980s, Irvine Company staff were working with state and federal ecologists on other regulatory matters, including mitigation for wetlands development. Irvine Company staff accordingly had many opportunities to interact with CDFG and FWS ecologists, who alerted them to the CSS issue and coached them on the habitat planning principles of conservation biology. TNC ecologists similarly sowed ideas about effective ways to preserve species. In conversations with Irvine Company executives and staff, these ecologists repeatedly emphasized the importance of preserving large blocks of unfragmented habitat. As Vice President Florian later recalled, "We started hearing the same thing over and over: first of all, you don't look at the species, you look at the habitats, and you do things as comprehensively as you can." Florian heard a clear and concise message: "It makes a lot more sense to preserve large areas of land in large tracts . . . than it does to save little pieces."[19] The conservation biology mantra was seeping into the highest levels of the Irvine Company.

By 1990, the ecological epistemic community was shaping Irvine Company planning processes, just as it was shaping agency planning processes throughout California, but the Irvine Company owned only a fraction of the remaining CSS, so it could not ward off future listings by itself. Many private and public actors were chipping away at CSS in the South Coast Bioregion, and they would have to be part of any long-term solution. If everyone continued business as usual, without setting aside land for a reserve system, the habitat would become increasingly fragmented, and the FWS would ultimately list one or more species, any one of which could shut down development. The costs of preservation would then be borne by those developers, like the Irvine Company, who were in no hurry to develop their land; by associated local governments, like Orange County, which would lose tax revenues; and by state and federal agencies, like Camp Pendleton Marine Base, which contained significant tracts of CSS. On the other hand, if landowners, developers, and public agencies designed and implemented a CSS reserve system to ward off future listings, then all would benefit from greater certainty regarding future land uses, without fear of lawsuits. In the absence of a federal listing, however, state and federal officials did not have a mechanism to encourage broad-based participation in habitat planning efforts. Therefore, Irvine Company executives asked the future governor of California to create a new program to preserve CSS before species were listed.

In 1990, U.S. Senator Pete Wilson was running for governor and was aligned with some developers in Southern California. He was not popular with the development industry per se, because he had previously championed growth control as mayor of San Diego, but he received large campaign contributions from some developers, including the Irvine Company. More significantly, several Irvine Company executives were closely tied to him. Pete Wilson's former mayoral press secretary and campaign manager, Larry Thomas, was an Irvine Company vice president; Jack Flanigan, also a former campaign manager, was the Irvine Company's vice president for government relations; and Donald Bren, a billionaire who owned 93 percent of the Irvine Company, was, as Pete Wilson acknowledged, "a friend."[20] Therefore, regardless of campaign contributions, the future Wilson administration would likely pay close attention to Irvine Company concerns.

Irvine Company executives basically wanted what all developers, landowners, and agency line managers wanted. They wanted certainty. They wanted to know what land they could develop, without worrying that the FWS would list species after investments had been made. Therefore, Irvine Company executives asked Pete Wilson to create a program that would encourage landowners, developers, and local governments to produce a cooperative plan to preserve CSS prior to any listings. They were willing to strike a deal that would lock away a large portion of

Irvine Company land from development, provided that others would do so voluntarily as well. This was a classic collective-action problem. The Irvine Company owned more land than other developers, but it did not own enough CSS to preclude listings through its own actions. Yet other landowners could free ride on the Irvine Company's preservation efforts by developing their own CSS in the meantime. They could also free ride on federal preservation efforts, because federal land would be an important component of any CSS reserve system. The Cleveland National Forest in Orange County and Camp Pendleton Marine Base in northern San Diego County contained some of the best remaining tracts of CSS. Indeed, Camp Pendleton line managers feared that the base would be treated as a "mitigation dumping ground" for development projects outside the base.[21]

Whatever assurances Pete Wilson gave to Irvine Company executives during his gubernatorial campaign, he supported their proposal after the November general election. He also appointed Wheeler, who strongly supported the idea of large-scale habitat planning, as Secretary for Resources. Wheeler, in turn, asked a like-minded colleague at the World Wildlife Fund and Conservation Foundation, Michael Mantell, to join him as Undersecretary. Shortly after Wheeler and Mantell assumed office, CDFG staff presented a rough sketch for the new program to them, which reflected ideas that CDFG and FWS staff, TNC staff, and Irvine Company executives had been discussing. The details, however, were not yet worked out. Mantell quickly assumed leadership for this program, which became known as NCCP.

The public first heard of NCCP on Earth Day, April 22, 1991, when Governor Wilson announced his statewide environmental agenda, "Resourceful California." At that time, the MOU on Biodiversity was not yet drafted, and it would not be signed until September 1991; the first bioregional meeting in the Klamath would not be held until October; and the Executive Council would not hold its first meeting until December. Months before these other cooperative efforts saw the light of day, NCCP was already underway. Of all the bioregional efforts conducted under the umbrella of the MOU on Biodiversity, this program was the flagship. More time and resources went into NCCP than into any other bioregional planning effort, and it attracted national attention.[22]

NCCP as an Ineffective Voluntary Program (1991–1993)

When Undersecretary Mantell assumed leadership of the program in early 1991, it had few supporters, no funding, no organizational structure, and no name. The briefing papers contained little more than an idea for multispecies planning. Supporters had not even agreed on the unit of analysis. Should the new program focus on habitats, natural communities, ecosystems, or something else? More important, what would drive participation? Could agency officials rely on the threat of future listings as an incentive? Or were listings necessary because they would trigger lawsuits?

Irvine Company executives did not want federally listed species. Vice President Florian even wrote a letter to Wheeler on April 2, 1991, in which she suggested that large landowners would not participate unless environmentalists dropped their petition to list the gnatcatcher as a federally endangered species.[23] Many environmental organizations—including the Natural Resources Defense Council, which filed the petition—rejected this implicit threat. Though environmentalists supported the goals of NCCP, most did not believe that developers, local governments, and public agencies would preserve CSS in the absence of federally listed species. Ironically, the Irvine Company had been sparked to action by staff ecologists working for TNC, CDFG, and FWS, none of whom had any intention of undermining the ESA.

The Irvine Company did not want state listings either. Undersecretary Mantell obliged by testifying at a hearing of the California Fish and Game Commission on August 30, 1991, regarding the commission's upcoming listing decision on the gnatcatcher. He asked the commissioners not to list the gnatcatcher because Governor Wilson's new program to preserve CSS would provide adequate protection for the gnatcatcher. Mantell later claimed he had been forewarned that the Fish and Game Commission would not list the gnatcatcher and lobbied against the listing to maintain support for NCCP (Thompson 1994: 195). Yet the Wilson administration lost credibility in the environmental community after Mantell's testimony. Regardless, a state listing would have been largely ineffectual because CESA was anemic compared to the federal ESA, and environmentalists wanted a federal listing.

The debate over listed species continued for two years, during which time NCCP withered on the vine because few participants enrolled in the program. In signing NCCP enrollment forms, landowners voluntarily placed their CSS under an 18-month development moratorium. The Irvine Company enrolled, as did other large landowners, because they had large tracts of land they could set aside for open space and habitat preservation. Smaller landowners and developers, by contrast, had less land with which to work, and they could free ride on the preservation efforts of large landowners and federal agencies. Thus, the threat of listed species was an insufficient incentive for many to sacrifice short-term gains. Meanwhile, CSS continued to disappear.

Despite slow progress on the ground, NCCP took administrative shape during this period. In 1991, the state legislature authorized and funded NCCP. This legislation was important from a federal perspective because the FWS needed a formal state program that it could recognize as a means for delegating authority and federal funding for species protection under Section 6 of the ESA. State officials also adopted regional and subregional components for NCCP, similar to the least Bell's vireo HCP. At the regional level, they established a Scientific Review Panel (SRP) to give the program credibility. The SRP included leading conservation biologists, like Reed Noss and Dennis Murphy. The SRP selected three target species for CSS planning purposes: the coastal California gnatcatcher, San Diego cactus wren, and orange-throated whiptail lizard. The SRP also established research protocols for habitat surveys of these target species, designed subregional planning boundaries, and issued conservation guidelines. These guidelines defined CSS preservation goals, established a research agenda, and provided specific biological guidance for preparing subregional plans. With so many local, state, and federal jurisdictions and private stakeholders involved, it would be difficult to develop a single regionwide plan for the entire 6000-square-mile area. Yet the subregional plans (and the subarea plans within the subregions) had to be scientifically credible; they had to provide corridors and linkages to existing reserve areas and adjacent plans. The planning process could not simply be a matter of setting aside a certain percentage of land; the set-asides and acquisitions had to form a preserve system with ecological integrity at regional, subregional, and subarea scales.

While NCCP bore similarities to the least Bell's vireo HCP in terms of its two-tiered planning structure, NCCP differed from HCPs in one very important respect. As initially conceived and implemented, NCCP operated in the absence of federally listed species. From 1991 to 1993, there were no listed species of fish or wildlife associated with CSS; hence, participation was not driven by the desire for incidental take permits or the imminent fear of lawsuits. NCCP was intended to ward off listed species, which meant there was no immediate regulatory threat to encourage proactive efforts to preserve CSS. As initially conceived, NCCP was based on the assumption that a large number of private and public actors would make voluntary sacrifices to protect habitat for the collective purpose of avoiding future regulatory threats. This was a big assumption, and an invalid one.

While the threat of future listings brought a few public and private actors together, this threat was insufficient to compel most actors to make sacrifices. If an effective reserve system could have been created by a few landowners and jurisdictions, such as the Irvine Company, Santa Margarita Company, and Orange County, events might have turned out differently. But with the cooperation of so many landowners, developers, local governments, and public agencies needed, too much opportunity for free riding existed. Small landowners and developers, in particular, hoped their larger counterparts would provide most of the necessary habitat, with public agencies setting aside the rest. As initially conceived, NCCP was a voluntary program, and as such, it could not overcome the free-rider problem.

Federal line managers also lacked a compelling reason to participate. If there had been federally listed species, their agencies would have to pursue Section 7 consultation with the FWS, which the FWS can tie to related conservation efforts like HCPs. Federal line managers also lacked discretion to enroll their lands in NCCP. They could sign cooperative agreements stating that federal lands would be managed in conformance with subregional NCCP plans, but they could not enroll federal lands in NCCP because enrolled lands would be managed by a nonfederal entity. This was important because federal agencies had some of the best remaining patches of CSS—particularly Camp Pendleton Marine Base, which some viewed as a core reserve and future national park.

NCCP also needed local government officials to participate because land-use authority on private lands in California is primarily granted to local governments. Implementing an effective reserve system would require changes in general plans to restrict growth in certain areas.

Counties were particularly important because unincorporated areas contained much of the remaining habitat. State and federal officials also depended on local government staff for assistance because they did not have enough staff to monitor private-sector compliance. With so many actors involved across 6000 square miles and numerous jurisdictions, developers could bulldoze a few acres of CSS here or there without anyone noticing, regardless of whether the property was enrolled in NCCP or not.

Therefore, to gather support from local governments, Governor Wilson appointed Carol Whiteside, a former mayor, to serve as liaison between local governments and the Resources Agency, which housed the CDFG. Unlike Mantell, Whiteside was not associated with the environmental community. Her role was to sell NCCP to local government officials. Governor Wilson also appointed John Sullivan as Deputy Director of CDFG. Sullivan, who was aligned with the building industry, took charge of NCCP implementation within CDFG. The die was being cast. The proponents of biodiversity in the Resources Agency did not control NCCP; instead, they shared implementation decisions with the proponents of development and devolution in the Wilson administration.

In this context, developers, local governments, and state agencies did little to preserve CSS, because there were no federally listed wildlife species. After two years as a voluntary program, NCCP was clearly not meeting its stated goals, though most participants would not admit this publicly. Several state and federal officials closely associated with NCCP stated forthrightly during interviews that NCCP did not work in the absence of federally listed species, but they did not want to be quoted on this point because the program's official purpose was to prevent species from being listed through proactive planning and management. Stating that NCCP did not preserve habitat without listings was an admission of failure, and it was not a message developers wanted to hear. According to one staff ecologist,[24]

> There is *no doubt* that the Natural Community Conservation Planning process was not working because it was voluntary. And this was not surprising—to me, anyway, it's not surprising—because when you're putting these planning efforts together, and when you have a listed species, then you are looking at people and saying "To get a permit [for incidental take] you gotta come up with a plan that protects the species, and that means you're gonna have to give up some of your property or [purchase] some property somewhere else. In other words, you're

> giving up some of your assets, hopefully to get permission to earn more." I strongly believe . . . that you're not gonna get very many people that will do this sort of thing voluntarily.

NCCP lacked regulatory teeth without a federal listing because it could not compel public or private actors to protect habitat—let alone cooperate in doing so.

For this reason, environmental groups continued to push for a federal listing, without which NCCP appeared to be a delay tactic while CSS continued to disappear. State officials had few enrollment agreements to show for their efforts, and they continued to make positive public statements about NCCP, which irritated many environmentalists. Mantell and Wheeler were stuck in the middle. They had spent much of their careers working for environmental organizations before joining the Resources Agency, where they were increasingly viewed by environmentalists in the South Coast Bioregion as puppets in the Wilson administration and servants of the development industry. To some, NCCP appeared to subvert the ESA, but to make that case stick one needs evidence that the FWS intentionally delayed a listing on the merits so that development of CSS could continue. Mantell did lobby against a state listing, but most thought state listings were largely inconsequential. Thus, a simpler interpretation than the subversion hypothesis is that NCCP as a voluntary program was a failed attempt to preserve habitat proactively before species' populations declined to the point at which they merited a federal listing. Mantell and Wheeler knew that listed species brought people to the table, but they could not publicly advocate a listing without political support.

NCCP as an Effective Regulatory Program (1994–2002)

Having made little headway after more than two years as a voluntary program, the Wilson administration abruptly changed course in late 1993 by asking the FWS to list the gnatcatcher as a threatened species. This represented a major change in the NCCP program. Previously, Wilson's appointees lobbied to keep species off the state and federal lists. They now reversed their position to make the regulatory hammer more immediate. In doing so, they approached FWS officials behind the scenes, without fanfare or official statements. This allowed them to deflect blame for new regulations to the federal government, while providing the necessary incentive for participation in NCCP.

The Wilson administration did not ask the California Fish and Game Commission to list the gnatcatcher, for reasons easy to surmise. First, the commission held public hearings prior to listing decisions, which would reveal to developers that the Wilson administration now supported the listing. Second, it was unlikely that the politically appointed commission would list the species when it had not done so before. Third, and most important, few believed that a state listing would provide the necessary incentive to protect CSS. Besides, CESA was largely irrelevant if the FWS listed a CSS-dependent wildlife species. As one of NCCP's designers and longtime supporters bluntly stated, CESA was "mostly inconsequential" with regard to NCCP.[25]

The Wilson administration worked out a clever arrangement with the FWS that hitched NCCP directly to the ESA. In this arrangement, the gnatcatcher was listed as threatened rather than endangered. The Natural Resources Defense Council had petitioned for an endangered listing, but the FWS could still opt for the less restrictive, threatened status. An endangered listing might lead to numerous HCPs rather than the comprehensive reserve system envisioned for NCCP. A threatened listing, on the other hand, left open a wider range of conservation and development activities, depending on the rule that accompanied the listing. Under Section 4(d), the FWS issues rules that specify regulations deemed "necessary and advisable" to provide for the conservation of threatened species. The FWS can also delegate enforcement responsibilities to states under Section 6. Thus, the FWS can issue a 4(d) rule for a threatened species that is as stringent as the prohibition on take for endangered species, but that delegates authority to a state program and includes specific guidelines or regulatory standards. Accordingly, the FWS listed the gnatcatcher as threatened and issued a 4(d) rule that hitched NCCP to the ESA.

The "Special Rule Concerning Take of the Threatened Coastal California Gnatcatcher" was published in December 1993.[26] Under this rule, development activities covered in an approved NCCP subregional plan would not be considered a violation of the ESA. Informally, FWS staff also let it be known that anyone choosing to develop a separate HCP, or find a Section 7 nexus with a federal agency, as alternatives to enrolling in NCCP, would have to demonstrate that this alternative was compatible with subregional NCCP plans. Thus, even if permit applicants did not participate in subregional NCCPs, they would likely be bound to these plans, so it behooved them to shape those plans by participating. With the federal listing, the regulatory hammer became imminent, and, with the 4(d) rule, NCCP became a means to meet legal obligations under the ESA without fear of lawsuits.

The 4(d) rule offered an additional incentive to encourage participation. It allowed participants to develop 5 percent of CSS in each subregion during the planning process. Thus, developers did not have to wait several years to complete their own HCP, or perhaps longer to complete a subregional NCCP, before they could bulldoze CSS. If they enrolled in NCCP, they were eligible to develop a limited amount of CSS without fear of legal retribution. This 5 percent interim take covered up to 20,920 acres and 116 known pairs of gnatcatchers across the entire planning area. The remaining land within each subregion would be set aside until the subregional plan was completed and the FWS issued a permit for activities covered by the plan. Environmentalists complained, as they had with the interim take permit for the Stephens' kangaroo rat, that this was a gift to the development industry, with no assurance that subregional NCCPs would be completed and a regional reserve system established.

Yet other incentives for long-term participation seemed sufficient to carry the process through. Most notably, NCCP covered numerous candidate species associated with CSS, not just the one listed species. Once the FWS issued an incidental take permit for a subregional plan, the permit would cover all activities specified in the plan, regardless of whether these candidate species were listed in the future. In other words, once a subregional plan was accepted and the incidental take permit issued, the habitat conservation planning process for CSS in that subregion would be over. While HCPs could also cover multiple species, the NCCP program was much more comprehensive, and the FWS made it known that separate HCPs had to be consistent with subregional NCCPs. Therefore, even though NCCP plans would likely take longer to complete than an HCP prepared by one applicant, NCCP provided more regulatory certainty by covering more species, and FWS officials implicitly threatened that HCPs would not be approved until subregional NCCPs were completed.

NCCP kicked into high gear with the federal listing and the 4(d) rule. It remained a state program but now relied on federal law to provide the incentive for participation. One year after the gnatcatcher listing, subregional planning processes were well underway in Orange and San Diego Counties, which contained most of the remaining CSS. Riverside County, which was still trapped in single-species planning for the Stephens' kangaroo rat, did not initially participate in NCCP, despite the concerted efforts of state and federal officials. It was not until June 1994 that a multispecies planning process was initiated in western Riverside County, with the intent being to graft it onto the Stephens' kangaroo rat HCP (Jasny, Reynolds, and Notthoff 1997). But the FWS did not issue a permit for the latter until May 1996, which meant that multispecies planning and NCCP were slow to start in Riverside County. Subregional plans were also underway in San Bernardino and Los Angeles Counties, but Los Angeles County contained so little CSS that it was of little practical relevance.

The first significant subregional plan was completed in Orange County.[27] The Orange County Central-Coastal NCCP Subregional Plan was approved in July 1996. It covers 208,000 acres, addresses 39 sensitive plant and animal species, and includes a 37,380-acre reserve system. With the Irvine Company in the lead and relatively few participants in the subregion, it was not surprising that the first significant subregional plan was completed in Orange County. A separate subregional plan for southern Orange County was still in the planning stage in 2002.

In December 1996, the San Diego Multiple Species Conservation Program was also approved. This plan covers 582,243 acres and 85 species (including 8 federally listed species) in southwestern San Diego County, and includes a 172,000-acre preserve system. This complex subregional plan includes numerous cities, some of which have not yet completed their subarea plans. A third significant subregional plan was also nearing completion in San Diego County in 2002. The San Diego Multiple Habitat Conservation Program covers 118,852 acres and numerous species in northwestern San Diego County.

It still remains to be seen, however, whether all of the subregional and subarea plans will be completed, whether they will be implemented as designed, and whether they will be effective. Nevertheless, the cooperative planning process itself has thus far been monumental.

What Happened to State and Federal Line Managers?

One of the notable features of NCCP is the relative absence of regional line managers. At the state level, NCCP was largely run by Governor Wilson's political appointees—both those favoring biodiversity and those favoring development. Line managers in some agencies were literally removed from the process, thereby collapsing the hierarchy. In the Klamath Bioregion, regional line managers in most agencies did not participate or encourage field staff to participate in the subregional groups, and agency directors in most agencies provided little or no leadership to pressure their line managers to support the groups. The BLM was an important exception, but other agencies were also needed, particularly the FS, which managed most federal land in the Klamath. In the South Coast Bioregion, line managers in some agencies were removed from NCCP decisions.

This was particularly notable in the FWS and CDFG. Political appointees in Sacramento and Washington wanted these regulatory agencies to speak with one voice, which was much easier to accomplish if fewer line managers were involved. Therefore, they removed regional line managers from the process so that FWS and CDFG field staff reported directly to them. CDFG field staff working on NCCP even had their own office space in San Diego, in a separate building from the CDFG regional office. As one participant put it, "The normal chain of command through Fish and Game has been completely side-stepped on this. This should be Region Five's project . . . but in fact it's been Carol Whiteside of the Resources Agency or Michael Mantell that make all the major and day-to-day decisions on how the process is run."[28] Regional line managers could not impede NCCP or slow it through indifference, neglect, or sabotage, because they were not responsible for it. In both the CDFG and FWS, the normal chain of command was eviscerated.

It may seem remarkable that political appointees representing both major parties cooperated in this regard. After all, why would a Republican administration in Sacramento and a Democratic administration in Washington both seek to eviscerate the chain of command on NCCP to aid developers, like the Irvine Company, that supported Governor Wilson? The short answer is that CSS-dependent wildlife species had a very large congressional delegation in Washington. In 1993–1994, the

ESA was up for reauthorization and was being hammered in Congress by property rights advocates for being ineffective and for taking private property. Interior Secretary Bruce Babbitt sought to bolster support for the ESA and ensure President Clinton's reelection in the densely populated South Coast Bioregion by avoiding "environmental train wrecks" like the Stephens' kangaroo rat and northern spotted owl.[29] This required proactive planning, interagency cooperation, and working directly with field staff, rather than through line managers.

Summary

The Klamath and South Coast bioregional efforts had one important element in common. Local stakeholders and public officials learned vicariously about the legal uncertainty that listed species created. In the Klamath, it was the northern spotted owl. In the South Coast, it was the Stephens' kangaroo rat. In both bioregions, participants wanted to avoid the uncertainty created by regulatory enforcement, particularly in courts. Further, in both bioregions, ecologists stepped forward with solutions, offering habitat planning and management principles from conservation biology as a means to maintain species' populations to prevent listings, enforcement, and lawsuits. The regulatory threat nevertheless had to be imminent for public and private actors to make collective sacrifices for habitat preservation. It took the gnatcatcher listing to make this threat imminent and credible in the South Coast Bioregion.

NCCP was initially intended to protect habitat proactively, rather than react once species were listed; to move away from species-by-species and site-by-site planning to habitat-based planning; to emphasize local solutions to regional problems; and to encourage responsible economic development during the planning process. Yet the program needed at least one federally listed species to provide the fundamental incentive for landowners, developers, and public agencies to sacrifice other land uses. With a listed species, NCCP became the functional equivalent of a state-sponsored HCP. It expanded the two-tier planning approach embodied in the least Bell's vireo HCP to encompass more species and a larger planning area. NCCP symbolized a remarkable move from single-species planning to broad-scale, multiple-species planning. While the Klamath Bioregion Project sputtered, NCCP took large strides forward in both interagency and public-private cooperation. In both bioregions, the ESA compelled actors to protect species and their habitats. Yet the Klamath Bioregion Project did not channel this incentive into a well-defined and attainable goal. NCCP channeled this incentive by making participation a means to get an incidental take permit, and it protected participants from future listings and lawsuits by covering dozens of candidate species as well.

One of the more interesting aspects of NCCP was the pronounced role of ecological knowledge. Participants accepted the planning principles of conservation biology even before the program had a name. There was little disagreement about ecological information, knowledge, or what needed to be done to maintain species' populations (Thompson 1994). The environmental and development communities were at odds over whether species should be listed, but they agreed on the need for a collegial body of scientists (the SRP) to oversee data collection and develop conservation guidelines. Thus, ecological knowledge and the planning principles of conservation biology were widely accepted by the participants. Knowledge was not disputed. The dispute was over the incentive to preserve CSS in the absence of a federally listed species.

In the San Joaquin Valley Bioregion, events played out differently. As we will see in the next chapter, knowledge was disputed and the bioregional effort was not tied directly to the ESA. County supervisors conspired against the bioregional effort, while leading agency officials to believe they might participate. Ironically, this bioregional effort was sponsored by the BLM, and Hastey asked county supervisors to lead it. Rather than leading, county supervisors undermined the effort, leaving the BLM, which had long been tied to rural counties, in the lurch. The San Joaquin Valley bioregional initiative would likely have failed anyway because it was not directly tied to the ESA, but the defection of county supervisors ensured its failure.

7

The San Joaquin Valley Bioregion: BLM's Co-optation Strategy Fails at the Bioregional Level

In 1993, BLM line managers and field staff attempted to form a bioregional council in the San Joaquin Valley Bioregion (see figure 3.1). They did not call it a "bioregional council," however, because the terms *bioregion* and *biodiversity* were politically controversial in the San Joaquin Valley. Therefore, they called it a "consortium." Regardless of its name, the bioregional effort followed directly from the principles laid out in the MOU on Biodiversity. The San Joaquin Valley Regional Consortium was arguably the purest example of what the framers of the MOU and the Executive Council were trying to accomplish at the bioregional level. It also reflected Hastey's efforts to co-opt county supervisors by inviting them to play a lead role in the Consortium. This co-optation strategy backfired, however, because county supervisors in the San Joaquin Valley intentionally sabotaged the effort.

As with the other bioregional efforts, the San Joaquin Valley Regional Consortium did not arise solely from the MOU on Biodiversity. Instead, it grew out of preexisting interagency efforts among professional field staff. Thus, as in previous chapters, we need to begin by examining these formative efforts. Indeed, several cooperative efforts to preserve biodiversity were already underway in the San Joaquin Valley. Some focused on wetlands (such as the California Central Valley Habitat Joint Venture, which was part of the North American Waterfowl Management Plan); others focused on riparian zones (such as the San Joaquin River Management Program). While important in their own right, these aquatic and riparian efforts were unrelated to the BLM-led bioregional effort, the roots of which were embedded in the drier, upland areas of the bioregion, where BLM land was concentrated.

Natural and Social History of the San Joaquin Valley Bioregion

The San Joaquin Valley constitutes the southern half of California's immense Central Valley, which stretches 400 miles from Red Bluff in the north to Bakersfield in the south. The San Joaquin Valley floor is nearly flat and more than 50 miles wide. It is surrounded by mountains on three sides—the Sierra Nevada to the east, the Coast Range to the west, and the Tehachapi Mountains to the south. Historically, much of the San Joaquin Valley was a seasonal flood basin, which captured winter and spring runoff from the mountains. In the northern part of the valley, this runoff flows into the San Joaquin River, which meets the Sacramento River in the center of the Central Valley. At the confluence of these two rivers was a large marsh, much of which has since been diked and drained for agriculture. Today, this area is known as the Delta, and through it flows the Central Valley's unused fresh water into the great saltwater estuary known as San Francisco Bay.

In the southern part of the San Joaquin Valley, most of the runoff flowed into seasonal lakes, marshes, and vernal pools, most of which no longer exist. Tulare Lake, for example, provided habitat for hundreds of species of birds, fish, insects, and plants. It covered 700 to 800 square miles during heavy-rainfall years, but it began to shrink around 1880 as settlers diverted water to irrigate the valley's fertile flood plains. It dried up completely for the first time in 1905 (Griggs 1992: 12), and with the lake went the plants and wildlife it once supported. The former lakebed is now highly productive agricultural land.

Visitors to the San Joaquin Valley today would find it hard to imagine the immense biodiversity that once existed. Its wetlands provided habitat for millions of wintering ducks and geese. The valley contained more than half a million acres of seasonal vernal pools, which provided habitat for amphibians and fairy shrimp before giving way to profuse wildflower displays as the pools evaporated in the spring (Williams, Byrne, and Rado 1992: ix). Large herds of elk, deer, and pronghorn roamed the uplands surrounding the valley floor. Salmon, trout, and other native fish were abundant in free-flowing streams and rivers. This great biodiversity supported one of the densest populations of Native Americans north of Mexico (Griggs 1992: 11). Yet it all disappeared rapidly after the Gold Rush. Tule elk were estimated to number more than half a million before the Gold Rush, but market hunters reduced the population to a few individuals by the 1870s (Griggs 1992: 12). Exotic livestock, which first arrived with Spanish settlers in the eighteenth century, replaced native ungulates. With the domestic livestock came the seeds of Mediterranean grasses, which were already adapted to a dry summer climate. These exotic grasses spread quickly, giving rise to the famed golden hills of California, while displacing native grasses.

Ecologists have identified more than a dozen natural communities that still exist in the San Joaquin Valley. These include aquatic communities, like freshwater marshes and vernal pools; riparian forests; oak savannas; sink scrub; and saltbush scrub. Sink scrub and saltbush scrub suffered from even lower public esteem than coastal sage scrub in the South Coast Bioregion. Sink scrub consists of low succulent shrubs dominated by alkali-tolerant plant species that grow in the alkaline clays of historic lakebeds, while saltbush scrub occurs on rolling alluvial fans and uplands (Anderson, Spiegel, and Kakiba-Russell 1991: 37–38). By the 1980s, these natural communities had been reduced to a small fraction of their former extent by plows, irrigation canals, cattle grazing, oil and gas extraction, road construction, and suburban development.[1]

The loss was particularly severe on the valley floor, which is now dominated by immense tracts of industrial agriculture. In the southern part of the valley, the soil is less productive for agriculture, but large oil and gas reserves lie beneath it. The valley produced roughly two-thirds of California's oil and gas resources in the 1990s, and included the nation's top three counties in gross farm sales. Cities in the valley were also growing rapidly, as California's expanding population sought affordable housing inland from the expensive coastal communities.

More than 90 percent of the valley was under private ownership, which meant that environmental protection ran headlong into private property rights. Local politics in the San Joaquin Valley were also conservative, which meant that biodiversity preservation was a politically divisive issue, particularly in the dry upland communities harboring uncharismatic species like the blunt-nosed leopard lizard, the giant kangaroo rat, the Fresno kangaroo rat, and the Tipton kangaroo rat. Agency officials faced an uphill battle gaining the support of county supervisors, who preferred to espouse the causes of local control and private property rights, while arguing that environmental laws and implementing agencies represented egregious examples of state and federal interference in local communities.

Staff Ecologists Develop a Natural Lands Inventory for the Southern San Joaquin Valley

The genesis of the San Joaquin Valley bioregional effort was in the drier, southern portion of the valley, where the energy industry dominated the economic landscape. Here, staff ecologists working for the California Energy Commission (CEC) took the first step toward regional cooperation by developing a natural lands inventory in the late 1980s—well before the CEC officially joined the Executive Council in 1994. In the southern part of the valley, the land is less fertile, particularly in the southern quarter of the valley. Yet beneath the surface lie immense fossil fuel reserves. Here, the energy industry rivals agriculture and ranching, with oil pumps dotting the landscape and pipelines crisscrossing the valley floor and foothills.

Ranching and agriculture had already transformed much of the Southern San Joaquin Valley, but energy projects were contributing to biodiversity loss in the 1980s. By 1986, the FWS had already listed seven species as either threatened or endangered. A few years later, this number rose to ten, representing "one of the highest concentrations of listed species in the continental United States" (Anderson, Spiegel, and Kakiba-Russell 1991: ix). With so many species nearing extinction, it was clear to staff ecologists that habitats were largely gone. Yet they had only a scattershot understanding of the distribution of upland species, habitats, and natural communities. A few species were well studied, but there was no inventory of natural lands in the area. Without a natural lands inventory, preservation efforts were haphazard, particularly when it came to mitigating the impacts of development projects.

Under the ESA and other environmental laws, public agencies can reduce the impacts of development projects when issuing permits in two basic ways. One is to require on-site mitigation, which can assume a wide variety of forms depending on the particular needs of a species, but usually entails limiting habitat alteration at the site. This might involve raising a pipeline so animals can move beneath it, limiting construction during breeding seasons, or setting land aside as a habitat preserve.

Public agencies can also require off-site compensation, in which developers purchase, restore, or otherwise preserve habitat at another site. On-site mitigation makes sense when the known range of a species is limited to the area near a development project. For widely dispersed species, however, there is no technical reason why the conditions of a permit should be confined to the vicinity of the project itself. In fact, all parties may prefer off-site compensation, especially if the best remaining habitat is in remote locations, far from development activities. Remote lands may have a lower market value and higher habitat value. By contrast, land near development projects may provide poor habitat for sensitive species, have higher market values, and be difficult to manage for conservation purposes. Therefore, off-site compensation is an important alternative, but it requires knowledge about the distribution and condition of natural communities across the landscape.

This was an important issue in the mid-1980s in the Southern San Joaquin Valley. Energy projects represented only one type of development, and they were not the most destructive. Most energy projects were exploratory, so they had limited impacts, while housing developments, shopping malls, and agriculture largely erased natural communities from the landscape. Yet the scale of energy development was immense. The BLM Caliente Resource Area Office, which managed 75,000 acres in the Southern San Joaquin Valley, received more than a thousand applications for drilling permits in the latter half of the decade (Sheppard 1992: 243). This is a remarkable figure because much of the land in the valley was private, not public; thus, it does not capture the entire scale of energy development. Yet it does suggest one of the reasons the BLM was interested in bioregional planning. The BLM was responsible for endangered species on its oil and gas leases; therefore, it had to consult with the FWS under Section 7 to ensure that drilling-related activities would not jeopardize listed species.

Energy developers on private land also had ESA obligations. To receive an incidental take permit, they could prepare HCPs individually or with local governments and other actors. Yet HCPs were more time consuming than Section 7 consultations, which could be completed in a few months. Therefore, private firms often sought a federal nexus so they could fulfill their legal obligations through Section 7 consultation, rather than pursue an incidental take permit under Section 10. For energy projects, finding a nexus was often easy because pipelines and related infrastructure often crossed one or more of the BLM's scattered parcels.

One might assume that BLM staff would have been reluctant to consult with the FWS on behalf of private firms, particularly since many federal agencies once routinely avoided their own Section 7 obligations. Yet these consultations gave BLM ecologists an opportunity to immerse themselves in regional planning issues because Section 7 regulations required federal agencies to examine the broader impacts of federal actions, not the immediate impacts on federal land. BLM ecologists could thereby extend their influence well beyond the agency's land holdings. In one consultation, for example, BLM ecologists in Bakersfield worked on a 250-mile pipeline that crossed only a few miles of BLM land. This allowed them to negotiate with private consultants and other agencies on endangered species far to the north, including the salt marsh harvest mouse in Martinez, where the BLM did not own any land.[2] The same was true for other types of projects, like canals and utility rights-of-way. These examples do not mean the BLM was necessarily a better steward of natural resources than other agencies, but they do suggest why BLM staff were at the forefront of bioregional planning and management in the San Joaquin Valley.

State agencies also issued permits that required energy developers to mitigate and compensate for the effects of their projects. The CDFG, for example, issued permits authorizing nonfederal actors to take state-listed species under CESA. The California Environmental Quality Act (CEQA) also required local and state agencies to conduct environmental reviews of their projects, and to consider alternatives and mitigation measures to prevent significant, avoidable damage to the environment.[3] Yet there was no regional data set that identified the location and condition of natural communities, which meant that on-site mitigation and off-site compensation were haphazard. To applicants, permit conditions were also inconsistent and seemingly arbitrary. Off-site compensation ratios, for example, were typically greater than 1:1—meaning that more than one acre of habitat was preserved off site for every acre a project destroyed—but the ratio was not consistent. Therefore, off-site compensation requirements were inconsistent, underinformed, uncoordinated, and possibly ineffective in preserving biodiversity.

So, in 1986, state ecologists initiated an inventory of natural lands in the Southern San Joaquin Valley. This inventory was undertaken by the CEC to identify compensation sites for thermal power plants. The CEC had been given statutory authority in 1975 to license thermal power plants of 50 megawatts or greater, a category that included cogeneration plants associated with oil and gas extraction. In licensing cogeneration plants, the CEC had to consider alternatives that would reduce environmental impacts under CEQA. The CEC also had responsibilities under the state and federal ESAs to limit the effects of licensed projects on listed species. Therefore, CEC ecologists wanted to know the distribution of listed species so they could require applicants to acquire specific sites to compensate for project impacts. The CEC inventory accordingly covered natural communities directly affected by energy development, including valley sink scrub, saltbush scrub, and nonnative grasslands, while focusing "on larger, high-quality areas offering a greater potential for long-term population viability and sustainability" (Anderson, Spiegel, and Kakiba-Russell 1991: ix).

A notable characteristic of the CEC inventory was its broad geographic scope. CEC staff could have taken a much more parochial perspective, collecting data on sites near potential projects, or simply requiring on-site mitigation. The agency was not obligated to undertake a regional inventory of natural lands—an inventory that would also benefit other agencies, local governments, and nonprofit organizations. Because it covered a broad geographic area, the CEC natural lands inventory provided the technical foundation for regional planning. To a large extent, the credit goes to the CEC representative on INACC, Richard Anderson, who hoped to convince his agency to expand the natural lands inventory from the Southern San Joaquin Valley to cover all of California.[4]

Despite its broad geographic scope, the CEC natural lands inventory was not a regional plan. It was intended to provide information to permit applicants and regulators so that everyone would be better informed about project impacts on listed species, and so regulators could target off-site compensation more effectively. It also provided a technical foundation for coordinated regional planning by depicting the condition and distribution of natural communities in several counties. Because the

CEC's natural lands inventory would be of general use, professional staff from other agencies contributed to it, along with TNC staff.

The findings were rather bleak. The CEC inventory found only 2.9 percent of the 2950-square-mile Southern San Joaquin Valley floor to be in good natural condition or better, with "good" being a 5 on a scale of 1 to 7; it found only 3.6 percent of the larger 5700-square-mile study area to be in good or better natural condition (Anderson, Spiegel, and Kakiba-Russell 1991: i) The data indicated that little natural land remained in the Southern San Joaquin Valley, which was consistent with the expanding list of threatened and endangered species. When the CEC initiated its study in the fall of 1986, the FWS had already listed seven species. When the study was completed, the number had risen to ten, with roughly two dozen additional species under consideration. Moreover, as the CEC report noted, "The overall trend will be continued loss of natural lands and continued reduction in quality of undeveloped natural lands" (Anderson, Spiegel, and Kakiba-Russell 1991: ix).

Despite continued loss of habitat during the study period, there were significant efforts to preserve biodiversity in the bioregion. Before the CEC completed its natural lands inventory, oil companies approached the BLM and TNC to help them acquire land to offset the impacts of oil development in Kern County. Their acquisition efforts focused on the Carrizo Plain.

The Carrizo Plain Acquisition: The BLM Builds Local Support for a Large Preserve

The Carrizo Plain lies southwest of the San Joaquin Valley, above the valley floor. It is not part of the San Joaquin Valley per se, but it contains significant tracts of natural communities once found in the valley. Due to physiographic similarities, INACC ecologists included the Carrizo Plain within the San Joaquin Valley Bioregion. Others had also recognized the Carrizo Plain as a prime location for off-site compensation before the CEC started its natural lands inventory because the land was isolated from development in the valley and appeared to be in relatively good natural condition.[5] The Carrizo Plain had been ranched but not intensively farmed. Therefore, the soil was relatively undisturbed compared to the San Joaquin Valley, and habitat restoration was feasible. This was important because the land already supported several endangered species, including the giant kangaroo rat, the blunt-nosed leopard lizard, and the San Joaquin kit fox.

The BLM owned scattered parcels in the vicinity, but most of the Carrizo Plain was privately owned. BLM ecologists knew about the area's natural values, and TNC staff were also interested in acquiring or managing the Carrizo Plain. Therefore, when energy companies approached the BLM and TNC to explore off-site compensation options for projects in Kern County, their attention turned west to the Carrizo Plain.[6] The BLM and TNC negotiated a cooperative management plan with the FWS and CDFG, and TNC negotiated the acquisition with private landowners. Because TNC could not foot the entire bill, and the permits did not require energy companies to acquire all of the land as compensation for projects in the Southern San Joaquin Valley, an acquisition bill was placed before Congress. Public ownership on the Carrizo Plain subsequently rose from 30,000 to 180,000 acres.

Local support for the Carrizo Plain acquisition was important because rural county supervisors generally do not favor removing lands from private ownership or economic production. Unlike urban areas, where the public often encourages elected officials to support public acquisition of private land in the name of open space and recreation, the Carrizo Plain was in the middle of nowhere. If local stakeholders and county supervisors did not support the acquisition, they might lobby against it. Therefore, Hastey gathered local support for this acquisition, and for regional preservation efforts elsewhere, because, as he put it, "I really felt that county involvement and county leadership in these things was the way to reduce the confrontation, and to make something like this work."[7] This included establishing a steering committee composed of energy companies, ranchers, environmentalists, and other stakeholders. This committee was chaired by a supervisor from San Luis Obispo County, in which most of the Carrizo Plain lies.

Participants and observers alike viewed the Carrizo Plain acquisition as a success story because it benefited all major interests. For the BLM, the acquisition consolidated the agency's scattered holdings in the area, making it easier to manage this land as habitat, while freeing up other BLM land for energy development. Public and private ecologists were pleased that a large portion of remaining natural lands had been preserved from future development. Ranchers were allowed to continue grazing cattle on the acquired lands. Ecologists even encouraged cattle grazing as a means of suppressing nonnative grasses because wild herds of native ungulates no longer played a similar ecological role. In this regard, the BLM issued grazing permits directly to TNC, allowing TNC to manage grazing in the area.[8] Local support for the acquisition gave Hastey a strong reason to believe that rural county supervisors would support similar efforts, so long as they were included in the process.

In sum, the Carrizo Plain acquisition was notable for more than its geographic scale. BLM line managers and staff gathered support from rural county supervisors and local stakeholders, and assembled some of the political infrastructure for bioregional planning and management. Perhaps more than other agencies, the BLM understood the benefits of building local support, which contributed to the agency's historic reputation of being captured by local interests. Yet BLM staff and line managers did not believe they could manage the agency's numerous, scattered holdings without cooperation from neighboring jurisdictions and private landowners. Moreover, they had a multiple-use mandate to fulfill. Therefore, cooperative management and planning was a sensible, if not necessary, strategy for achieving traditional missions and complying with new legal mandates to protect species and habitats.

From Data to Planning: The San Joaquin Valley Biological Technical Committee

The CEC natural lands inventory was simply a survey of the distribution and content of scrublands and grasslands in the Southern San Joaquin Valley. It was not a planning document. Nor was it the only data on natural communities in the San Joaquin Valley. The BLM also collected data, as did the CDFG. At the local level, HCPs were underway in Kern, Tulare, and Fresno Counties; these planning processes also assembled data on the location and distribution of specific listed species. The CEC study was arguably the largest in scope, and represented the most complete inventory of natural lands in the Southern San Joaquin Valley, but it was not the only repository of ecological data in the bioregion, and it was not yet integrated into regional or subregional planning processes.

Therefore, to address regional planning concerns and integrate existing data sets, a new interagency group of ecologists formed after the CEC completed its natural lands inventory in 1991. The San Joaquin Valley Biological Technical Committee included two authors of the CEC study, but most participants came from other agencies, with the FWS, CDFG, and BLM well represented.[9] In some respects, this committee was similar to the bioregional INACCs that emerged elsewhere in the state, though it was not officially recognized as such, nor did a bioregional INACC emerge in the San Joaquin Valley. Nevertheless, the committee included agency ecologists in field offices throughout the San Joaquin Valley, along with representatives from nongovernmental organizations such as TNC. Like the North Coast INACC and other bioregional INACCs, the San Joaquin Valley Biological Technical Committee worked toward the long-term goal of developing a regional conservation strategy that would encompass more than reserved natural areas, like the Carrizo Plain.

The San Joaquin Valley Biological Technical Committee extended the CEC study in at least three respects. First, it was a cooperative group of professional staff in several agencies, not a single-agency program. Second, its work was based on data from several sources, including the CEC inventory. Third, and most important, the committee produced a regional planning document, titled *A Biological Framework for Natural Lands and Endangered Species in the Southern San Joaquin Valley*, released as a draft in May 1993. This 25-page document—plus appendixes and map—was intended to guide planning in the southern half of the San Joaquin Valley Bioregion. Like the CEC inventory that preceded it, the *Biological Framework* did not cover wetlands, so it did not address all listed and candidate species in the bioregion. It also did not cover the northern half of the valley, primarily because comparable data sets did not then exist to identify lands with high conservation value. Despite these limitations, the *Biological Framework* provided interagency guidelines for regional planning based on ecological data and the principles of conservation biology.

In developing this document, the San Joaquin Valley Biological Technical Committee was driven by several regional planning concerns. One was recovery planning under the ESA, a legal obligation the FWS had been unable to meet due to staff and funding constraints. The ESA required the FWS to develop recovery plans for listed species, but

Congress and the President did not appropriate sufficient funds for the agency to meet all of its mandates under the ESA. Underfunding was particularly prevalent in California, which contained numerous listed and candidate species. FWS staff in California were consumed by, and backlogged with, the decision-making processes associated with listing species, consulting with federal agencies, and approving HCPs, so they completed few recovery plans. In the San Joaquin Valley, recovery plans were completed for the blunt-nosed leopard lizard and the San Joaquin kit fox in the early 1980s, but these plans were not implemented and they needed to be updated in the 1990s. Moreover, recovery plans for other listed species—not to mention future recovery plans for species that would likely be listed—still had to be written and implemented. The *Biological Framework* provided a significant step in this direction by painting a picture of recovery needs for a suite of co-occurring species.

A second concern was the compatibility of HCPs in the valley. Local governments were then developing several HCPs, and more would likely be initiated in the future.[10] Though some of these HCPs covered large areas, like the 3000-square-mile Kern County Valley Floor HCP, these HCPs did not cover entire habitats or natural communities—let alone the entire bioregion. The Biological Technical Committee believed a regional planning guideline was needed so that HCPs would be consistent. This included identifying large reserve areas for natural communities, smaller reserve areas for particular species, and corridors connecting these reserves. Because committee members participated in some of these HCPs, they had reason to believe the *Biological Framework* would guide these planning processes.

A third concern of the Biological Technical Committee was to develop consistent on-site mitigation and off-site compensation requirements for development projects. The lack of consistency between permits led to uncertain outcomes for applicants, unequal treatment among applicants, and distrust of state and federal agencies. Though on-site mitigation measures remained largely site-specific, the committee agreed by consensus to recommend a standard compensation ratio of 3:1, or three acres of habitat preserved off site for each acre of habitat lost to development.[11] These compensation lands could then be given to a public agency or nonprofit organization (such as TNC), or placed under a private conservation easement.

The San Joaquin Valley Biological Technical Committee acknowledged social, political, and economic factors, but left these issues for line managers, elected officials, and other actors to address. Like the state-level INACC and the North Coast INACC in the Klamath Bioregion, the committee was a professional group of agency ecologists; it did not include line managers, so its members did not have authority to commit agency resources to an interagency preserve system. The group also did not include local government representatives, so it could not serve as a forum for building local support for regional conservation efforts. Local support would be crucial in the San Joaquin Valley Bioregion because most of the land was privately owned and local governments had zoning authority over private land. It would be difficult to implement a regional preserve system based solely on mitigation and compensation requirements for development projects.

The San Joaquin Valley Regional Consortium: The BLM Attempts to Form a Bioregional Council

The BLM managed more acres of public land in the San Joaquin Valley Bioregion than any other state or federal agency, with numerous parcels scattered throughout the foothills surrounding the valley floor. Though virtually absent from the valley floor, BLM lands contained habitat for many listed and candidate species once widespread on the valley floor. This was a major reason why BLM staff participated actively in these interagency efforts. In fact, the San Joaquin Valley Biological Technical Committee was largely organized and coordinated by BLM Wildlife Biologist Larry Saslaw, who also worked on the Carrizo Plain acquisition and contributed to the CEC natural lands inventory.

Saslaw was the BLM's top wildlife biologist for the agency's Bakersfield District, which included lands spread across roughly one-quarter of California, including the San Joaquin Valley Bioregion. He believed that BLM line managers and staff increasingly participated in public-private and interagency efforts to preserve biodiversity because BLM lands were intermixed with other jurisdictions and contained some of the best remaining habitat. Though habitat transformation had occurred primarily on private land, BLM land was important for maintaining and recovering species. This was not simply the view of professional ecologists. Saslaw believed "the whole BLM management team" saw the endangered species component of public lands management "as one of the best things we can do for local economies."[12]

Indeed, BLM line managers looked positively on the work of the San Joaquin Valley Biological Technical Committee. Not only was Hastey paying attention, word traveled all the way up the hierarchy to Interior Secretary Bruce Babbitt in Washington, D.C.[13] Hastey and other BLM line managers believed the committee had made a good effort to reach consensus on the technical requirements for conservation, but they now wanted to move beyond the *Biological Framework*, to initiate a second phase focused on building broader support for regional conservation efforts. This was not a task for which the committee was well suited, so Hastey sought to create a bioregional council for this purpose.

The committee itself recognized that public relations and education would be important for implementing the conservation strategy in the *Biological Framework*. As Saslaw noted,

> We recognized from the beginning that you can't separate the biology from the social, economic, and cultural aspects of natural resources management; but . . . we figured that the biology needs to be on the table first because you run a risk if you don't have the biology, which oftentimes doesn't have the same political weight as jobs, economies, land ownership. . . . It's easier to get tweaked around [if you don't have a biological picture of] where you want to go in the long run; and we felt that if we put a strategy on the table for other people to work with we might have a better chance of actually accomplishing something.

Indeed, the committee hoped the *Biological Framework* would provide the fundamental strategy for a future bioregional council.[14]

Yet the BLM-sponsored effort to form a bioregional council faced an uphill battle from the beginning. Resistance came primarily from county supervisors, who Hastey hoped would ultimately lead the effort. But it was so threatening to county supervisors that agency officials did not even refer to it as a bioregional council. Instead, they called it a "consortium"—a synonym of *partnership*. Unlike the term *bioregional council*, *consortium* did not contain the politically loaded and divisive *bio-* prefix. As one staff ecologist put it, the word *bioregion* upset county supervisors.[15]

Many county supervisors in the San Joaquin Valley openly held the state and federal agencies in disdain, particularly on the issue of biodiversity. They rejected the idea that species were in decline, disputed agency data, and were suspicious of agency ecologists, particularly in the FWS and CDFG. Kings County Supervisor Nick Kinney said biodiversity was particularly contentious

> because none of us believe what the federal and state government tells us is true. We think they lie to us. We believe that they come in and they own 50 percent of the state now and they're not happy with that. They would like to own 100 percent; and that way they can control all the species, and control all land use, and all other types of decisions, because they feel that they are more qualified [to act on behalf of the public than the public itself].[16]

Kinney's statement indicated the huge chasm separating county supervisors and agency officials because staff ecologists viewed Kinney as one of the most open-minded supervisors on biodiversity issues.[17] Kinney also sat on the state-level Executive Council, where he represented the San Joaquin Valley Regional Association of County Supervisors. If he did not trust the agencies, then it is likely few supervisors—if any—did.

The FWS was probably the most distrusted agency because of the large number of federally listed species in the valley, and the perceived havoc that listed species wreaked on the local economy. One county supervisor referred to the FWS as "the KGB," noted Hastey, because the agency listed species with little advance warning and no public input.[18] In part, this was simply a communication problem. The FWS had developed a poor reputation for communication within the bioregion because staff did not inform local stakeholders of impending decisions that might affect them. FWS regulatory requirements also differed from those of the CDFG, and sometimes conflicted with them, putting county planners in the uncomfortable position of reconciling regulatory requirements. Because Hastey wanted county supervisors to lead the Consortium, and because the Consortium would be the forum within which state and federal agencies, local governments, and other stakeholders would develop regional plans to preserve biodiversity, these communication problems had to be resolved.

From Hastey's perspective, county supervisors were the obvious choice to lead the Consortium. Hastey believed supervisors best represented local interests and had been the key to success with the Carrizo Plain acquisition. Hastey had also co-opted county supervisors at the state level by inviting them to sit on the Executive Council, without requiring them to sign the MOU on Biodiversity. They were a logical choice, too, if agency officials hoped to influence county planning processes and address the highly contentious issue of private property rights.

Hastey assigned Patty Gradek to be the full-time BLM coordinator for the outreach effort to build support for the future Consortium. Gradek was a hazardous materials specialist who had just completed the BLM's management training program. She was not a member of the San Joaquin Valley Biological Technical Committee and was not associated with the *Biological Framework*. Hastey thus sent an implicit message to other agencies and to local governments that the Consortium would not be wedded to the *Biological Framework* or bound by biological objectives, leaving more room for political compromise. Nevertheless, the outreach effort was an interagency project, and several members of the Biological Technical Committee worked with Gradek on it.

This new group dubbed itself the San Joaquin Valley Interagency Team. Gradek chaired the team, which was composed primarily of agency ecologists—an unlikely group to be accepted by county supervisors. This was an important issue because the supervisors were expected to lead the Consortium, and the long-term purpose of the Interagency Team was to provide staff support for the Consortium. Line managers were also largely absent from the Interagency Team. The only line manager who participated regularly worked for the BLM.[19] In other words, the Interagency Team was strikingly similar to the Biological Technical Committee, but now with BLM management support and leadership.

With line managers from other agencies absent, and with Gradek chairing the Interagency Team, BLM's presence and influence loomed large. Participants from other agencies acknowledged the BLM's prominent role and credited BLM officials with making the outreach effort happen, but they still considered it an interagency project.[20] Therefore, BLM officials played down their leadership role and avoided taking credit for it. When asked who was leading the outreach effort, Hastey said, "We all are."[21] Hastey understated his own influence because BLM line managers and staff had much to gain from cooperative planning and management, and he did not want to derail the outreach effort by seeming pushy or preventing others from taking ownership of it. This put Gradek in an awkward position. Within the BLM, Hastey pressured her to move the outreach effort along as fast as possible, while staff in other agencies sought to slow the process down.[22] They were pleased that the BLM provided administrative support, and that Hastey raised the profile of their work with other agency directors, but field staff in other agencies did not want to be led by the BLM.

The Interagency Team's primary medium for public outreach was a traveling slide show. This slide show presented stories about species-specific problems in the San Joaquin Valley and elsewhere in California, and introduced audiences to the MOU on Biodiversity and other cooperative efforts as effective means for addressing these problems. The basic message was that working together on legal requirements would be more productive than working independently. During the summer and fall of 1993, the Interagency Team presented the slide show to city and county planning directors, county boards of supervisors, and a wide range of interest groups, including local chapters of the Farm Bureau, Sierra Club, and Audubon Society. Because the Consortium was intended to represent all interests, the Interagency Team sought to inform them of the potential benefits of cooperation and to encourage them to participate.

The Interagency Team emphasized that the future Consortium would be built from the bottom up, with broad-based public involvement and county leadership. State and federal agency officials would participate, but the Consortium would decide as a group how to meet regulatory requirements and what it wanted the San Joaquin Valley to look like in the future. In these respects, the Interagency Team introduced the *Biological Framework* as a potential guide for regional planning and management, not as a blueprint. According to Gradek, the purpose of the outreach effort was to "drum up some grassroots enthusiasm" for the Consortium:[23]

> It was not to go out and say: "We as government have a plan and we want you to buy into this." It was to say: "We as government have at least gotten ourselves together; we think we agree on a number of things; and we would like to now get together with you . . . and talk about the issues that you see that need to be addressed in the valley, in terms of biological diversity and your concern with your economy and how we make both of these things happen together."

The *Biological Framework* provided the technical foundation for the outreach effort, but the Interagency Team played it down in the slide show. As Gradek noted, "It was very difficult to describe it in a way that doesn't give people the impression that we've already got a plan—'We want you to bless it'—especially when you've got maps that already show places we've got certain ideas about."

Yet the mere existence of the *Biological Framework* troubled audiences, so the Interagency Team revised the slide show, removing all references to it, before hitting the road again in early 1994. Agency ecologists were not pleased by these revisions, but Gradek believed the Interagency Team was losing support whenever it announced the existence of the *Biological Framework*, even if offered as one of several alternatives for regional planning. Nevertheless, BLM officials and the Interagency Team expected the Consortium to work within the bounds of the ESA. As Hastey argued, the outreach effort and Consortium were intended "to get the public, through the counties, to better understand the Endangered Species Act; and recognize that it's not gonna be changed drastically, and we're always gonna have the Endangered Species Act, and we gotta figure out ways to work within it and . . . at the same time still have some assurance about some kind of social and economic development." The Consortium would provide a forum for discussing means to meet legal obligations under the ESA, not to avoid them. Agency officials hoped the Consortium would produce or adopt a regional conservation strategy, with locally crafted plans at the county level.

In addition to the Interagency Team's traveling slide show, agency directors—including Hastey and Wheeler—made their own pitch to county supervisors at a semiannual meeting of the San Joaquin Valley Regional Association of County Supervisors. The supervisors did not reject the idea of the Consortium at this meeting, giving Hastey reason to be optimistic they would participate.[24] Five agency directors also signed a statement of intent to support the San Joaquin Valley Regional Consortium.[25] Among other points in this agreement, the signatories agreed to designate managers and staff to assist with the formation and operation of the Consortium, empower staff to make decisions on behalf of the agencies, and cooperate in efforts to seek grants and other funding sources to assist the Consortium. The signatories also agreed to evaluate existing regulatory processes to identify opportunities for developing coordinated, consistent, and streamlined permitting processes. The latter point was of great interest to local governments and stakeholders, and the Interagency Team soon followed through on it.[26]

While not legally binding, this agreement seemingly evinced a more serious commitment by agency directors to support the San Joaquin Valley bioregional initiative than the Klamath Bioregion Project. In the Klamath, subregional groups were skeptical that the agencies would accept any resource management plans they developed, and agency officials did not sign an agreement stating they would. The Statement of Intent to Support the San Joaquin Valley Regional Consortium listed the ways the agencies would support the Consortium, but it did not state that the agencies would necessarily accept plans developed by the Consortium. After all, these plans would have to be consistent with state and federal laws.

County Supervisors Sabotage the Consortium

Hastey hinged the Consortium's success on the participation and leadership of county supervisors. Yet there was little evidence to suggest they would cooperate without a radical rethinking of the Consortium's purpose. The Kern County Board of Supervisors, in particular, had assumed an overtly reactionary stance against state and federal environmental laws. Kern County supervisors—and many supervisors in other counties—were much more interested in weakening the state and federal ESAs than in complying with regulatory requirements, so they attacked these laws on several fronts.

In 1992, just one year before the Interagency Team began its outreach effort, the Kern County Board of Supervisors petitioned the California Fish and Game Commission to remove the Mohave ground squirrel from the state list, on the grounds that it had been erroneously listed. While the squirrel's habitat was in the Mojave Bioregion, in eastern Kern County, the petition clearly indicated that Kern County supervisors did not accept the existing regulatory regime. Moreover, the petition received a great deal of media attention because the commission voted—over the objections of professional staff—to delist the squirrel in 1993, marking the first time a nonextinct species had been removed from the state list. Kern County Supervisor Roy Ashburn, who led the petition, hailed the decision as a victory for the people of Kern County, who he argued "have suffered under the most outrageous form of heavy-handed treatment over these last years."[27] The delisting decision was short lived, however, because the Office of Administrative Law found that the Fish and Game Commission violated state law by failing to allow sufficient time for public comment.

The petition to delist the squirrel was one of many indications that Kern County supervisors sought to change the state and federal regulatory regime rather than work within it, which suggested they would be unlikely partners in any effort to implement the laws. Indeed, the Interagency Team quickly realized they confronted this dilemma because supervisors in other counties followed Kern County's lead.[28] In particular, the supervisors sought to weaken the ESA, but viewed cooperative planning and management as a means to meet regulatory requirements as long as the ESA was in effect. The Interagency Team was well aware of this dynamic. According to one FWS official,[29]

> The county supervisors are only talking with us because they absolutely have to, because there is a law. They would like to ignore it, hoping that it's gonna die and go away, and then they won't have to do anything anyway. They really don't want to have to talk about the Endangered Species Act at all. Other than a few individuals that I've heard speak, most of 'em really look at the Endangered Species Act as at best an inconvenience, and at worst an absolute infringement on property rights. . . . I think some of it's posturing for their constituency because their constituency really feels that way.

This was not simply the perception of agency officials. County supervisors themselves readily admitted as much.

Kings County Supervisor Kinney, who represented the San Joaquin Valley Regional Association of County Supervisors on the Executive Council, held relatively moderate views, but he believed most supervisors agreed with Kern County Supervisor Ashburn on the legitimacy of state and federal regulations: "Roy Ashburn thinks that all federal and state government [agencies] should leave the State of California, and Roy should be able to run his own county without their interference. And we all tend to agree with that. But it's *not gonna happen*!"[30] Kinney believed the existing regulatory regime was fixed, which led him to explore strategies for coping with the laws. He was open to the idea of a Consortium as a forum for reconciling state and federal regulations, streamlining permitting processes, and developing regional plans for listed species. The competing strategy, promoted by Kern County Supervisor Ashburn, rejected the legitimacy of state and federal environmental laws. Rather than working with state and federal agencies to implement the laws, this strategy emphasized actions to weaken or overturn the laws.

Kinney and Ashburn both held formal positions in the San Joaquin Valley Regional Association of County Supervisors, which enhanced the debate. Ashburn served as President of the Association, so his opinions carried significant weight. Kinney represented the Association on the Executive Council, which indicated he saw value in working with the agencies.[31] Because Kinney was one of the few supervisors open to the Consortium idea, the Interagency Team focused their attention on him, hoping he would sell the idea to his colleagues.

Kinney indeed tried to sell the idea, but the supervisors did not immediately bite. Instead, they established a Biodiversity Task Force in late 1993 to discuss what role the supervisors might play in the Consortium and how it would be structured, operated, and funded. The formation of this task force gave the Interagency Team a reason to believe the supervisors might eventually participate in the Consortium. Yet the Supervisors' Biodiversity Task Force did not invite the Interagency Team to its meetings, and they did not meet with agency officials for several months, so it was not easy to discern their intent. Indeed, there was no small irony in the supervisors using the word *biodiversity* in the name of their task force.

The Supervisors' Biodiversity Task Force subsequently held five monthly meetings with agency officials and local government planners from April to August 1994, during which time participants drafted and discussed a partnership agreement and a work plan for the Consortium.[32] The meetings and documents gave the appearance that county supervisors might eventually participate in the Consortium, but these activities were little more than a ruse on the part of the supervisors. As a group, they never intended to participate in the Consortium—at least not as agency officials conceived it. Unbeknownst to the Interagency Team and BLM officials, the supervisors had already decided before these meetings to put their efforts behind legislative reform.

This decision was made on March 24, 1994, at the semiannual meeting of the San Joaquin Valley Regional Association of County Supervisors—one month before the Supervisors' Biodiversity Task Force first met with the Interagency Team. While this meeting was technically open to the public, it was held at the Rio Bravo Country Club in the foothills east of Bakersfield, with no signs posted at the facility to identify the meeting room and no public participation.[33] Agency officials knew about the meeting but did not attend. The day before this meeting, Gradek met with the Supervisors' Biodiversity Task Force to present a revised proposal for initiating the Consortium. The next day, the San Joaquin Valley Regional Association of County Supervisors discussed this proposal and formally agreed on a course of action. They decided, by consensus, to meet with the Interagency Team in the short run, but to put their long-term efforts behind reforming, if not repealing, the federal ESA. Kern County Supervisor Mary Shell, who sat on the Supervisors' Biodiversity Task Force, made the case for this position:

> I think that the supervisors should meet [with the agencies]. But, in really thinking this through, the major problems with the Endangered Species Act have to be resolved at the legislative level. I asked [agency officials] the question: "Just how much can they change under the current law?" They can't change very much. They can change attitudes, which I think they should because we've had some horror stories, but the real problem is in the legislation itself and the power that it has over private property rights. . . . So, I think if you're looking for what you want, you're not gonna get it by meeting with them. Although, I think we ought to meet with them, and I think that they ought to know the problems that people are encountering in dealing with this act and in dealing with the people who are implementing it. But you're not gonna get what you really want, and that is a change in the law; and I think you ought to pursue that as vigorously as possible.[34]

Kinney concurred, and there was no dissent.

Pursuing legislative reform rather than cooperative planning was a potentially effective strategy at that time because several bills to amend the ESA were then before Congress. The ESA was overdue for reauthorization, and several members of Congress were pushing for reform. Congressman Richard Pombo held a series of hearings in the San Joaquin Valley to build support for ESA reform. Environmentalists believed these hearings were intended to gather ESA "horror stories" from farmers and other economic interests because few environmentalists were invited, the time and location of the hearings were announced to the public a few days before the hearings, and environmentalists who showed up were given little time to speak. Kern County Supervisor Shell was also then working with the California State Association of Counties on a survey of county grievances to compile ESA "horror stories."

In the short run, the supervisors had nothing to lose by meeting with the Interagency Team, other than a few days of their time, because the meetings would not conflict with their pursuit of ESA reform in the long run. Moreover, if they refused to meet with agency officials, it would appear as if they were making no effort to allow the ESA to work in practice. For the supervisors, meeting with the Interagency Team served a public relations purpose. The meetings provided an opportunity to air local grievances against state and federal laws and agencies, without the supervisors committing themselves to implementing those laws or working with the agencies.

The supervisors accordingly scheduled five meetings with agency officials and local government planners. To agency officials, the compressed time frame indicated the supervisors did not want the bioregional effort to succeed. Kings County Supervisor Kinney claimed that the limited time frame was due to resource constraints: "We're limiting the time frame because we don't have a lot of extra money and stuff. . . . So either we get this set up real quickly, in the next three to four months, to where we have a working task force that's going to be able to do something that is to our liking and everybody else's liking, or we're gonna walk away."[35] Agency officials believed the Consortium would reduce the costs of ESA compliance in the long run through regional planning and streamlined permitting. Moreover, the agencies were covering most of the organizational costs. Therefore, the cost issue raised by Kinney was probably a red herring. Most of the supervisors wanted to reform the ESA, not reduce the costs of implementing the law. The supervisors had already decided to walk away from the Consortium before they met with the Interagency Team because they knew the agencies could not give them what they wanted. For the supervisors to participate in the Consortium, the Interagency Team would have to make all of the concessions. Compromise and deliberation seemed unlikely.

After the last meeting with the Interagency Team, the San Joaquin Valley Regional Association of County Supervisors met again for their semiannual meeting in October 1994, at which they voted against participating in the Consortium and formed a committee to advocate reforming the state and federal ESAs. While not a surprising outcome in light of their previous meeting in March, agency officials did not attend the March meeting, so they had reason to believe the supervisors might actually join the Consortium. The supervisors had now officially turned their backs on the agencies and walked away. Kinney subsequently resigned from the Executive Council, and the Executive Council seat reserved for the San Joaquin Valley Regional Association of County Supervisors was vacant for a year.

The supervisors' decision underscored how far apart the agencies and supervisors were in terms of goals and strategies. Kinney emphasized to his colleagues before the vote against the Consortium that participation would not interfere with their efforts to reform the laws.[36] Indeed, there is no evidence the supervisors believed a conflict existed between these strategies. The Kern County Board of Supervisors also distorted the history of agency efforts to work with the supervisors. For example, they adopted a resolution that stated, in part, that "despite the efforts of County Supervisors to have State and Federal agencies define the purpose, funding and goals of this proposal, there has been no clear statement of the goal to be accomplished nor a commitment by State and Federal agencies that they will make changes in their regulatory or enforcement policies if such a consortium were established."[37] This claim is simply inaccurate, and the Kern County Board of Supervisors must have known so because some of them had been meeting with agency officials. Moreover, five state and federal directors had already signed the Statement of Intent to Support the San Joaquin Valley Regional Consortium, which laid out the purposes and goals of the Consortium. The Executive Council also offered $10,000 to fund initial expenses, and the BLM assigned a full-time staff member to coordinate the effort. In addition, the Interagency Team was working with the Supervisors' Biodiversity Task Force to draft a work plan and partnership agreement that specified the mission, principles, and goals of the Consortium, and it was standardizing the process for complying with the state and federal ESAs.[38]

Yet standardization and streamlining were not what the Kern County Board of Supervisors had in mind when they adopted the resolution. They wanted the agencies to weaken ESA regulations, not implement them. They wanted more than an efficient, expedited process for complying with the laws. They wanted the agencies to subvert the state and federal ESAs by allowing local governments and other stakeholders to develop projects without fully complying with the laws. Because the agencies would not do this, the Kern County Board of Supervisors adopted a resolution that implied the agencies, not the supervisors, refused to cooperate.

The Subversion Hypothesis Revisited

This case offers a good test of the subversion hypothesis. The strong form of the hypothesis argues that agency officials initiated the MOU on Biodiversity, the Executive Council, and the bioregional efforts to lull elected officials and the public into believing that the laws could be weakened because the agencies were now working together to preserve biodiversity. The weak form of the hypothesis argues that agency officials initiated the cooperative efforts in good faith as a means for implementing the laws, but others subsequently used these efforts as a pretext to lobby legislators to weaken the laws and convince political appointees and agency officials to relax enforcement.

This case does not support the strong version of the subversion hypothesis because the evidence indicates that the BLM line managers and staff who initiated the outreach effort to build the Consortium sought to work within the ESA's regulatory framework. If the strong form of the subversion hypothesis were correct, the Interagency Team would likely have conceded ground to county supervisors by relaxing regulatory enforcement. Yet neither side conceded anything of substance with regard to implementing the federal ESA. The Interagency Team sought to streamline permitting procedures, not weaken enforcement. If the Consortium was intended to subvert the ESA, it is likely that BLM officials would have found a way to entice the supervisors to participate. Instead, the supervisors walked away from discussions and established their own committee to reform the ESA. In so doing, they led Hastey to believe they might participate in the Consortium as a means for implementing the ESA, when their actual intent was to reform the law itself. Hastey misjudged their intent; there is no evidence he sought to subvert the ESA through the Consortium.

The case does, however, support the weak form of the subversion hypothesis. Rather than make a good-faith effort to work with the agencies through the Consortium to implement state and federal laws, the supervisors met with agency officials to put a good spin on their lobbying efforts for regulatory reform, and to delay or weaken regulatory enforcement in the meantime. Though the agencies conceded nothing in this regard, the meetings helped the supervisors deflect blame for implementation failure onto the agencies. While the evidence clearly indicates that Kern County supervisors fit this profile, it is much less clear whether it fits Kings County Supervisor Nick Kinney, who sat on the Executive Council. Kinney may have sincerely hoped to work with the agencies, but could not gather support from other supervisors in the bioregion.[39]

Summary

The proximate cause that derailed the San Joaquin Valley Regional Consortium is clear. Hastey wanted county supervisors to lead the Consortium, but they did not want to participate because they sought to weaken the ESA, not implement it. The supervisors led the agencies on a yearlong wild-goose chase, allowing Hastey and the Interagency Team to believe they might join the Consortium while never intending to participate in the way agency officials conceived. Hastey overestimated the willingness of county supervisors in the San Joaquin Valley, particularly in Kern County, to work within the existing legal regime. The Consortium was not part of a grand scheme to subvert the ESA. To the contrary, agency ecologists largely developed the underlying idea as a means for implementing the state and federal ESAs more effectively through proactive planning and management. They sought to build public support for regional planning and were grateful that BLM line managers supported their efforts.

Yet cooperation at the staff level among agency ecologists made large strides, and BLM line managers attempted to bump professional cooperation up to the managerial and political levels. Following the CEC's natural lands inventory in the Southern San Joaquin Valley, agency ecologists forged the Biological Technical Committee and drafted the *Biological Framework*, a regional planning guideline for upland natural communities. Hastey supported the Biological Technical Committee's work and sought to build public support for bioregional planning and management. He assigned a BLM staff member to work full time coordinating the Interagency Team's outreach effort to forge a bioregional council from the bottom up. This team, composed primarily of agency ecologists, spent more than a year trying to gain the support of county supervisors and local stakeholders, but their effort failed because the supervisors wanted Congress to dismantle the ESA. The supervisors did not want to work with the agencies to implement the ESA and other environmental laws.

It is easy to dwell on the role of county supervisors in this case because Hastey hinged the success of the Consortium on them, but it is important to recognize that interagency cooperation progressed quickly in the San Joaquin Valley Bioregion. In this respect, the case highlights two central themes of this book. One theme is the role of professional ecologists in developing the technical infrastructure for cooperation. Agency ecologists gathered data identifying the location and condition of species and natural communities, and disseminated the planning principles of conservation biology. In so doing, they provided the underlying logic for interagency cooperation in preserving biodiversity. The county supervisors, however, rejected this logic because they had other goals in mind.

The second theme is the role of BLM line managers in expanding cooperative efforts beyond the professional level. These managers confronted a different set of incentives than line managers in other agencies. Generally speaking, line managers sought to maintain their autonomy because autonomy reduces uncertainty, which increases the ease of managing their organizations. BLM line managers likewise sought to reduce uncertainty, but autonomy was not a viable option for them. More so than other agencies, the BLM was inextricably intertwined with its neighbors because BLM lands were highly dispersed and fragmented. As one BLM official in the San Joaquin Valley stated, "The reality is that we can't hide behind a national park or a national forest boundary and say: 'The hell with what happens outside of that boundary, we know what's best here, we have a large staff, we have lots of information in our boundaries, come on up and visit and we'll tell you how the world is.' We *can't* operate that way."[40] Reducing uncertainty therefore required managing interdependence. This could be done in two ways. One strategy was to consolidate parcels, which BLM line managers did in the

Carrizo Plain acquisition, thereby reducing their dependence on neighbors in a specific geographic area. Another strategy for reducing uncertainty involved cooperative planning and management, which BLM officials pursued throughout California.

In the San Joaquin Valley Bioregion, the BLM managed numerous parcels scattered throughout the hills surrounding the valley floor. These parcels were highly permeable to activities occurring on neighboring property. This was particularly a problem with endangered species because BLM parcels contained some of the best remaining natural lands, and endangered species on BLM land were affected by activities occurring within neighboring jurisdictions. Even if authorized activities on BLM land had no effect on endangered species, these species still suffered from impacts on other lands, so the BLM would ultimately be responsible for protecting their populations on BLM land. Moreover, some external impacts crossed BLM land, as in the case of oil and gas pipelines, power grids, and the California Aqueduct. As a multiple-use agency, the BLM could not avoid entanglement with these activities, so interdependence was thrust on BLM line managers. Interagency, intergovernmental, and public-private cooperation was a means for managing this interdependence by preserving biodiversity on a regional basis.

If not for BLM line managers, cooperation may not have expanded beyond the professional level because there is little indication that line managers in other agencies would have stepped forward in their place. Although five agency directors signed the Statement of Intent to Support the San Joaquin Valley Regional Consortium, Hastey gathered the signatures, and the other directors were not subsequently put in the position of having to encourage their regional line managers to participate in the Consortium. In the Klamath, CDFG regional managers were particularly unsupportive, so there is little reason to expect their counterparts in the San Joaquin Valley Bioregion would have participated, let alone encouraged their staff to participate. This is an important point because line managers had the authority to commit agency resources to interagency processes, and their participation would have been crucial if the Consortium had formed. As in the Klamath Bioregion, line managers in most agencies would likely have resisted participation, so agency directors would have had to find a way to circumvent line managers to make the Consortium work, as they did with NCCP in the South Coast Bioregion. Other field staff would likely have participated to maintain the socioeconomic stability of their communities, as they did in subregional groups in the Klamath, but we cannot know for sure because the Consortium did not emerge, so long-term participation by nonecologists at the field level was not a factor.

As for county supervisors, they might have participated in the Consortium had it been tied directly to state and federal permitting processes. In the South Coast Bioregion, participation in NCCP was weak and ineffectual until state and federal officials hitched NCCP to the ESA. Enrolling in NCCP became a means for receiving incidental take permits. An NCCP-type approach could have been attempted because several HCPs were underway in the San Joaquin Valley Bioregion, and more were expected. The agencies could have drawn up rules requiring HCPs and Section 7 consultations to be consistent with a regional plan, such as the *Biological Framework*. Such consistency requirements would have given county supervisors a reason to view the Consortium as a better means for meeting legal obligations, so long as the ESA remained intact.

8

Explaining Interagency Cooperation: Or, Why the BLM Cooperates More Than the NPS

The case studies in chapters 3 through 7 sketched the scope, strength, and duration of interagency cooperation in preserving biodiversity in California, and explored the causal antecedents of cooperation at the local, regional, and state levels. This chapter ties the case studies together by reviewing the organizational and institutional incentives for cooperation among line managers, professionals, and field staff. Table 8.1 lists these incentives in the order presented in this chapter and indicates whether their impacts vary over time or across agencies. In each section, the incentives are discussed in the form of hypotheses that address the book's central question: "Under what conditions do individuals in different public agencies cooperate with one another?" To provide additional focus, I also pose a specific question: "Why did cooperative efforts tend to emanate from the BLM rather than the NPS?" The NPS had a much clearer mandate to preserve natural resources than the BLM. Yet BLM line managers and staff often led interagency efforts to preserve biodiversity, while NPS participation was the exception rather than the rule.

Why Cooperation Expanded during the 1990s

Agency officials expressed differing views about the causes of cooperation, but they largely agreed the agencies cooperated more in the 1990s than they had ever cooperated before on resource management issues. Notably, they expressed this belief even though local CRMPs had been thriving for decades and INACC had largely folded in 1991. So why did they believe that interagency cooperation was on the rise? First, state and federal directors—both political appointees and high-level line

Table 8.1
Explaining interagency cooperation: Time-series and cross-sectional variables

I. Time-series variables—or, why cooperation expanded during the 1990s:
   a. The regulatory hammer—court enforcement of environmental laws changed task requirements for some agencies.
   b. Consensual ecological knowledge—conservation biologists and their ecological allies offered consensual advice that cooperative habitat planning and management was the most effective way to preserve biodiversity.

II. Cross-sectional variables—or, why cooperation varied across agencies:
   a. Likelihood of litigation due to the presence of federally listed species.
   b. Likelihood of litigation due to relationship between an agency and environmentalists.
   c. Size and fragmentation of agency landholdings.
   d. Frequency of job rotations within agency.
   e. Discretion in standard operating procedures and reporting relationships.
   f. Travel time to attend interagency meetings.

managers—were now involved at the state level. Second, cooperation was emerging and expanding at the local and regional levels under banners other than CRMP and INACC. These changes occurred due to two significant changes in the organizational environments of public agencies. First, court enforcement of environmental laws compelled some agencies to protect species and habitats. Second, consensual ecological knowledge infused public agencies with the idea that protecting species and habitats is a collective-action problem. Together, the regulatory hammer and consensual ecological knowledge provided a two-pronged interactive effect that motivated cooperation beyond the ecological epistemic community.

The Regulatory Hammer

Line managers and staff in most agencies would have done relatively little to preserve biodiversity absent court enforcement of environmental laws. Biodiversity preservation was not a priority for local, state, and federal agencies, or for county supervisors, private landowners, and developers. Even the NPS, a longtime ally of environmental groups, had long manipulated ecosystems and developed infrastructure for recreation and tourism at the expense of biodiversity. Absent court enforcement of environmental laws, particularly the ESA, staff ecologists would have continued to gather information about the decline of biodiversity, but their planning efforts to preserve biodiversity would not have been supported by the line managers who controlled agency resources necessary for designing and managing reserve systems. Court enforcement also gave field staff, local government officials, and local stakeholders a reason to change their patterns of resource use as well.

Because public agencies have discretion to emphasize one use of natural resources over others, and because most agencies were primarily oriented toward human uses, biodiversity protection received haphazard attention on public lands until court enforcement constrained agency decisions. In places where management autonomy once predominated, the courts drew lines in the sand delimiting the bounds of managerial discretion, particularly where species balanced on the brink of extinction. Line managers increasingly feared to cross these lines because they might lose most or all of their autonomy to the cause of species protection, but no one knew precisely where they stood due to technical uncertainties. Because court decisions were feared and technical uncertainty was great, line managers and field staff increasingly perceived a looming regulatory hammer that could drop with shattering impact on management prerogatives and the socioeconomic stability of local communities. Of all the environmental laws they feared, none stood out more than the federal ESA.

Public and private actors could remove themselves from the shadow of this regulatory hammer in two basic ways. One set of strategies put a premium on weakening or repealing the ESA. This included lobbying Congress to eviscerate the law or finding agency officials willing to subvert it. The other set of strategies emphasized maintaining viable populations of species so the ESA's provisions would not kick into gear. Simply following the letter of the law was risky because it did not sufficiently ward off lawsuits, so minimal compliance faded as a preferred option.[1] These two sets of strategies—weakening the ESA and maintaining viable populations—were not mutually exclusive, but individuals tended to prefer one to the other, depending on ideological convictions.

At the local level, county supervisors in the San Joaquin Valley Bioregion lobbied Congress to weaken the ESA, even as planners in these counties were preparing HCPs under Section 10 of the ESA. Rather than seeking ways to live within the ESA's legal regime, these supervisors sought to weaken or repeal the law itself. They did not support the BLM-led effort to develop a Consortium (or bioregional council) in the San Joaquin Valley. Instead, they conspired to undermine it by acting as if they were amenable to the idea, and then pulling out without allowing sufficient time for the Consortium to form and accomplish its objectives. While BLM line managers and field staff sought to live within the spirit of the law in the San Joaquin Valley Bioregion, county supervisors undermined them by pursuing a different strategy.

At the state level, Governor Wilson's political appointees were also of two minds about biodiversity issues. Some supported and sponsored efforts to preserve biodiversity, while others lobbied Congress and federal agencies to loosen the ESA's regulatory strictures. Given this split personality, some observers believed the Wilson administration sought to subvert the legal regime by claiming that existing cooperative efforts sufficiently protected biodiversity, while simultaneously impeding these cooperative efforts from accomplishing their stated goals. While it is clear that some county supervisors and political appointees in the Wilson administration sought to overturn or weaken the state and federal ESAs, and sometimes used the existence of cooperative efforts for rhetorical purposes to aid them in this cause, there is no evidence that agency officials initiated these cooperative efforts for this purpose.

Instead, most agency officials sought ways to comply with the existing legal regime rather than overturn or subvert it. Minimal compliance, however, was dangerous—like walking near the edge of a cliff in a dense fog. Therefore, the risk-averse strategy for avoiding lawsuits was to maintain viable populations of species in natural communities, rather than allowing them to balance on the brink of extinction. This required proactive thinking and new strategies to maintain or increase the populations of listed and unlisted species before the ESA ran its course in the courts. Yet, without at least one federally listed species, there was no strong incentive for public or private participation. In the South Coast Bioregion, the original goal of NCCP was to avoid listings altogether, but participation was limited until a federally listed species made the regulatory threat credible and imminent for those who otherwise hoped to free ride on the voluntary contributions of others.

Consensual Ecological Knowledge

Ecologists played a dual role in this legal context. On the one hand, they front-loaded the decision-making process by providing information demonstrating that species were in decline, thereby placing public and private actors in the shadow of the regulatory hammer. They also offered these actors strategies to remove them from the hammer's shadow. Conservation biologists were in a particularly strong position in this regard because, as an epistemic community focused on the preservation of biodiversity, they offered a common set of management practices for maintaining viable populations. Their influence resulted not from their absolute numbers or line authority within the agencies, but rather from their consensual knowledge. The management principles of conservation biology offered synergistic possibilities for joint gains that spread throughout the larger epistemic community of ecologists, and were gradually understood and accepted by line managers and field staff as well.

The effect on cooperation was interactive because court enforcement and ecological knowledge would have prompted only limited cooperation in the absence of either. If the ESA had been perceived as a paper tiger, like the CESA, cooperation would have been limited to members of the ecological epistemic community. Court enforcement of the ESA prompted nonecologists to devote more of their time and resources to managing listed and candidate species. Yet, by itself, court enforcement was insufficient to spur cooperation because line managers, field staff, and other stakeholders needed a reason to believe that collective action was an important component of species protection. For line managers facing increased uncertainty and diminished autonomy, the functional postulates of conservation biology held the possibility for joint gains on this newly imposed task, which would allow them to pursue the traditional tasks for which they were responsible. For field staff and local stakeholders, the functional postulates of conservation biology provided a means for maintaining the socioeconomic stability of their communities by ensuring continued access to natural resources and development opportunities on public and private lands.

Conservation biology provided the strategic logic for collective action by suggesting that public and private neighbors could gain through joint planning and management within the ESA's legal regime. This was not simply an issue of reducing operating costs by sharing data, staff, and other administrative resources. It was a technical issue driven by the concept of "minimum viable populations" and the habitat required to sustain them. Because most habitats were interwoven with multiple private parcels and public jurisdictions, it was not possible to maintain viable populations for most species within a single parcel or jurisdiction. Therefore, effective reserve systems would have to encompass multiple jurisdictions and ownerships. Even if an effective reserve system could be theoretically designed within a single jurisdiction, public officials could still gain through cooperation by designing and implementing a multijurisdiction plan that spread the burden of habitat protection and reduced the risk of court actions. In other words, rather than each agency or landowner preserving habitat for species without regard to the activities of others, they could coordinate their activities by jointly managing adjacent land, designing wildlife corridors between habitat patches, swapping parcels to create larger and more contiguous habitat patches, and developing consistent regulatory standards and management guidelines. By developing coordinated reserve systems, line managers could save more of their jurisdictions for other uses, and even design a more inclusive reserve system for a suite of candidate species, thereby preventing further population declines and obviating the need for the FWS to list more species.

Cooperation was not only an effective means to maintain viable populations; it was also an effective means for enhancing management autonomy and the socioeconomic stability of local communities. Agency officials were no longer simply protecting their own turf, but they were also creating and protecting cooperative turf. The relative utility of cooperative turf varied, depending on other factors discussed below, but the operational capacity of the interagency system to produce synergistic benefits for line managers and field staff changed significantly when the ESA was enforced *and* they learned about the functional postulates of conservation biology. Though line managers generally seek autonomy, they may pool their resources when external forces threaten an even greater loss of autonomy should they attempt to operate independently. Under these circumstances, cooperation provides more stability for managers than does autonomy.

As the 1980s drew to a close, this interactive effect gathered momentum. Conservation biology quickly became a distinct epistemic community. Some of the leading academics were based in California, including Michael Soulé, who founded the Society for Conservation Biology. A cadre of professional conservation biologists, allied with the larger epistemic community of ecologists, emerged within the agencies. The knowledge base was in place, but other agency officials needed a reason to pay attention. In this regard, an enormous event occurred in 1991 that shook traditional sensibilities. U.S. District Judge William Dwyer issued an injunction on timber sales on federal land in the range of the northern spotted owl. With a single decision, Judge Dwyer largely shut down timber operations for three years on federal land in the Pacific Northwest, including the Klamath Bioregion in northwestern California, thereby impeding a traditionally strong agency—the FS—from carrying out its primary mission.

The loss of autonomy by FS line managers was so pronounced that it prompted others to consider proactive strategies to protect biodiversity before this happened to them. The injunction was an ominous demonstration of what could happen in courts if line managers ignored the advice of professional staff regarding the ecological impacts of agency actions. The ESA had already caused numerous smaller headaches for public officials in California regarding other species—including the Stephens' kangaroo rat in the South Coast Bioregion—but no one expected a court decision like Judge Dwyer's, in which the technical core of a powerful agency was shut down for several years and its primary mission challenged, diverted, and impeded over a significant portion of three states because line managers failed to meet legal obligations. Even more ominously, the injunction came in response to a lawsuit filed under NFMA and NEPA, not the ESA, which suggested that litigants could also achieve major victories on behalf of biodiversity through other environmental laws. For agency officials in California, the owl was a harbinger of things to come and a specter of worst-case scenarios. If it could happen to a powerful agency like the FS, it could happen to any agency. Moreover, with more than a hundred listed species and many more candidate species in California in the early 1990s, plenty of "spotted owls" loomed on the horizon.

The logging injunction was an opportunity for vicarious learning because individuals who were not directly affected witnessed a scenario they had not previously considered. In such circumstances, individuals may engage in different types of learning, such as single-loop or double-loop learning (Argyris and Schön 1978; Clark 1997). Single-loop learning simply involves a cybernetic self-correcting cycle, much like a thermostat stabilizes the temperature inside a house (Steinbruner 1974). It is similar to an iterative game, in which individual preferences and goals do not change, but strategies might change based on what participants learn about other players in previous stages of the game. In double-loop learning, individuals may change their preferences and goals in addition to changing their strategies.

Peter Haas (1990: 61) argues that an "even more sophisticated mode of learning" may occur, in which "policymakers adopt entirely new patterns of reasoning." They may make this leap either because the authority of an epistemic community's knowledge base sways them, or because "they may recognize that the context in which policy is made has changed, and alter their reasoning process accordingly." In the former situation, they become members of the epistemic community; in the latter, they become strategic adherents of the causal mechanisms, not the value premises, espoused by the epistemic community. In California, most line managers were not concerned about biodiversity per se, at least not enough to override traditional agency missions and tasks. Most did not see an imminent ecological crisis. Instead, they became strategic adherents of the causal mechanisms to deflect court enforcement of environmental laws.

Ideas—such as the planning principles of conservation biology—do not simply become policy. They must attach themselves to political problems. As Kingdon (1984: 181) argues, "Solutions float around in and near government, searching for problems to which to become attached or political events that increase their likelihood of adoption. These proposals are constantly in the policy stream, but then suddenly they become elevated on the governmental agenda because they can be seen as solutions to a pressing problem or because politicians find their sponsorship expedient." Adler (1992: 124) offers a similar argument in his research on epistemic communities, noting the existence of a "political selection process" in which epistemic ideas are selected and turned into policy because they "best fit the interests of policymakers," not because they are the best-fitted ideas.

In California, agency ecologists had been cooperating for years—within professional networks, across agency boundaries—trying to develop plans and gather agency resources to implement these plans. Yet just because interagency coordination may be a technically effective or efficient means for preserving biodiversity does not mean it will occur. State and federal agencies had multiple missions, many of which were conflicting, and most of which were intended to satisfy human wants rather than species' needs. Line managers and staff needed a strong incentive to support these efforts. Court enforcement of the ESA and other environmental laws created a problem for line managers because it threatened management autonomy, and it created a problem for field staff and other local stakeholders because it threatened the socioeconomic stability of the communities in which they lived. Conservation biology provided functional postulates and best management practices to preserve biodiversity, but line managers and field staff did not follow them until the regulatory hammer loomed in court. Line managers and field staff saw a different problem than professional ecologists, but they all drew on the same solution. Only when line managers became strategic adherents of the causal mechanisms did the marginalized ecologists achieve prominence in agency decision-making processes.

Why Cooperation Varied across Agencies

Court enforcement of environmental laws and consensual ecological knowledge provided a joint incentive for cooperation during the 1990s. Yet this interactive effect primarily explains the emergence of cooperation over time, not variation in participation across agencies. BLM officials participated in cooperative efforts more routinely than did FS officials, even though the FS suffered the brunt of the logging injunction. Cooperative efforts also emanated more from the BLM than the NPS, even though the NPS had a much stronger mandate to protect natural resources. In the remaining sections, I focus on the BLM and NPS to demonstrate the effects of incentives that varied across agencies.

Likelihood of Litigation Due to the Presence of Federally Listed Species

Because line managers viewed court enforcement of the ESA as the biggest threat to their management autonomy, one or more listed species had to be present to prompt cooperation beyond the ecological epistemic community. California had more than a hundred federally listed species in the 1990s, with nearly a thousand species under consideration for listing. Thus, it is not surprising that cooperative efforts emerged throughout the state. Yet these efforts were not evenly distributed across the landscape. They were concentrated in areas with listed species, particularly species whose habitat sprawled across several jurisdictions. The more such species were found in a bioregion, the more likely interagency cooperation to preserve biodiversity emerged in that bioregion.

Variation in participation also existed across agencies, with the BLM often in the lead. For example, of those directors who contributed time, voice, and agency resources to cooperative efforts, none stood out more than BLM State Director Ed Hastey. Some referred to Hastey as the godfather of biodiversity in California because he brought agency directors to the table on this issue. Not only did he provide leadership in bringing the MOU to fruition, but he worked hard to implement its principles through the state-level Executive Council and the San Joaquin Valley Regional Consortium. These actions surprised many observers because Hastey and the BLM had not previously developed reputations for protecting biodiversity. To the contrary, Hastey had been trained as a logging engineer in a traditional forestry school, and the BLM had long been the target of environmental activism because its multiple-use mandate often produced adverse environmental impacts. Yet BLM line managers and staff led these interagency efforts because their jurisdictions contained a large number of listed and candidate species. At the time, the BLM managed 17 percent of the surface area of the state, including parcels in all bioregions, with 78 listed species and 361 candidate species residing somewhere on BLM's 17.1 million acres in California (Bureau of Land Management 1992).

Additional evidence of the impact of listed species can be found at the local and regional levels, where habitat planning and management largely occurred. In the South Coast Bioregion, NCCP floundered until there was at least one federally listed species to provide the basic incentive for participation. Public and private actors would not make sacrifices for a collective reserve system when the regulatory hammer did not lurk immediately in the background. Minimally, the prospect of a listing had to be imminent and credible. Cooperative planning and management occurred in the shadow of the law, not in the absence of it. For the ESA to cast a significant shadow, at least one species had to be listed.

Likelihood of Litigation Due to Relationships between an Agency and Environmentalists

The mere presence of listed species within a bioregion or jurisdiction explains only part of the variation in participation among agencies. Public officials also had to believe the ESA would be enforced against them. Generally speaking, the magnitude of the perceived threat depended on the probability that an agency's activities would prompt a lawsuit in the foreseeable future. Without the impending threat of a lawsuit that might constrain management autonomy, the presence of listed species did not prompt cooperation from line managers.

For BLM line managers, the probability of a lawsuit was high because environmentalists routinely targeted the agency on many issues, including biodiversity. For NPS line managers, the probability was low because the agency was seldom sued, so they were much less concerned about the presence of listed species within the parks. Environmental groups routinely turned the other cheek with the NPS because it is a traditional ally of the environmental movement. This does not mean that activities within national parks necessarily complied with the ESA or other environmental laws; it simply means that NPS line managers were not concerned that the ESA would be enforced against them. Thus, NPS line managers demonstrated little enthusiasm for interagency efforts to preserve biodiversity. The NPS Regional Director attended six of the first thirteen Executive Council meetings, but his participation at those meetings was minor, and he was not a forceful advocate for the MOU on Biodiversity within his agency. In the Klamath Bioregion, the Redwood National Park superintendent did not support staff ecologists working on the North Coast INACC. None of the NPS officials I interviewed mentioned listed species or court enforcement as an incentive for cooperation, even though many listed species occurred in the parks.

Ironically, many public officials and private actors believed their common antagonist to be the FWS, not the environmental groups that prodded the FWS into action. Interagency cooperation did not result from one or more agencies threatening the others. The FWS was slow to list species and hesitant to compel others to comply with the ESA's provisions. The FWS was the only agency with regulatory authority for nonmarine species under the ESA, but it was not strategically situated to challenge the prerogatives of other agencies.[2] Environmentalists prompted the FWS and other agencies to act by filing numerous lawsuits to speed up the listing process, and to compel local, state, and federal agencies to fulfill their duties for listed species.[3] Legal standing under the ESA was important because environmentalists could raise the regulatory hammer on behalf of listed species in court. This meant that agency officials did not wield control of the hammer itself. Instead, the hammer loomed in the background—an ominous and unwieldy threat to management autonomy and the socioeconomic stability of local communities. Had environmental advocates not been able to sue on behalf of listed species, the threat posed by the ESA would have been significantly diminished.

Size and Fragmentation of Agency Landholdings

Even when agency officials were motivated to protect species, either because of epistemic beliefs or to avoid lawsuits, the size and shape of landholdings provided an additional incentive for cooperation. Generally speaking, agencies responsible for large blocks of consolidated land depended less on neighboring jurisdictions than agencies managing small parcels dispersed across the landscape did. As parcels become larger and more consolidated, agencies depend less on the condition of habitat in neighboring jurisdictions. Thus, technical interdependence in the preservation of biodiversity arises in part from the geometry of land ownership. Yosemite National Park, for example, is large, well rounded, and borders on relatively few neighbors. The park is also consolidated, meaning there are no private inholdings within its boundaries. The park's large size and consolidated shape allow its managers the luxury of believing—rightly or wrongly—that they control sufficient habitat within their jurisdiction to manage associated species independently of their neighbors.

The traditional view of national parks is that they are biological islands largely insulated from outside disturbances. This metaphor fell on hard times, however, as academics and environmentalists increasingly documented external impacts on parks, such as air pollution and development. Yet, with regard to biodiversity, many NPS line managers did not confront a pressing technical imperative to coordinate their planning and management practices with neighboring jurisdictions because most national parks are consolidated and well rounded. This was particularly true of Yosemite National Park, where public officials in neighboring jurisdictions expressed concern about the absence of NPS participation in regional planning processes. Moreover, some NPS officials believed the island metaphor still influenced management decisions. As one Yosemite ranger commented in 1994,[4]

> My perception is that the Park Service culture is beginning to change, and the rhetoric has changed, and certain individuals have changed radically, but the dominant culture is still Freemuth's [1991] "island" mentality. . . . I think it's partly sustained [by] a large segment of the environmental community that still holds that notion. Although I think that's breaking down more rapidly within the environmental community than it is within the Park Service because it's institutionalized in the Park Service by the autonomy that superintendents have, and the fact that most superintendents now have come up through the Park Service over years and years and years, and have been steeped in that model of the park, and so that's how they govern their little fiefdom.

Yet the NPS is not the best test case for assessing the effect of parcel size and consolidation on cooperation because NPS line managers were not concerned about lawsuits threatening their management autonomy. Therefore, we need to examine an agency that also managed large, consolidated jurisdictions and that was likely to be sued by environmentalists. The Department of Defense (DoD) managed several large bases in California that provided habitat for listed species. In addition to potential lawsuits, DoD base managers also faced uncertainty from potential base closures, which led some base managers to expand their missions to justify the continued existence of their bases. Ironically, some military bases had become de facto nature preserves, prized by environmentalists as future national parks, because much of the land on the bases remained undeveloped.

Camp Pendleton Marine Base, for example, contained some of the largest remaining patches of natural communities in the South Coast

Bioregion. Camp Pendleton managers therefore had functional and symbolic reasons for becoming good stewards of the land, and they spent millions to study and maintain viable populations of species on the base. Yet they distanced themselves from regional habitat planning efforts such as NCCP because they did not depend heavily on the condition of habitat in neighboring jurisdictions to maintain viable populations within the base. Base managers were hesitant to become part of a larger regional reserve system because they worried that Camp Pendleton might become a "mitigation dumping ground," with restrictions placed on base activities to compensate for habitat damage caused by development outside the base. Having little to gain from cooperation because of the size and shape of the base, and potentially much to lose in compensation deals for development outside the base, they saw little reason to participate in regional planning efforts.

Unlike the NPS and DoD, the BLM managed numerous fragmented parcels scattered throughout California. BLM line managers long recognized they had insufficient staff to manage these highly fragmented parcels independently of their neighbors. As one BLM line manager put it,[5]

> We recognize it's impossible that this little office here—thirty people, more than a thousand parcels of land scattered over a ten-million-acre planning area—it's impossible to manage [this land]. Not improbable, impossible! So we had to look at doing cooperative things—local stewardship, administrative jurisdictional transfers, disposals of land to consolidate through exchanges—and that means you're dealing with everybody.

BLM line managers and staff have been entering into partnerships for years, largely with local landowners, to manage the agency's highly fragmented lands. These public-private partnerships contributed to the perception that the agency is captured by local interests, but partnerships also provided forums within which BLM officials pursued nontraditional missions, such as the preservation of biodiversity.

Due in part to this fragmentation, Hastey was a strong supporter of CRMP, encouraging his line managers and field staff to reach out to neighbors and develop consensus-based plans. The NPS was not even a signatory to CRMP, let alone a participant. Moreover, Hastey was one of the first directors to participate in natural areas planning. Not only did he bring the MOU on Biodiversity to fruition in 1991, he was the first director to sign the INACC agreements in 1983 and 1989. The northern spotted owl focused the attention of other directors, but Hastey had been spurred to coordinated action long before the logging injunction.

The BLM also lacks mandated boundaries like the FS and the NPS. BLM officials have much more discretion to buy, sell, and trade real estate, which means they can consolidate landholdings in desired locations, discard undesirable parcels, and move into entirely new geographic areas.[6] As Hastey put it, "We have an advantage over some of the other agencies because we don't have any boundaries; we can go anywhere we want."[7] The FS and NPS can only acquire private inholdings within their designated boundaries. BLM officials have therefore been leaders in many cooperative efforts in California, working closely with counties and local stakeholders on plans in which developers purchase habitat in remote areas to swap it in compensation deals for developing habitat elsewhere. Such swaps can increase the absolute number of acres set aside for habitat and can provide better habitat for species because it is more remote from development and is contiguous with other protected parcels. The ability to trade land has made the BLM a desirable partner for TNC and similar organizations that seek to protect private lands from development. Ironically, if the BLM continues to consolidate its holdings in this fashion, it may gradually lose an important incentive for cooperating with neighboring jurisdictions. In the extreme, its jurisdictions could become large, well rounded, and consolidated like national parks.

Frequency of Job Rotations within Agency

Personnel systems that encouraged individuals to rotate from one position to another, or from one location to another, created a subtle but powerful incentive against cooperation. In some agencies, like the BLM, individuals routinely stayed in a local, regional, or state office for significant portions of their careers. In other agencies, individuals were either required to rotate or encouraged to rotate if they wanted better jobs. In the FS, employees believed that promotion depended on rotation. According to Regional Forester Ron Stewart, "It was not a national policy that was ever on paper that I know of, but it was pretty well understood that if you were gonna get ahead in the Forest Service you were expected to be mobile."[8] In addition to rotations from one location to another, incentives also existed for rotation from one project to another within an agency. In the FWS, for example, recovery planning for endangered species was a lower priority than Section 7 consultations. The ESA required the FWS to do both, but the agency was understaffed, so short-term consultations took priority over long-term recovery planning, with the latter usually handled by temporary staff who sought more stable jobs in the agency.

Whether mandatory or incentive-based, personnel rotation systems have significant impacts on long-term cooperative processes. When an individual is required to rotate or desires to rotate, their participation in an interagency planning or management process comes to an end and the intellectual and social capital they have accrued go with them. Months—sometimes years—may pass before a replacement arrives, learns about the cooperative effort, and decides to commit time and resources to it. In the meantime, the replacement is still learning her job and may have other priorities than a project initiated by others. Moreover, even if she commits herself to the project, she still must familiarize herself with the details, learn about the interests of others, come to trust in their ability to deliver resources to the group, and decide how much she can deliver herself. This takes time and repeated interactions.

Repeated interaction allows individuals to build interpersonal trust (Thomas 1998). Repeated interaction also provides an opportunity to monitor and punish others for egregious behavior (Axelrod 1984; Ostrom 1990). Because interagency planning and management processes tend to last years, repeated interaction is important for maintaining cooperative relationships, particularly when a single person represents agencies at meetings. Cooperative efforts, such as CRMPs, stumbled when familiar faces disappeared and new ones arrived because participants learn about and come to trust one another over the course of many meetings. Continuity in attendance matters greatly in the long run.

Personnel rotation policies are thus an effective means for inhibiting cooperation. In part, the FS previously used its rotation system for this very purpose because agency leaders feared that forest rangers would be captured by local interests if they stayed in local offices too long. Kaufman (1960) once praised the FS for its ability to pursue its mission single-mindedly, in spite of the agency's geographic decentralization. He attributed the agency's success in part to its personnel system, which socialized individual behavior by regularly moving employees to new locations before they could develop personal relationships within local communities. Though the FS did not officially state that centralized control was a reason for the rotation policy, FS staff widely believed it to be the intent.[9]

Ironically, what was valued for agency success several decades ago became increasingly dysfunctional in a new era of ecological interdependence. As Kaufman (1956) himself once argued, the demands for bureaucratic change in any given era result in part from successful implementation of the demands of an earlier era. The FS was also compelled to rethink its rotation policy in the 1990s due to the high cost of resettling families. According to Regional Forester Stewart, "Employees aren't so willing to move any more, and so you don't get the same numbers of people applying for positions requiring mobility. [There are] dual careers and spousal placement issues, and then the cost of moving has run $50,000–80,000 to move a family nowadays—that's a year's salary, at least!"[10] Yet "mobility for line officers is still expected," as Stewart noted, which is another reason to expect less cooperation from FS line managers. Indeed, Stewart himself left the regional office in 1994 after three and a half years as Regional Forester, while Hastey served as BLM State Director from 1975 to 1979, and again from 1982 until he retired in 1999. BLM staff and line managers were more likely to stay in the same positions much longer than their counterparts in the FS and NPS.

Discretion in Standard Operating Procedures (SOPs) and Reporting Relationships

In addition to personnel rotation policies, other organizational factors provided disincentives to cooperation. SOPs and hierarchical reporting relationships limited individual discretion by circumscribing acceptable behavior within agencies. Interagency cooperation depended on the breadth of discretion given to staff and lower-level line managers to make decisions for and contribute resources on behalf of their agencies. As Bardach (1996) argues, participants must be able to leverage resources for the interagency collaborative. If they do not have the authority to do so, the scope, strength, and duration of cooperation will necessarily be limited.

SOPs also inhibited interagency cooperation because recalcitrant individuals invoked them as a convenient excuse for not participating. The BoR was a signatory to several interagency efforts in California, but some BoR officials read directly from the agency's multivolume codebook at interagency meetings to delimit what they could not or would not do. As one observer noted, "This is a major stumbling block. Reclamation goes by the book. I mean, one of the first things they ever gave us . . . was that blue set of all their laws. I've had meetings where people come in and they quote this thing like *The Bible*."[11] This behavior is similar to what Bardach and Kagan (1982) call "going by the book," but in this example it is done to maintain agency independence from cooperative relationships, rather than as a choice of regulatory style. In another form of this going-by-the-book mentality, individuals sometimes called on agency lawyers to find legal constraints. The TTF Ad Hoc Committee, for example, was nearly derailed by a CDFG official who threatened to take a draft of the MOU on Biodiversity to agency lawyers for review because he was dissatisfied with its language. This did not happen, however, because Hastey terminated debate at this point and personally took the MOU to agency directors for their signature.

This example also suggests the role of agency hierarchy in interagency relationships. When groups decide on a course of action, participants often need clearance from higher-level managers before proceeding. Like SOPs, which constrain what participants can say and do as representatives of their agencies, hierarchical reporting systems constrain participants by requiring them to clear decisions with superiors. The more levels of clearance, the longer it will take to receive permission and the less likely it will be given. Hastey's leadership in the TTF Ad Hoc Committee circumvented some of these barriers because he personally approached the other agency directors and sold them on the MOU on Biodiversity before lower-level line managers and attorneys reviewed the document in those agencies. This also occurred with NCCP, in which state and federal political appointees removed FWS and CDFG line managers from the decision-making process to reduce the layers of clearance between field staff and political appointees.

Interagency cooperation stumbled as well when participants believed they had authority to speak for their agency, or led others to believe they were speaking for their agency, when in fact they needed clearance. Such misinterpretations were particularly disruptive if an interagency group had to reinitiate deliberations once they had identified the appropriate person. If the person without decision-making authority continued to represent the agency at meetings, the group's trust in that individual's ability to meet commitments and deliver agency resources declined. This problem routinely occurred with local CRMPs because agencies were typically represented by field staff with limited authority to speak for their agencies. Private landowners spoke for themselves at meetings, but agency staff usually had to get clearance from line managers, making the agencies comparatively unreliable partners at the local level.

The BLM was different in this respect from most other agencies. Line managers and field staff in local BLM offices had significant decision-making authority. The BLM also had fewer SOPs than agencies like the FS, BoR, and NPS. This allowed Hastey's presence to loom large. Hastey pressured his line managers and field staff to work with individuals in neighboring jurisdictions, and they followed his lead. To a large degree, Hastey's leadership substituted for SOPs within the agency. The contrast with other agencies is remarkable. In the NPS, the regional director had little authority over individual park units. When the NPS regional director signed the MOU on Biodiversity and distributed it to park superintendents, the document essentially disappeared into an organizational void because he lacked the authority and desire to pressure his line managers to adhere to its principles. Regardless, NPS line managers had little desire to participate because they were not threatened by lawsuits, and because most parks were large and consolidated.

Travel Time to Attend Interagency Meetings

Travel time is important for explaining the duration of cooperation.[12] When individuals had to drive or fly more than a couple of hours to attend meetings, their participation waned, even in the presence of other motivating factors. Agency directors were more likely to attend the quarterly meetings of the Executive Council if their offices were in Sacramento, where most of the meetings were originally held (see table 4.1). Travel time was even more important at the local and regional levels, where most interagency planning and management activities occurred. In the Klamath Bioregion, travel time proved to be a determining factor for regular participation. The effort to form a bioregional council in the

Klamath failed in large part due to the size of the region, far-flung offices and living arrangements, and winding mountain highways. At the second bioregional meeting, participants divided themselves into four subregional groups. While they did this in part because they saw more common interests within the subregions than between them, they were also not prepared to drive several hours to meetings in different parts of the bioregion. One of the subregional groups even organized itself around a freeway, which allowed participants to drive more readily from one end of the subregion to another. Yet this also meant that agencies with far-flung local offices, like the BLM, SCS, and UC/DANR, were more likely to participate in local cooperative efforts.

Summary

Several factors shaped the emergence, scope, strength, and duration of cooperation in preserving biodiversity in California. Preeminent among these was an interactive effect combining court enforcement of environmental laws with consensual ecological knowledge. Ecologists did not need the ESA as an incentive for cooperation, but court enforcement gave other agency officials an incentive to protect species, habitats, and ecosystems, while the functional postulates of conservation biology offered a set of solutions to their collective dilemma. Within this context, coordinated action held out the potential for synergistic benefits. As Weiss (1987) showed in her research on cooperation among school districts, public officials are more interested in cooperation when coordinated action solves problems or produces financial benefits. Court enforcement of the ESA threatened line managers with a loss of autonomy, and it threatened field staff with socioeconomic instability in their communities. Coordinated planning and management solved these problems because it was understood to be an effective means for avoiding lawsuits. Cooperation was driven by the potential for joint gains among line managers, professionals, and field staff, not by the ability of one agency to punish others for defecting from cooperative efforts. Participants were influenced instead by an external threat, in which environmental groups were strategically positioned to file and win lawsuits.

Within this legal milieu, variation in cooperation across agencies was determined by each agency's dependence on the actions of its neighbors.

The more neighboring agencies depended on one another, the greater the potential for joint gains and the more likely cooperation would emerge. Some bioregions contained more listed and candidate species than others, and some agencies managed more habitat for these species than others. Ownerships and jurisdictions also had to be intermixed to create a collective-action problem. The BLM owned highly fragmented parcels that harbored a large number of listed species, and environmentalists routinely targeted the agency. The NPS, by comparison, was seldom sued and its parcels were less interspersed with other ownerships. Therefore, autonomy was a viable strategy for park managers. Park managers were also relatively autonomous of the NPS regional office, which lacked the necessary line authority to provide a potential source for integration. Unlike BLM managers, park managers viewed the logic of interagency cooperation within the parochial context of their park boundaries.

While these factors largely determined where cooperation would occur and which agencies would take the lead, other factors shaped cooperation on the margin. In particular, cooperative efforts were more productive if participants stayed in their jobs for long periods of time, had the authority to speak for their agencies, and traveled short distances to attend meetings. Agencies developed poor reputations for cooperation if their employees moved routinely from one job to another, needed clearance from superiors before committing agency resources to interagency processes, or had to travel more than a couple of hours to attend meetings. In short, job stability, delegated authority, and a large network of local offices promoted interagency cooperation; staff rotations, centralized authority, and centrally located staff impeded cooperation, particularly at the local and regional levels within California.

These factors explain most of the variation in participation. The BLM and NPS fell at opposite ends of the spectrum on most of these variables. These organizational and institutional incentives allowed the ecological epistemic community to have a much larger impact on BLM operations, producing a surprising outcome: BLM staff and line managers took the lead in cooperative efforts to preserve biodiversity in California, while NPS participation was largely evanescent. This does not mean that biodiversity was better protected under the BLM, but it does mean that BLM officials were trying much harder to protect biodiversity than popular stereotypes and the academic literature have led us to expect. While the

NPS probably did a better job of protecting biodiversity than the BLM, the NPS largely did so by default because it had a dual-use mandate to protect natural resources while providing access for tourists, not a multiple-use mandate that included resource extraction. Moreover, the BLM was pursuing one of the fundamental tenets of conservation biology—interagency coordination—to a much greater degree than the NPS. Ironically, the Clinton administration heightened this disparity in 1993 by transferring NPS scientists to the newly created National Biological Service, further emaciating the agency's ecological core. While ecologists were assuming prominence throughout the BLM—not just in California, but also in neighboring states like Nevada and Oregon—they were disappearing from the NPS. The logic of cooperation in preserving biodiversity was evaporating within the NPS as it burgeoned within the BLM.

BLM employees in California presented a complex profile. On the one hand, the BLM retained a large number of employees representing an earlier generation of resource managers, many of whom were derided by employees of the FS, NPS, and FWS for lacking professional training. On the other hand, the BLM was also populated by a significant number of professionals with ecological predilections, who demonstrated an entrepreneurial, proactive spirit with regard to environmental issues. They were enthusiastic about developing cooperative relationships, were encouraged by the BLM's culture and leadership to do so, and were relatively unconstrained by SOPs and hierarchical reporting relationships. Though Downs (1967) argued that we should expect to find older agencies like the BLM dominated by conservers who resist change, several BLM employees I interviewed better fit Downs's description of zealots.[13] This image is surprising because popular stereotypes also suggest that environmental activists will likely be found in agencies charged specifically with protecting biological resources, like the NPS and FWS, not the BLM.

Yet the BLM still maintained its ties to ranchers, private landowners, and the local governments representing them. Hastey did not intend to sacrifice this clientele to the preservation of biodiversity. Interagency cooperation was simply a means for pursuing traditional multiple-use goals when confronted by lawsuits that mandated new tasks. As with the FS, cooperation on the externally imposed task of species preservation became a prerequisite for pursuing traditional missions. Local stakeholders nevertheless feared that the agencies were developing regional preservation plans at their expense. So, to assuage their concerns, Hastey co-opted them by inviting county supervisors (and city council members in Southern California) to join the Executive Council, without requiring them to sign the MOU on Biodiversity, which had been drafted by members of the ecological epistemic community. While a few county supervisors with ecological predilections regularly attended Executive Council meetings, most local government representatives simply wanted to know that the agencies were not challenging their prerogatives, and stopped attending the meetings once satisfied this was the case.

Co-optation, of course, has its costs. As Selznick (1949: 13) argued, co-optation is "the process of absorbing new elements into the leadership or policy-determining structure of an organization as a means of averting threats to its stability or existence." Co-optation may therefore change an organization's goals. The ecologists who drafted the MOU on Biodiversity hoped agency directors would bring top-down pressure within the agencies to support their efforts to preserve biodiversity. To some extent this occurred in agencies like the BLM, but not in agencies like the NPS. Yet, once county supervisors were co-opted, it was less likely the Executive Council would adopt the radical planning principles in the MOU on Biodiversity. The scope of cooperation laid out in the MOU was determined by the staff ecologists who wrote it, but the scope of cooperation in the Executive Council was determined largely by the participants, few of whom were ecologists, and many of whom were local government officials. If staff ecologists had any hope the Executive Council would design a statewide strategy to preserve biodiversity, recommend consistent statewide goals for the protection of biodiversity, and recommend consistent statewide standards and guidelines to meet those goals, as stated in the MOU on Biodiversity, these hopes evaporated after local governments were co-opted into the policy-determining apparatus in 1992. The failure of the co-optation strategy rang loudest in the San Joaquin Valley Bioregion, where Hastey tried to put county supervisors in the driver's seat, and they refused to board the bus.

Appendix A: Research Methodology

Why Qualitative Methods?

As a discipline, political science has become increasingly enthralled with numbers. The contents of the *American Political Science Review* and the *American Journal of Political Science*, two of the most prestigious journals in the United States, lead readers to believe that concepts that cannot be quantified, hypotheses that cannot be tested with multivariate equations, and theories that are not amenable to mathematical proofs are of little importance. Unfortunately, this bias in favor of quantitative and formal methodologies gives short shrift to concepts, theories, and entire fields of study that are less amenable to mathematical and statistical techniques.

The study of public organizations is a case in point. Relatively little can be learned about public organizations through numbers, data sets, and equations. Very few agencies generate useful data sets, in contrast to Congress, where the moves of members and committees are tracked statistically like baseball players and teams. Regulatory agencies produce enforcement data, but this is only one indicator of agency behavior, leaving much to be described and explained. Principal-agent models relying on enforcement data can tell us something about the political determinants of a particular type of activity in a discrete set of agencies, but if we want to understand agency behavior more broadly we have to go well beyond existing or manufactured data sets.

Interagency relationships are particularly difficult to quantify, and even more difficult to model in game-theoretic terms. Game theory is helpful in modeling situations with relatively few actors, such as nuclear deterrence and congressional committees, but interagency relationships involve many individuals in many agencies, who are driven by diverse motives and respond to a broad range of organizational and institutional incentives. In writing the case studies for this book, I simplified very complex relationships observed in the field. It is difficult to imagine how these relationships could be modeled in formal terms without committing gross oversimplification. There are simply too many individuals, too many agencies, too many alternatives, and too many exogenous forces, most of which would have to be ignored to make formal models tractable.

The dependent variable—cooperation—is also not easily quantified. There is no central repository of interagency agreements, such as memoranda of understanding; researchers would have to rummage through agency files just to find such agreements, let alone sample them in a statistically meaningful way. Moreover, even if we had a sample of these agreements, we could not assume that each represents a case of cooperation because the text does not necessarily reflect the underlying purpose or resulting actions. Case studies are therefore necessary to decide whether an observed interagency agreement represents cooperation or some other type of relationship. Cooperation can also exist in the absence of formal agreements. Hence, qualitative research must precede quantification.

Given these problems, I decided against using quantitative and formal methodologies, and relied instead on field observations, interviews, and case studies. Rather than testing hypotheses, I searched for hypotheses. In doing so, I was inspired by E. O. Wilson (1992: 5), who states in the opening passages of *The Diversity of Life*:

> The best of science doesn't consist of mathematical models and experiments, as textbooks make it seem. Those come later. It springs fresh from a more primitive mode of thought, wherein the hunter's mind weaves ideas from old facts and fresh metaphors and the scrambled crazy images of things recently seen. To move forward is to concoct new patterns of thought, which in turn dictate the design of the models and experiments. Easy to say, difficult to achieve.

It is odd that political scientists increasingly accept the notion that mathematical models and statistical relationships represent the best social science, when eminent natural scientists like Wilson do not believe the best science consists of such techniques.

Beyond these opening comments, I do not see a need to justify the choice of qualitative over quantitative methodologies. Nevertheless, quantitative reasoning can inform qualitative methodology (King, Keohane, and Verba 1994). Readers should also be informed about the means by which qualitative researchers gather and interpret information. The following sections accordingly describe the qualitative methodologies I followed.

Case Selection

I introduced the logic of case selection in chapter 1, but did not fully articulate the research design. I noted, for example, that I chose cases related to biodiversity because agency tasks are interdependent, and that I selected California because the state contains a large number of jurisdictions and a large number of endangered species. As alternatives, I could have compared cases in California with cases in other states, or cases in other issues areas. These alternatives would have increased the number of potentially confounding explanatory variables. Therefore, I decided to hold these variables constant by comparing cases within the same issue area and the same state—which proved to be a significant challenge in its own right.

The case studies in chapters 3 through 7 are not historically isolated phenomena, so each chapter includes historical discussions of related cooperative efforts, each of which could be considered a case study in its own right. The three bioregional cases focus on a specific region-wide cooperative effort, the genesis of which is tied to preexisting activities at the local level and within lower levels of agency hierarchies. Comparing bioregions is logical because bioregional activities were advocated by state-level actors who planned their activities at the bioregional scale. Bioregional comparisons also allow us to examine the relationships among different sets of local, state, and federal agencies, as well as the political tensions between resource use and preservation in each bioregion. By allowing ecological characteristics and land ownership patterns to vary across bioregions, we can look for similarities across habitat types and human uses of the land. If the fundamental characteristics of interagency cooperation are similar regardless of bioregional variation, we should find only idiosyncratic differences between cooperative efforts in the Klamath, San Joaquin Valley, and South Coast Bioregions.

In one sense, I can be accused of selecting cases on the dependent variable because I have chosen bioregions in which rapid resource extraction and development have left only tiny fractions of native habitat. Therefore, it is much more likely that agency officials would have been concerned about particular species and habitats in these bioregions and would have attempted to cooperate in managing them. I acknowledge this shortcoming, and justify case selection based on the assumption in chapter 1 that technical interdependence is a necessary condition for cooperation. Unless habitat is significantly diminished and fragmented, it is unlikely that more than a few people will care about losses, and it is unlikely that agency officials would interpret the losses as a problem to be addressed. Because the bioregions I selected demonstrate little variation in this regard, I cannot attempt to answer the question "How much interdependence is necessary for interagency cooperation to occur?"

Open-Ended Interviews

Because the literature indicated that relatively little was known about interagency cooperation, I relied on open-ended interview questions as a means for identifying the important dynamics that underlie cooperative behavior. I interviewed 102 individuals in the field—ninety-five public officials and seven individuals from the private and nonprofit sectors. Of the latter, five worked for public interest groups and two for private firms. All five interest-group respondents worked for environmental groups. Individuals working for industry associations declined my requests for interviews. The other two nongovernment respondents worked for private developers in the South Coast Bioregion; thus, the sample of nongovernment officials is not entirely biased toward the environmental community. Because I interviewed nongovernment representatives only as checks on the validity of self-reports by government officials, I was not concerned about this bias in the sample.

Of the public officials, forty-six worked for federal agencies, thirty worked for state agencies, and nineteen worked for local governments. At the local level, the respondents included county supervisors, city council members, resource conservation district representatives, and professional staff employed by cities, counties, and local government associations. For each state and federal agency, I attempted to interview the director or associate director, and at least one line manager, one professional staff member, and one field staff member to see how individual incentives and perspectives differed throughout the agency. I cannot list the respondents by position and affiliation because this would compromise the anonymity that many requested. Relevant resource management agencies included the BLM, FS, NPS, FWS, EPA, CDF, CDFG, CSLC, CDPR, UC/DANR, and the California Resources Agency. This list is not inclusive, however, because the interagency network for terrestrial biodiversity extended to some additional agencies that were relatively minor players.

The sampling technique was nonrandom. I relied on agency-produced contact lists and snowball sampling, with the aim of producing a final sample representing a wide array of local, state, and federal agencies, and variation within agencies by hierarchical level. I also sought respondents who were not participating in cooperative efforts, along with those who were. In other words, I interviewed participants in cooperative efforts, individuals who could have been participating but were not, and those who stopped participating. The interview format was loosely structured. Throughout, I relied on Dexter's (1970) suggestions for elite interviewing.

The 102 interviews were conducted from October 1993 to October 1994. All of the interviews were tape-recorded. Most took place in the respondent's office or conference room. A few were conducted in homes, in restaurants, or at conferences, depending on the respondent's preferences. At the beginning of the interviews, I informed respondents that I wanted to record the interview and that the tapes would only be heard by myself in preparing academic manuscripts. No one objected to the use of a tape recorder, though some asked me to turn the recorder off when they answered what they thought was a sensitive question.

I also offered the respondents anonymity. If they requested anonymity, any quotes I used in the text are cited in a footnote as "Interview #__." Political appointees, elected officials, and line managers typically waived anonymity, while staff often requested anonymity. In listening to the tapes, it was not obvious to me that individuals who requested anonymity were more candid than individuals who waived it. In all cases, the quotes were transcribed verbatim from interview tapes, with the exception of filler words like *um* and *uh*. I used italics sparingly to indicate significant vocal inflections on the tape. The choice of punctuation necessarily required personal interpretation.

In addition to the 102 on-site interviews, I conducted numerous phone interviews from 1992 to 2000. Unlike the on-site interviews, the phone interviews did not have uniform structure, and were used primarily for informational purposes and to supplement specific on-site interviews.

Other Sources

Formal agency and interagency documents provided useful background for specific cases, but these documents were much less informative than the interviews, so I only used them to supplement and support information reported by respondents. Some respondents gave me full access to their files, which provided more insight into interagency processes than documents released to the general public. Because complete access to files was relatively rare, and certainly not random, I did not rely on these files to construct stories. Instead, the files primarily served to corroborate stories told by respondents. In all of the case studies, I triangulated the evidence to support significant points. Thus, for example, I did not quote respondents if their version of events was not corroborated by others, or by official documents.

In addition to these primary sources, I also used Web sites, newspaper articles, and journal articles to round out the case studies.

Appendix B

MEMORANDUM OF UNDERSTANDING*
California's Coordinated Regional Strategy to Conserve Biological Diversity

The Agreement on Biological Diversity

September 19, 1991

I. Preamble

California is one of the most biologically diverse areas in the world. The state's rich natural heritage—vegetation cover and distribution, wildlife and fish habitat, recreation and aesthetic values, water and air quality—provides the basis for California's economic strength and quality of life. Sustaining the diversity and condition of these natural ecosystems is a prerequisite for maintaining the state's prosperity.

Public agencies, private organizations, and individual citizens have long shared a commitment to conserving the natural environment of their state. Laws, policies, and programs already in place protect many of the elements of California's natural heritage. That experience, and a growing body of scientific research, demonstrate the need to move beyond existing efforts focused on the conservation of individual sites, species, and resources. Californians now recognize the need to also

* The text has been retyped verbatim for legibility. Signatures were scanned from the original document.

protect and manage ecosystems, biological communities, and landscapes.

These broader systems represent an important component of the state's biological diversity—the full variety of living organisms in California, the genetic differences among them, and the communities and ecosystems in which they occur. These ecological systems appear throughout the state across a variety of ownerships and jurisdictions. To effectively conserve California's biological resources and maintain social and economic viability, public agencies and private groups must coordinate resource management and environmental protection activities, emphasizing regional solutions to regional issues and needs.

II. Purpose

This Memorandum of Understanding establishes an Executive Council to develop guiding principles and policies, design a statewide strategy to conserve biological diversity, and coordinate implementation of this strategy through regional and local institutions.

III. Policy and Principles

This memorandum recognizes the following set of policies and principles.

A. The signatory parties agree to make the maintenance and enhancement of biological diversity a preeminent goal in their protection and management policies. Furthermore, they agree to work with the Executive Council to develop and adopt a coordinated regional strategy that ensures protection of biological diversity and the maintenance of economic viability throughout California.

B. The basic means of implementing the strategy are to be improved coordination, information exchange, conflict resolution, and collaboration among the signatory parties. In addition, the signatories agree to pursue the development of local and regional institutions and practices necessary to conserve biological diversity. These tools may include the establishment of mitigation and development banks, planning and zoning authorities, land and reserve acquisition, incentives, alternative land management practices, restoration, and fees and regulation.

C. Community and public support are vital to the success of a bioregional program. Human communities, local economies, and private property are important regional attributes to be maintained. As a consequence, signatories will develop procedures and guidelines to facilitate public education, dialogue and participation, and to minimize the disruption of human communities and expectations. Public lands are to be given first preference as reserves and conservation areas. Impacts on private lands will be minimized to the degree possible.

D. Biological diversity is to be viewed as an attribute of natural processes operating at the landscape, ecosystem, species, and genetic levels. These processes are dynamic varying over time and space. A recognition is made that these processes are altered by both human and natural factors. While the focus of the agreement is on biologic factors, abiotic elements are also recognized as important components of natural systems. The signatories agree to pursue the establishment of measurable baselines and standards of diversity as a means of conserving biological resources over time.

E. Given the changing characteristics of both the biological and social environment, the signatories agree to an adaptive approach in the development of bioregional strategies. Such an approach will place substantial emphasis on monitoring, assessment, and research programs. These programs will help determine if strategies are accomplishing their intended objectives, maximize the opportunities to learn from experience, and enhance the flexibility in the face of new knowledge.

IV. Authority

This Memorandum does not modify or supersede existing statutory direction of the signatories.

V. Organization

A. Statewide Executive Council—The Executive Council is to be chaired by the Secretary of the Resources Agency of California and made up of the principal signatory agencies. The Council will set statewide goals for the protection of biological diversity, recommend consistent statewide goals for the protection of biological diversity, recommend consistent statewide standards and guidelines, encourage cooperative projects and sharing of resources, and cooperate in the following program areas:

1. Biodiversity-related policies and regulations;
2. Land management, land use planning, and land and reserve acquisition and exchange;
3. Private landowner assistance;
4. Educational outreach, public relations, and staff training;
5. Monitoring, inventory, and assessment;
6. Restoration; and
7. Research and technology development.

The Council will seek adequate funding to implement regional strategies and to develop necessary state and regional institutions, such as trading and mitigation banks. Further, the Council will cooperate with regional representatives to define the boundaries of bioregions and to help establish Bioregional Councils.

The Council will meet quarterly to review progress in accomplishing its mission. Representatives of other state and federal agencies and sponsors will be invited to participate in the meetings of this group. The Council will produce and distribute to the public regular summaries of its activities.

B. Sponsors—A sponsor may be any special interest group or organization that supports the purpose and intent of this Memorandum of Understanding. Sponsors will be expected to promote the development and adoption of biodiversity strategies and principles through their membership and activities. Sponsor representatives are to be invited to attend and participate in any Executive Council meeting or activity. Sponsorship should help enhance consensus and participation in the adoption of bioregional strategies.

C. Bioregional Councils—Regional administrators of signatory agencies will develop regional memoranda of understanding with the purpose of establishing Bioregional Councils. Participation of additional organizations specific to each region, such as county governments and local environmental and industry groups, will be encouraged. The Councils will develop regional biodiversity strategies that incorporate the policies, principles, and activities listed above under the mission of the Executive Council. Regional solutions to regional issues and needs will be encouraged, consistent with statewide goals and standards. The Councils are to work with regional and local authorities to implement biodiversity policies. In addition, Bioregional Councils will actively encourage the development of watershed or landscape associations to assist in implementing regional strategies.

D. Watershed and Landscape Associations—Local staffs of signatory agencies will encourage the participation of local public, landowner, and private organizations in the formation of watershed or landscape associations. These associations will be encouraged to develop specific cooperative projects that help to achieve regional and statewide objectives. Use of a Coordinated Resource Management Planning process will be encouraged. The local associations are to be a primary forum for the resolution of local issues and conflicts related to biodiversity concerns.

VII. Modifications

This agreement is to remain into effect until modification by the parties in writing; it is negotiable at the option of any one of the parties.

Douglas Wheeler
Secretary
The Resources Agency

Ed Hastey
California State Director
USDI Bureau of Land Management

Peter Bontadelli
Director
California Department of
Fish and Game

Ronald Stewart
Regional Forester
Pacific Southwest Region
USDA Forest Service

Richard A. Wilson
Director
California Department of
Forestry and Fire Protection

Marvin L. Plenert
Regional Director
USDI U.S. Fish and Wildlife
Service

Henry Agonia
Director
California Department of
Parks and Recreation

Stanley Albright
Regional Director
Western Region
USDI National Park Service

Charles Warren
Executive Officer
State Lands Commission

Kenneth R. Farrell
Vice President, Division of
Agriculture and Natural Resources
University of California

Pearlie Reed
State Conservationist
Soil Conservation Service
U.S. Department of Agriculture

Henry Voss
Director
California Department of Food and Agriculture

John Smythe
State Executive Director
USDA Agricultural Stabilization and Conservation Service

Roger Patterson
Mid-Pacific Regional Director
Bureau of Reclamation
U.S. Department of the Interior

Ed Heidig
Director
California Department of Conservation

David Kennedy
Director
California Department of Water Resources

Charles A. Pritchard
California Association of Resource Conservation Districts

Charles R. Imbrecht
Chairman
California Energy Commission

Peter Douglas
Executive Director
California Coastal Commission

Felicia Marcus
Regional Administrator
U.S. Environmental Protection Agency

James Strock
Secretary
California Environmental
Protection Agency

Michael Fischer
Executive Officer
California Coastal Conservancy

James W. van Loben Sels
Director
California Department of
Transportation

John D. Buffington
Acting Regional Director
National Biological Service

John Conomos
Acting Regional Director
U.S. Geological Survey

John D. Morgan
Associate Director
Information and Analysis
U.S. Bureau of Mines

Appendix C

STATEMENT OF INTENT TO SUPPORT THE AGREEMENT ON BIOLOGICAL DIVERSITY*

California Counties have long recognized the importance of maintaining productive, healthy natural resources and ecosystems which in turn provide the setting of lifestyles, scenery, recreation, and the diversity of natural life systems, while providing resources for raw materials to produce products, jobs, and community stability.

California Counties support the Agreement on Biological Diversity in the context of balanced and wise use of natural resources. To alleviate the difficult task of allocating uses of these resources, Counties support the idea of coordinated and cooperative planning efforts of multiple jurisdictions, species, and ecosystems. These efforts should be conducted with strong local leadership and the participation of everyone concerned with natural resource use and management and implemented consistent with existing local, state, and federal laws and regulations.

With the active participation of locally elected leaders, land managing agencies, and locally affected publics, we believe the Agreement can help conserve California's rich biological diversity for future generations to enjoy and promote responsible development as we strive to meet the future needs of California's citizens.

* The text has been retyped verbatim for legibility. Signatures were scanned from the original document.

San Joaquin Valley Regional Association of County Supervisors

North Coastal Counties Supervisors Association

Mike Fluty, Chairman
Sacramento–Mother Lode Regional Association of County Supervisors

Northern California County Supervisors Association

Southern California Regional Association of County Supervisors

Fred Keeley
Central Coast Regional Association of County Supervisors

Regional Council of Rural County Supervisors

South Central Coast Regional Association

Approved: Chairman, Executive Council

11-13-92
(Date)

Southern California Association of Governments	San Diego Association of Governments

Approved: Chairman, Executive Council

September 13, 1993

(Date)

Notes

Chapter 1

1. See, for example, the literatures on congressional oversight (Aberbach 1990), the administrative presidency (Durant 1992; Golden 2000), and political appointees (Heclo 1977). On the relationship between regulators and regulatees, see La Porte and Thomas 1995 and Bardach and Kagan 1982.

2. On subgovernments from an agency perspective, see Knott and Miller 1987. For a congressional perspective, see Fiorina 1989. On the relationship between bureaucratic power and clienteles more generally, see Rourke 1969 and Clarke and McCool 1996.

3. For reviews of the early implementation literature, see Stoker 1991, Hasenfeld and Brock 1991, Ingram 1990, and Goggin et al. 1990.

4. Some researchers prefer to use the term *collaboration* (Wondolleck and Yaffee 2000; Bardach 1996, 1998; Gray 1989) rather than *cooperation*. I prefer the latter because it is more commonly used in the political science and organization theory literatures. *Collaboration*, however, has the interesting connotation of working with the enemy.

5. The number of local governments in California is not stable over time. Rather than present the most current figures, these figures represent the era in which events discussed in this book occurred. The number of counties is stable, but the number of cities is growing, while the number of special districts fluctuated between 5000 and 5200 between 1982 and 1992. The number of cities (468) comes from California Office of State Controller (1994). The number of special districts (4995) comes from California Office of State Controller (1993).

6. Figures for state and federal agencies are from Kreissman 1991. These figures do not account for all agencies or for all holdings by the agencies mentioned. Figures for the NPS also precede passage of the California Desert Protection Act in 1994, which significantly expanded the agency's holdings at the expense of the BLM. Rather than report current data, I report data from the period in which the case studies occurred.

7. For technical discussions of the links between species viability and habitat, see Harris 1984; Soulé 1986, 1987; Morrison, Marcot, and Mannan 1992; Noss

and Cooperrider 1994. For breezier treatments, see Wilson 1992 and Quammen 1996.

8. In their review of the literature on ecosystem decline in the United States, Noss, LaRoe, and Scott (1995: 2) state that "biologists agree that the major proximate causes of biotic impoverishment today are habitat loss, degradation, and fragmentation." Also see the data in Kareiva et al. 1999: 28–31.

9. This definition is from Jensen, Torn, and Harte (1993: 5), who similarly used italics for emphasis. Noss and Cooperrider (1994: 5) provide a more complex definition of biodiversity, emphasizing processes in addition to scale: "Biodiversity is the variety of life and its processes. It includes the variety of living organisms, the genetic differences among them, the communities and ecosystems in which they occur, and the ecological and evolutionary processes that keep them functioning, yet ever changing and adapting." For other definitions, see Grumbine 1992: 22–28, and U.S. Office of Technology Assessment, 1987. See Takacs 1996 for in-depth history and meaning of the term *biodiversity*, Golley 1993 for a similar treatment of *ecosystem*, and Worster 1994 on *ecology*.

10. For philosophical discussions regarding the value of preserving species and ecosystems, including biocentric and anthropocentric arguments, see Leopold 1949; Ehrenfeld 1978; Callicott 1987, 1989; Norton 1987, 1986; Devall and Sessions 1985; Nash 1989; Eckersley 1992; Heal 2001.

11. Unless otherwise indicated, all data on federally listed species were downloaded from the FWS Web site (<http://www.fws.gov>) during the years indicated in the text. The figures reported in this sentence were posted by the FWS in March 1994 and March 1995.

12. The Endangered Species Act of 1973 (as amended), Section 4(b)(1)(A).

13. In part, this reluctance was due to the FWS being a relatively weak agency, as suggested by the indicators developed by Clarke and McCool (1996). Yet, even arguably stronger federal regulatory agencies have found it difficult gaining the compliance of their sister agencies (Durant 1985; Wilson and Rachal 1977). Nor has it been a simple matter for federal regulatory agencies to gain compliance from their state counterparts (Scicchitano and Hedge 1993). The Clean Water Act, for example, gives the Environmental Protection Agency (EPA) authority to set water-quality standards when states fail to do so, but the EPA struggled for years to get California to set its own standards.

14. See, for example, Agee and Johnson 1988; Salwasser, Schonewald-Cox, and Baker 1987; Keystone Center, 1991; California State Assembly Office of Research, 1991. Even the NPS, whose parks were long viewed as islands of preservation surrounded by human development, did not escape this critique. According to a 1988 workshop convened to enhance biological diversity in the parks, "Inadequate research and management coordination between parks and their neighbors pose immediate threats to the integrity of biodiversity within the parks" (Dottavio, Brussard, and McCrone 1990: 3).

15. Section 6, P.L. 94-588, 90 Stat. 2949.

16. Section 101, 43 U.S.C. 1701.

17. National Park Service Organic Act, August 25, 1916.

18. Freemuth (1991), who uses the island metaphor, places the blame for problems within the parks on external actors, while overlooking the "internal threats" wrought by park managers identified by Abbey (1968), Chase (1987), Seligsohn-Bennett (1990), Hess (1993), Wagner, Foresta, and Gill (1995), and Sellars (1997). Taking a somewhat different tack, Lowry (1994) places the blame largely on elected representatives in Washington, D.C.

19. 50 CFR 17.3. The Supreme Court upheld this regulation in *Babbitt v. Sweet Home Chapter of Communities for a Great Oregon*, 515 U.S. 687 (1995). Species listed as "threatened" are protected under Section 4(d), which requires the FWS to promulgate regulations deemed "necessary and advisable to provide for the conservation of such species." In practice, the FWS has extended the prohibition on take to threatened species, except where otherwise authorized by a special regulation.

20. Section 7(a) (2). See Rohlf 1989: 105–171 for discussion of these terms, the procedural aspects of interagency consultation, and substantive protections under Section 7.

21. Interview #1. All citations identified by "Interview #" indicate the respondent requested anonymity. See appendix A for research methodology.

22. No clear biological standards compelled placing a species on one list or another. To a large extent, administrative and political factors, rather than biological factors, governed the listing process (Tobin 1990: 113–117). See also U.S. General Accounting Office, 1993. Nevertheless, Wilcove, McMillan, and Winston (1993: 91) found that vertebrate animals listed as endangered between 1985 and 1991 did have significantly fewer individuals (median = 407.5) than those listed as threatened (median = 4161). For plants, the median numbers were 99 individuals for endangered species and 2944.5 for threatened species.

23. See, for example, Stevens 1996; Himmelspach 1994; Gorman 1989.

24. I use *interagency* rather than *interorganizational* or *intergovernmental* because *agency*, *organization*, and *government* are not synonymous. An organization is a cooperative system composed of the activities of human beings (Barnard 1938: 73,77). Public agencies are formally constituted organizations at the local, state, and federal levels. Informal organizations also exist within and between local, state, and federal agencies (Chisholm 1989). Therefore, the set of government organizations includes public agencies, formally constituted interagency committees, and informal systems of cooperation both within and among agencies. I use *interagency* broadly to include relationships both within and among the three levels of government. While federalism scholars might prefer a finer conceptual distinction between *inter*governmental and *intra*governmental activities, the findings from my field research suggest little reason to make such a distinction for the purposes of this study. In referring to individuals within agencies, I use the terms *public official*, *agency official*, and *bureaucrat* interchangeably, but tend to avoid the latter because of its pejorative and stigmatizing connotations.

Chapter 2

1. See Bendor 1990 for a review of the budget-maximization hypothesis in public administration. See O'Toole 1988 for an application to the FS. See Durden 1991 for evidence that Niskanen 1971 is now a "classic" of social choice theory.

2. O'Toole (1988), for example, provided extensive documentation of the FS planning process, and claimed that his data support the budget-maximization hypothesis. Yet the evidence only indicates that FS planners respond to an incentive-based planning system; it does not support the assumption that FS officials are inherently budget maximizers. Sigelman (1986), on the other hand, directly tested the assumption by analyzing surveys of state administrators that asked them how much they would like their budgets increased. He found that they favored only modest increases in their budgets, suggesting that the budget-maximization hypothesis "fails as a descriptive generalization" (Sigelman 1986: 57).

3. Niskanen also relaxed his original assumption in this collection (Blais and Dion 1991: 18), arguing that "bureaucrats act to maximize their bureau's discretionary budget, defined as the difference between the total budget and the minimum cost of producing the output expected by the political authorities."

4. Lebovic (1994) recognized the possibility of such constraints, but nevertheless analyzed defense agency budgets from 1981 to 1993, a period in which the defense budget was increasing. Although he concluded that defense agencies actively sought future budgetary resources, he hedged his conclusions about their general behavior by noting that "the services are shown here acting under the exceptional conditions of the 1980s" and "might act differently in times of austerity" (Lebovic 1994: 849).

5. The quote is from *Webster's New Twentieth Century Dictionary*, 2nd ed. (William Collins and World Publishing Co., 1978), p. 1970.

6. I want to thank Gene Bardach for pointing out this important distinction. Bardach (1996: 177) defines *turf* as "the exclusive domain of activities and resources over which an agency has the right, or prerogative, to exercise operational and/or policy responsibility." Unlike most authors, he does not limit his discussion of turf-related behavior to higher-level managers: "Employees at all levels of the organization may have their own stakes in protecting the agency's turf." Also see Bardach 1998.

7. As cited in Cawley and Fairfax 1991: 428.

8. On epistemic communities and international cooperation, see the special issue of *International Organization* (Vol. 46, Winter 1992) devoted to the topic. For a critique of this special issue, see Jacobsen 1995. Susskind (1994: 73–76) provides an additional critique of the concept. See Ernst Haas (1990) on the role of epistemic communities in organizational learning. Thomas (1997) explores the utility of epistemic community theory for explaining cooperation within the United States.

9. Gruber (1994: 21–22) argues that political capital includes individual contacts, political debts, and constituents, all of which can be used to sway voters

and legislators. For another treatment of social, intellectual, and political capital, see Innes et al. 1994.

10. The Society for Conservation Biology was formed in 1985 and 1986 (Grumbine 1994: 99). Its journal is *Conservation Biology*. Exemplars of conservation biology from a policy perspective include Soulé 1985; Noss and Harris 1986; Grumbine 1992, 1994; Noss and Cooperrider 1994; Noss, O'Connell, and Murphy 1997; Soulé and Terborgh 1999.

11. Nash (1989) places Leopold at the vanguard of the movement in contemporary Western culture to extend ethics from humans to the rest of nature. For additional praise of Leopold's legacy, see Callicott 1987. For an alternative interpretation, see Worster (1994: 289), who argues that Leopold "never broke away altogether from the economic view of nature."

12. While this circularity troubled Kuhn and his critics, it is of little concern here because I do not seek to explain the emergence of new paradigms and epistemic communities. I merely want to be clear about terms.

13. Gifford Pinchot (1910: 42), the Progressive Era's preeminent conservationist and first Chief of the FS, argued that "the first principle of conservation is development, the use of the natural resources now existing on this continent for the benefit of the people who live here now. There may be just as much waste in neglecting the development and use of certain natural resources as there is in their destruction." Grumbine (1992) labeled this anthropocentric philosophy "resourcism." Even national parks were established to preserve monumental natural wonders for aesthetic consumption, with their borders drawn to exclude areas containing other commercial values (Runte 1987).

14. Noss and Cooperrider (1994: 72–80) provide a useful overview of the traditional training instilled by professional schools of forestry, range management, and wildlife management. They argue that these schools trained resource managers to think primarily in terms of the efficient production of commodities rather than ecological principles, resulting in numerous deleterious effects on biodiversity.

15. An earlier literature on FS employees explored organizational identification and the conformance of individual attitudes with agency values (Kaufman 1960; Schiff 1966; Hall, Schneider, and Nygren 1970; Twight 1983).

16. Interview with Carl Rountree, Chief, Biological Resources Branch, California State Office, U.S. Bureau of Land Management; Sacramento, California, December 10, 1993.

17. Several books also spurred individuals to action by trumpeting the impending biodiversity crisis and proposing means to prevent it. See, for example, Grumbine 1992; Jensen, Torn, and Harte 1993; Noss and Cooperrider 1994.

18. As Leopold (1953: 146–147) argued, "If the biota, in the course of aeons, has built something we like but do not understand, then who but a fool would discard seemingly useless parts? To keep every cog and wheel is the first precaution of intelligent tinkering."

19. Conservation biologists disagree, however, over whether a single large reserve is preferable to several smaller ones of equivalent area. Noss and Cooperrider (1994: 140) note that "the literature on this debate, which came to be known by the acronym SLOSS (single large or several small), is perhaps larger than on any other topic in the history of applied ecology." Given that the answer appeared to depend on the species of concern, the two sides resolved the debate by agreeing that large reserves *and* multiple reserves are both important. If planners need to make a trade-off when designing a reserve system, they should consider the needs of the relevant species. See Quammen 1996 for a historical account of the SLOSS debate.

20. See Noss and Cooperrider 1994 for a lengthy and detailed exposition of the links between empirical evidence, hypothesized ecological relationships, and principles for designing natural reserve systems. They also offer a concise statement of the normative postulates guiding their work (1994: xxvi), and suggest the benefits of drawing the social sciences into their interdisciplinary mix of applied and theoretical science (1994: 84).

Chapter 3

1. The California Interagency Wildlife Task Group, for example, developed and managed the Wildlife Habitat Relationship program, which included common methodologies for describing habitat and the effects of management practices on habitat. This program supplied a central repository of information on the management status, distribution, life history, and habitat requirements of terrestrial vertebrates in California, as well as a predictive model, which could be used by biologists, land managers, and planners.

2. For guidelines see Coordinated Resource Management and Planning, 1990, a joint publication of the cooperating agencies in California. This handbook lays out a 12-step approach to consensus-based planning. For discussions of CRMP in other states, see Cowart and Fairfax 1988: 425–426 on Nevada, and Anderson and Baum 1987 on Oregon. CRMP dynamics are similar in all three states.

3. Federal MOUs on CRMP were signed in 1971, 1975, 1980, and 1987. The 1987 version of the federal MOU provided guidance for California's 1987 MOU on CRMP, but there are notable differences. For example, the federal MOU dispensed with the word *planning* in 1987, so it became Coordinated Resource Management, or CRM. According to a memo sent to regional foresters on October 5, 1987, from the Associate Deputy Chief at FS Headquarters, *planning* was dropped "to avoid the appearance of imposition of another planning process or of conflict with formal planning processes of agencies."

4. This assertion is based on participant impressions rather than a rigorous count of attendance at CRMP Executive Council meetings. Suffice it to say that BLM State Director Ed Hastey, a CRMP booster and past chair of the CRMP Executive Council, once commented that the only agency director who attended the annual Executive Council meetings was the chair. (From the minutes of the August 5, 1992, meeting of the Executive Council on Biological Diversity.)

5. Interview with Leonard Jolley, State Range Conservationist, U.S. Soil Conservation Service; Davis, California, March 29, 1994.

6. In 1994, the agency received a new name, the Natural Resources Conservation Service, to emphasize that its mission included conservation of all natural resources, not simply soil. The agency nevertheless remained the technical delivery arm for conservation in the U.S. Department of Agriculture.

7. For various interpretations of the BLM's relationship with clientele groups, see Foss 1960; Calef 1960; McConnell 1966; Dana and Fairfax 1980; Culhane 1981; Clarke and McCool 1996.

8. Interview with Leonard Jolley, State Range Conservationist, U.S. Soil Conservation Service; Davis, California, March 29, 1994.

9. Interview with Kent Smith, McCollum Associates (formerly with the California Department of Fish and Game); Sacramento, California, May 5, 1994.

10. Assembly Bill 1039; California Fish and Game Code 1930.

11. These figures are from Hoshovsky 1992: 21. Also see Hoshovsky 1990 and CDFG's Web site for more current information.

12. As stated in the California Fish and Game Code §1932, the CDFG "shall, after consultation with federal, state, and local agencies, education institutions, civic and public interest organizations, private organizations, landowners, and other private individuals, identify by means of periodic reports those natural areas deemed to be most significant." To reduce unnecessary duplication of effort, the CDFG also "shall provide coordinating services to federal, state, local, and private interests wishing to aid in the maintenance and perpetuation of significant natural areas."

13. Interview with Marc Hoshovsky, Biodiversity Protection Planner, Natural Heritage Division, California Department of Fish and Game; Sacramento, California, November 10, 1993.

14. Interview with Chris Unkel, Director of California Wetlands, The Nature Conservancy (formerly with the California Department of Fish and Game); Sacramento, California, November 16, 1993. See Pearsall 1984 for a discussion of the Tennessee Protection Planning Committee.

15. Excerpts from INACC minutes, August 14, 1986.

16. According to Hoshovsky, the INACC minutes no longer exist as a full set. The following discussion draws from the minutes of seven quarterly meetings between 1986 and 1988, a relatively active period in INACC's history.

17. Experiment stations and regional offices are distinct units within the FS. At the state level, they are essentially two different agencies because the directors report directly to the Chief Forester in Washington, D.C. The vast majority of FS line officers in the regional offices were professionally trained foresters, not research ecologists. Stewart was one of the rare regional foresters who moved to that position after serving as an experiment station director. In 1994, Stewart subsequently became an assistant to Chief Forester Jack Ward Thomas, another research ecologist who bypassed traditional promotion routes.

18. For further details, see Sahagun and Stein 1989.

19. The NPS published these figures to draw attention to its plight (National Park Service, 1980: 36). Yet the agency's personnel profile did not change significantly in the 1980s. According to a report by the National Research Council (1992: 73), the NPS "maintains a smaller research staff" than most other federal land management agencies. The report also suggested that even these relatively small numbers might be exaggerated because the agency combines scientists and resource managers in a single statistical category (1992: 61).

20. Comment by Rich Baker, as recorded in the INACC minutes, August 14, 1986. Baker was involved with the NPS Gene Pool Task Force at the University of California, Davis, which was studying the agency's role in protecting genetic diversity. The NPS was suffering from similar research shortfalls throughout the country. According to a NPS report (National Park Service, 1980: 35), "Very few park units possess the baseline natural and cultural resources information needed to permit identification of incremental changes that may be caused by a threat. The priority assigned to the development of a sound resources information base has been very low compared to the priority assigned to meeting construction and maintenance needs."

21. The extant minutes from seven quarterly meetings between 1986 and 1988 show eight people attending at least half of the meetings: Marc Hoshovsky and Chris Unkel, CDFG ecologists; Jim Barry, Senior Ecologist in the technical report division of the California Department of Parks and Recreation; Dan Cheatham, Field Representative for the University of California's Natural Reserve System; Lynn Lozier (and later Leslie Friedman) from TNC; Gail Kobetich, endangered species recovery planner for the FWS; Alex Young from the BLM; and Dave Diaz from the FS Regional Office in San Francisco. An NPS representative usually attended the meetings, but no one came on a routine basis. Diana Jacobs, staff ecologist for the California State Lands Commission, also began participating in 1987, two years before the director of her agency signed the revised INACC agreement in 1989. Several of these individuals later assumed prominent roles in implementing the MOU on Biodiversity and related efforts around the state. Ed Hastey credited Dan Cheatham with providing the original impetus behind INACC.

22. Biogeographers use the term *province* to divide the planet into locations containing unique assemblages of plants and animals. *Bioregion* is more anthropomorphic because it also takes into account human communities and administrative jurisdictions. Therefore, I rely on the latter, but the term *province* nevertheless appears in some quotes because several people I interviewed referred to INACC's "Klamath Bioregion" as the "Klamath Province."

23. In some versions of the bioregional map, the Sierra Bioregion is divided into northern and southern bioregions, and the Klamath Bioregion is labeled "Klamath/North Coast."

24. Interview with Ed Hastey, California State Director, U.S. Bureau of Land Management; Sacramento, California, November 8, 1993.

25. Public Resources Code §4800 et seq.

26. Interview #39. As stated in *The Report of the California Timberland Task Force* (Timberland Task Force, 1993: 18), "Frequent charges were inadequate review of timber harvest plans filed under the authority of the California Environmental Quality Act (CEQA) and the Z'Berg-Nejedly Forest Practice Act. Court rulings in several of these cases made it clear that CDF and the board needed to improve their ability to assess the singular and cumulative effects of private and state forestry operations on wildlife species and their habitats."

27. According to Fortmann (1990: 372), "There is no question that the agency has better relations with the timber industry than with other groups. Agency staff are professional foresters who have been socialized to believe that the best use of trees is for production as long as environmentally sound practices are observed."

28. U.S. District Court, Western District of Washington, "Memorandum Decision and Injunction," *Seattle Audubon Society v. Evans*, No. C89-160WD, May 23, 1991, pp. 33–34, 20 (as cited in Yaffee 1994: 134, note 46). For extensive background on litigation over the owl, see Sher 1993 and Yaffee 1994.

29. The desert tortoise, for example, occurred on several million acres of BLM land in California, primarily in the Mojave Desert. Tortoise management had enormous implications for other uses of BLM land because studies suggested that each tortoise population required nearly 100,000 acres of undeveloped land to remain genetically viable. (Interview with Alden Sievers, Project Manager, Barstow Resource Area, U.S. Bureau of Land Management; Barstow, California, January 4, 1994.)

30. Private and state forestlands in the Klamath were also directly affected by litigation over the northern spotted owl, though state agencies were not as severely hampered as the FS. The state forestry agencies nevertheless began a habitat conservation plan for the owl so they would be eligible for an "incidental take" permit (under Section 10 of the ESA), which would allow the CDF to approve timber harvest plans. The Board of Forestry also adopted specific rules for private and state forestlands to comply with the ESA.

31. Interview with Ron Stewart, Regional Forester, Pacific Southwest Region, U.S. Forest Service; San Francisco, March 16, 1994.

32. Interview with Harley Greiman, Regional Forester's Representative, U.S. Forest Service; Sacramento, California, November 8, 1993.

33. The California Environmental Quality Act (CEQA) requires local and state agencies to submit environmental impact reports to the CDFG if their projects will significantly affect the environment and involve either public funds or a discretionary permit from a local or state entity. In this capacity, CDFG staff reviewed timber harvest plans. According to one estimate by a former CDFG employee, the agency received over 17,000 environmental impact reports in 1989 or 1990, but could review only 4400 of these. To help fill the gap, volunteers from organizations like the California Native Plant Society and Audubon Society acted "as a shadow resource management department, reviewing a fraction of the many environmental documents produced." (Testimony by Deborah Jensen, *The Second Annual Natural Diversity Forum*, Senate Committee on Natural

Resources and Wildlife, California Legislature, Los Angeles, November 26, 1990, p. 5.)

34. The nonagency representatives on the TTF were Dan Taylor, Audubon Society, and Gil Murray, California Forestry Association.

35. Letter to Governor George Deukmejian from Jon Kennedy (signed by Harley Greiman), September 21, 1989. The letter reads, in part: "Perhaps the most important notion in this legislation is the recognition that resource problems cross administrative and property boundaries; and responsible interests and agencies must cooperatively seek resolutions to these important issues."

36. As stated in the California Public Resources Code §4802, TTF's duties were to (1) develop a coordinated base of scientific information on the location, extent, and species composition of timberland ecosystems in California; (2) design and contract for studies to validate wildlife habitat models, evaluate the effectiveness of alternative mitigation measures designed to minimize significant adverse environmental impacts of timber harvesting, and develop and evaluate alternative management programs designed to maintain or develop the physical characteristics of wildlife habitats; (3) identify critical habitat areas necessary to maintain and restore viable populations of species dependent on specific timberland habitats, focusing initially on species dependent on old-growth forests along the North Coast and then proceeding to other regions of the state; and (4) identify species that are or may become endangered, threatened, or of special concern as the result of management activities on private and public timberlands.

37. Interview with Robert Ewing, Chief, Strategic Planning Program, California Department of Forestry and Fire Protection; Sacramento, California, November 15, 1993.

38. Interview with Ed Hastey, California State Director, U.S. Bureau of Land Management; Sacramento, California, November 8, 1993.

39. "Revised Work Plan of the California Timberland Task Force," April 9, 1990, p. 4.

40. This unit subsequently experienced at least two name changes during the 1990s, first becoming the Strategic Planning Program and then the Fire and Resource Assessment Program.

41. Interview with Marc Hoshovsky. Elsewhere, Hoshovsky (1992: 23) wrote, "The major challenge in making bioregional planning work is to gain the support of regional administrators, such as forest supervisors or county supervisors. Hopefully the statewide biodiversity agreement will provide the 'top-down' direction needed to enable at least state and federal agencies to work together at a regional level."

42. Interview with Kent Smith, McCollum Associates (formerly on staff of CDFG's nongame programs); Sacramento, California, May 5, 1994.

43. The second draft of the MOU on Biodiversity includes all of the TTF agencies, with the exception of the University of California. It also lists the California Department of Conservation, California Department of Water Resources, California Fish and Game Commission, California State Board of

Forestry, California State Lands Commission, California Energy Commission, California Department of Food and Agriculture, Governor's Office of Planning and Research, California Department of Transportation, U.S. Soil and Conservation Service, U.S. Bureau of Reclamation, and U.S. Department of Defense.

44. Interviews with Joan Reiss, former Regional Director, Wilderness Society; San Francisco, April 11, 1994; and Wayne White, State Supervisor for Ecological Services, U.S. Fish and Wildlife Service, Sacramento, California, December 16, 1993.

Chapter 4

1. Attendance is a reliable measure of cooperation because the minutes specify which directors or alternates were present at each meeting. The measure is not perfectly reliable, however, because there are some errors in the minutes. For example, on at least two occasions a director was not listed as present on the first page of the minutes even though comments were later attributed to him in the text of the minutes. In these cases, I counted the director as present and assumed he arrived late. In light of such discrepancies, it is possible that table 4.1 undercounts attendance, but there is no reason to believe the undercounting is systematically biased or substantively important.

2. The rank order of the agencies is not sensitive to changes in the ordinal scale. In other words, if the scale is expanded to capture greater differences in organizational hierarchy (e.g., by increasing the ordinal range from 0–3 to 0–7), then variation in the scores listed in the right-hand column increases, but the order in which the agencies are ranked changes little. The methodology therefore underestimates variation between agencies in terms of hierarchical status, but is unbiased in ranking the agencies.

3. The first nine meetings were convened in Sacramento because most of the signatories were based there. When the directors began holding meetings outside of Sacramento at the end of the Executive Council's third year, they did so to reach out to field offices and local stakeholders in the bioregions. This increased travel time for most directors because the last four meetings analyzed in tables 4.1 and 4.2 were held in Morro Bay, El Portal, San Diego, and Napa. None of the directors were based in these cities.

4. Wayne White attended seven meetings as Plenert's alternate, during which time he served as Field Supervisor and then State Supervisor of Ecological Services. In the latter capacity, he was the highest-ranking FWS official responsible for implementing the ESA in California.

5. "'Bio's' Threaten California Agriculture," a comment by President Bob L. Vice, California Farm Bureau Federation. Released by the California Farm Bureau Federation, August 17, 1992.

6. Fortmann (1990: 374–375) argues that Cooperative Extension's county-based farm advisors were closely aligned with large-scale farmers and ranchers. More generally, Fortmann notes that extension offices throughout the United States

have historically been closely aligned with, if not willingly captured by, prosperous Caucasian farmers, particularly members of the Farm Bureau. Also see Selznick 1949.

7. Having taken extensive notes at several meetings, I am aware of the disjunction between the minutes and actual discussion. My interpretation of the minutes was also supplemented by notes taken by a third party, and by interviews with numerous participants.

8. Although the minutes are not inclusive, there is no readily apparent reason to believe they are biased either in favor of or against any one director. Therefore, the total number of comments attributed to each director is probably underestimated, but not systematically biased. The minutes do, however, include instances of passive paraphrasing, in which the person preparing the minutes did not attribute comments to their sources. Agency staff or members of the public likely made some of these comments.

9. Interview #31.

10. Though Hastey won this battle, he may have lost the war, because the EPA subsequently signed the MOU in 1994. The Executive Council nevertheless shunned aquatic diversity issues until Warren left the CSLC in 1994, leading Warren to state in a survey of Executive Council members that "the major failure of the Council has been its refusal to expand its agenda to include aquatic diversity and agencies responsible therefor." (Memorandum to Ruth Mazur, Strategic Planning Program, CDF, from Charles Warren, CSLC Executive Officer, in response to a membership survey of the accomplishments of the Executive Council on Biological Diversity, September 22, 1993.)

11. The CSLC actively traded school lands with the BLM and FS to combine its holdings into more manageable units. In managing the school lands, the agency served as a fiduciary trustee for the State Teachers' Retirement System, and, in that capacity, was required to generate a revenue stream.

12. According to federal court interpretations of common law, states have an "affirmative duty" to take the public trust into account in managing sovereign lands. Public trust uses include commerce, navigation, fisheries, recreation, and the preservation of lands in their natural state to protect wildlife habitat. Public trust law is old, but is increasingly interpreted in novel ways. See Coggins, Wilkinson, and Leshy 1993: 323–332 and citations therein.

13. Warren was a member of the California state legislature from 1963 to 1977, where he had been instrumental in passing several pieces of environmental legislation. He later chaired the federal Council on Environmental Quality in the Carter administration from 1977 to 1979, and served as a member of the California Coastal Commission.

14. The annual *Governor's Budget* shows the CSLC budget hovering around $16 million during Wilson's first three years in office, and then declining precipitously to about $8.6 million in fiscal year 1994–95. Staff size similarly declined, from a high of about 250 in the early 1990s to 105.5 in fiscal year 1994–95.

15. Interview with Charles Warren, Executive Officer, California State Lands Commission; Sacramento, California, February 14, 1994. It is not clear from the tape-recorded interview whether Warren was referring to the state or federal ESA. Many local and state officials believed the state ESA was relatively anemic, and were much more concerned about the federal ESA.

16. See, for example, McHugh 1994. Governor Wilson's position became more strident in 1995, as he sought to shore up his funding base among major donors in the agricultural and development industries who claimed to be burdened by environmental laws. Shortly before officially announcing his presidential candidacy, Wilson declared a five-year moratorium on enforcement of the California ESA in cases where individuals or public agencies could claim they harmed species to prevent or mitigate an emergency or natural disaster. See McHugh 1995. Wilson also released a plan to overhaul the California ESA, which included dropping all currently listed species in five years and requiring a stricter review for new listings. See Gunnison and Lucas 1995.

17. Interview with Joan Reiss, former Regional Director, Wilderness Society; San Francisco, April 11, 1994.

18. NPS staff did play a role in some bioregional efforts—particularly in the Sierra Bioregion, where Yosemite, Kings Canyon, and Sequoia National Parks are located. It should also be noted that NPS participation in the Staff Committee increased in 1994 with the arrival of Sarah Allen as NPS Endangered Species Coordinator in the regional office.

19. Interviews #31, #48, #51, and #52.

20. The mission of CDFG's Natural Heritage Division specifically focused on the identification and conservation of California's sensitive plants, animals, and species. The mission of CDF's Forest and Rangelands Resources Assessment Program was more general, aiming to provide the CDF with the capacity to examine, evaluate, and assess the goals of the agency's programs, including fire protection and resource management.

21. Interview with John Hopkins, Sierra Club; Davis, California, March 29, 1994.

22. Interview with Joan Reiss, former Regional Director, Wilderness Society; San Francisco, April 11, 1994.

23. "'Bioregional' Strategy Adopted for Resources Conservation," a press release issued by the Resources Agency of California, September 19, 1991.

24. See, for example, "State, Federal Land Managers to Cooperate in Regional Goals," 1991; "State, Federal Officials Announce New Conservation Strategy," 1991. For a somewhat earlier perspective, see Bowman 1991b.

25. I first heard Wheeler deliver this quip in opening remarks to a conference of state and federal resource managers in Sacramento on January 25, 1993. The conference was convened to promote interagency coordination and cooperation in preserving biodiversity in California.

26. For more background, see Duane 1998: 23–29.

27. Paraphrased comments from the five workshops—held in Quincy, Placerville, Mariposa, Visalia, and Bishop—can be found in *The Sierra Nevada: Report of the Sierra Summit Steering Committee*, Resources Agency of California, July 1992.

28. Interview #51.

29. See Zea 1991.

30. As quoted in Blackburn 1992.

31. See Bancroft 1992.

32. Interview with Robert Ewing, Chief, Strategic Planning Program, California Department of Forestry and Fire Protection; Sacramento, California, November 15, 1993.

33. See Rose 1992.

34. From the minutes of the Executive Council on Biological Diversity, February 28, 1992.

35. See Bowman 1991a.

36. Interview #51.

37. The minutes reveal a large number of critical comments from the audience at this particular meeting. As one observer later commented, "There were hot words at the second meeting of the Executive Council in early 1992, challenging the authority of the group to convene or to make decisions" (Jensen 1994: 276).

38. The MOU on Biodiversity is not clearly worded in this regard. It states that the Executive Council will "recommend consistent statewide standards and guidelines," leaving open the possibility that the agencies might promulgate new standards and guidelines rather than simply reconciling existing ones. There is no evidence that agency officials seriously considered promulgating new rules. While some agency staff expressed interest in stronger regulations, this was not the position of the Staff Committee, whose Implementation Plan (second draft, February 28, 1992, p. 2) stated, "The Council's approach is to seek ways to more effectively coordinate the implementation of existing laws, policies, and direction among the widest array of regulatory and land management agencies and bodies."

39. Interview with Ed Hastey, California State Director, U.S. Bureau of Land Management; Sacramento, California, November 8, 1993.

40. Although Hastey had previously argued that the Executive Council might become too large if the three regulatory agencies suggested by Charles Warren were added, he now believed that "the advantages of including the full seven county organizations . . . outweigh any disadvantages that may result from increasing the size of the Executive Council." (From "Proposed Membership of Executive Council on Biological Diversity," signed by Ed Hastey (BLM), Richard Wilson (CDF), and Jon Kennedy (FS), undated.)

41. One county supervisor (Laurence "Bud" Laurent, San Luis Obispo County) signed for two of the regional associations (Southern California and South Central Coast), so the actual number of signatories was seven.

42. Interview with John Hopkins, Sierra Club; Davis, California, March 29, 1994.

43. Interview with Art Baggett, Supervisor, Mariposa County; Sacramento, California, March 23, 1994.

44. Two additional supervisors also played important roles. Siskiyou County Supervisor Patti Mattingly aided Hastey's effort in 1992 to bring county supervisors on board. She signed the Statement of Intent for the Northern California Counties Association, but passed away in December 1992 at the age of 41 from a pulmonary hemorrhage before having the opportunity to participate on the Executive Council or forthcoming initiatives in the Klamath. Modoc County Supervisor Nancy Huffman subsequently represented this association in an active manner through the latter half of the 1990s.

45. Interviews with Maura Wiegand, Council member, City of Encinitas, Encinitas, California, May 13, 1994, and Tom Sykes, Council member, City of Walnut, Walnut, California, May 16, 1994.

46. Interview #18.

47. The FS Pacific Southwest Experiment Station was the only exception. The Experiment Station did not officially join INACC until 1989 when Ron Stewart signed the revised natural areas agreement. In 1990, Stewart became the regional forester, in which capacity he signed the MOU on Biodiversity in 1991.

48. "'Bio's' Threaten California Agriculture," Farm Bureau Comment by Bob L. Vice, President, California Farm Bureau Federation. Released by the California Farm Bureau Federation, Sacramento, August 17, 1992.

49. Interviews #31, #34, #35, and #53.

50. Interview with Doug Wheeler, Secretary for Resources, California Resources Agency; Sacramento, California, May 3, 1994.

51. RCDs do not cover the entire state. They do not typically encompass cities, nor do they occur in some rural areas, particularly where single landowners have very large parcels, as in the southern portion of the San Joaquin Valley where large corporations farm most of the land. Some rural communities are also inhospitable to the creation of new layers of government, and therefore do not establish RCDs or let existing RCDs go fallow.

52. Interview #31.

53. I found memos sent by four directors in the original ten agencies: Ed Hastey (BLM), Ron Stewart (FS), Boyd Gibbons (CDFG), and Kenneth Farrell (UC/DANR). Staff Committee minutes also indicate that Charles Warren (CSLC) and Donald Murphy (CDPR) sent memos to their line managers. The Resources Agency is not germane because it is the superstructure housing the other state agencies. In the NPS regional office, Deputy Director Lew Albert reported that copies of the MOU were sent to individual park units. I could not account for the FWS or CDF.

54. Memorandum to Regional Forester Ron Stewart from Attorney Jack Gipsman regarding the California Biodiversity Memorandum of Understanding, October 1, 1991.

55. Hoshovsky, for example, referred to national parks as the "fiefdoms" of park superintendents.

56. Interview with Lew Albert, Deputy Director, Western Regional Office, U.S. National Park Service; San Francisco, February 22, 1994.

57. Interview with Al Wright, Associate State Director, Bureau of Land Management; Sacramento, California, November 16, 1993.

58. Interview with Carl Rountree, Chief, Biological Resources Branch, California State Office, U.S. Bureau of Land Management; Sacramento, California, December 10, 1993.

59. The author(s) of an unsigned, undated "Membership Subcommittee Report" to the Executive Council believed that dues were necessary to cover the increasing costs of operations, but in a clear indication of their pessimism they wrote,

> If agencies are unable to provide additional funding, then agencies must be willing to devote a specific amount of their staff's time to the Council's activities. In the past, agency staffs have participated as time and circumstances have allowed. The Council can [no] longer afford to operate in this fashion if it is to meet the demands watershed groups and others are placing on it. *The Subcommittee, therefore, recommends that member agencies designate specific members of their staff to devote a predetermined amount of time to the Council's activities. These staff members will serve on the Council's Staff Committee.* [Italics in original.]

60. INACC had been discussing these issues since the 1980s. According to INACC minutes, the group agreed at its September 15, 1987 meeting that there was a "need for a statewide GIS, coordinated between agencies, or at least systems that are intercompatible."

61. To see the wealth of information available from CERES, go to <http://ceres.ca.gov>. (Ceres was the Roman goddess of agriculture, and is also a small farming community in California's Central Valley.)

62. Online versions of the newsletter can be read at <http://ceres.ca.gov/biodiv/newsletter.html>. This newsletter was supplemented in 2001 by a large, glossy brochure titled *Decade of Diversity*, which celebrated the council's first ten years.

63. This project emerged from a CARCD proposal to the Resources Agency following the Sierra Summit, which the CARCD drafted before joining the Executive Council. The purposes of this project are laid out in the Memorandum of Understanding for a Coordinated Strategy for Resource Conservation in the Sierra Nevada, signed by Doug Wheeler (Resources Agency), Ed Heidig (Department of Conservation), Charles Pritchard (CARCD), and Pearlie Reed (SCS) on November 9, 1992.

64. The directors included Doug Wheeler (California Resources Agency), Boyd Gibbons (CDFG), Donald Murphy (CDPR), Ron Stewart (FS), and Kenneth Farrell (DANR). Farrell's appearance is particularly notable because he never attended Executive Council meetings. This meeting differed from Executive Council meetings, however, because only public officials were invited. Given the

backlash against the MOU and Executive Council, the organizers wanted participants to feel relaxed rather than exposed to the scrutiny of interest groups and the press. In his speech, Farrell suggested DANR could participate in research, the development of inventories, and ecological monitoring; Cooperative Extension offices in the counties could participate as facilitators. Other speakers included Robert Ewing (CDF), Al Wright (BLM), and Wayne White (FWS).

65. Interview #39.

66. Expecting field staff to be concerned about the lack of support at the regional level, state-level staff planned to emphasize at the local workshops that locally developed projects were supported from above, which meant telling field staff "We'll take the heat at the regional level." (Quote excerpted from "Notes from the meeting of the Regional Orientation/Teambuilding Planning Workgroup (established at the Granlibakken workshop), June 20, 1994," by Dennis Pendleton, convener, p. 2.)

67. Quote from a memorandum to Executive Council Members from the Interim Committee on Watershed Group Liaison, June 3, 1994, p. 2. In responding to the inertia from regional managers, this memo makes ten recommendations for linking state offices directly with local groups. All ten recommendations were adopted by the Executive Council at its June 16, 1994 meeting, but one of the recommendations—to form "regional managers' committees" within each bioregion to coordinate the relationship between the agencies and the local groups—was abandoned later that summer. The first forum for regional managers was not convened until October 1995. Approximately 40 line managers attended this regional meeting in Redding, near the Executive Council's original timberlands focus in the Klamath Bioregion.

68. In 1996, the council's Web page posted several definitions of *biodiversity*, but the council did not adopt any of these definitions. For current definitions posted at this site, see <http://ceres.ca.gov/biodiversity/bioregions.html>.

69. This tension was particularly pronounced in the CDFG, where Director Boyd Gibbons regularly ceded management authority to Deputy Director John Sullivan, who had much closer ties to the Governor's constituents in the building industry. One CDFG staff member, for example, believed the Wilson administration purposely impaired the agency's enforcement capability by installing a weak leader, allowing the deputy director to run the agency by default. (Interview #89.) Among other actions, Sullivan offered the state legislature a bill that would have significantly weakened the California ESA. Neither Wheeler nor the Executive Council endorsed this bill. For more on this leadership tension within CDFG, see Woody, 1993.

70. For current membership, see <http://ceres.ca.gov/biodiv>. As of 2002, there were four additional state agencies (the California Conservation Corps, the Native American Heritage Commission, the San Francisco Bay Conservation and Development Commission, and the State Water Resources Control Board). At the federal level, three agencies departed because they were either folded or merged with other agencies (the U.S. Agricultural Stabilization and

Conservation Service, the U.S. National Biological Service, and the U.S. Bureau of Mines). They were replaced by four additional agencies (the Monterey Bay National Marine Sanctuary, the National Marine Fisheries Service, the Western Ecological Research Center of the U.S. Geological Survey, and the U.S. Marine Corps). There were no changes at the local level.

71. Memorandum to California Biodiversity Members from Douglas P. Wheeler regarding the formation of the CBC Executive Committee, February 27, 1995.

Chapter 5

1. The U.S. Office of Technology Assessment (1992: 15) found that revenue-sharing payments "account for a large portion, up to 80 percent, of operating budgets in some Pacific Northwest counties." Counties received additional payments under the Payment in Lieu of Taxes Act of 1976, which provides a minimum payment of 75 cents per acre of federal land to local governments, regardless of revenues from timber sales or other developments on these lands.

2. For additional background on timber conflicts at that time in the Klamath, and throughout the Pacific Northwest, see Lipschutz 1996: 81–125; Nechodom 1994; Yaffee 1994; Raphael 1994; Grumbine 1992; Dietrich 1992.

3. This account of Forsman's early interest in the owl is from Dietrich 1992: 47–48. The northern spotted owl is one of three spotted owl subspecies in the United States. The others are the California spotted owl and the Mexican spotted owl. Under the ESA, a geographically isolated population of a species can be listed even if the species itself is not threatened or endangered.

4. As one of the litigants later noted, "This set of lawsuits has in just over five years invoked more provisions of the ESA—and uncovered more violations of that law by federal agencies—than any previous litigation effort on behalf of any species or groups of species" (Sher 1993: 42). For case law, see Sher 1993. For associated agency behavior, see Yaffee 1994 and Grumbine 1992: 143–155. Also see Bonnett and Zimmerman 1991.

5. The Endangered Species Act of 1973, as amended, Section 4(b)(1)(A).

6. Section 6(g)(3)(B) of NFMA requires the FS to "provide for diversity of plant and animal communities based on the suitability and capability of the specific land area in order to meet overall multiple-use objectives."

7. On June 6, 1994, Judge Dwyer accepted the Clinton administration's plan (popularly known as "Option 9"), allowing limited logging to resume, but withheld final approval of the plan on its ecological merits. At that time, the Clinton administration believed it would take another two years to reach one million board feet, or one-fifth of the timber harvested in the 1980s. See Cushman 1994.

8. Figures from William F. Delaney & Associates, "Feasibility of Alternative Forest Products in Trinity RC&D and Hayfork Adaptive Management Area," March 1994.

9. Memorandum to Redwood National Park Superintendent and Deputy, Chief, R&RM, Supervisory Botanist, Ecologist, Fish and Wildlife Biologist from Environmental Scientist Lee Purkerson, December 19, 1990. Interview with Lee Purkerson, retired Environmental Scientist, Redwood National Park, U.S. National Park Service; Eureka, California, August 29, 1994.

10. "Examples of Coordinated Strategies to Conserve California's Biological Diversity," an undated attachment to a press release issued by the Resources Agency of California on September 19, 1991, titled "'Bioregional' Strategy Adopted for Resources Conservation." Also see Timberland Task Force, 1993: 12–13, which links this bioregional effort directly to the northern spotted owl.

11. Figures from a letter from Jerry Moles to participants in the Klamath Bioregion Project, February 14, 1992.

12. "Summary Klamath Report," prepared by Tim Wallace for Robert Ewing, March 8, 1993.

13. Interview with Tim Wallace, Klamath Bioregion Project Coordinator, Division of Agriculture and Natural Resources, University of California; Berkeley, California, November 1, 1993.

14. Interview with Lee Purkerson, retired Environmental Scientist, Redwood National Park, U.S. National Park Service; Eureka, California, August 29, 1994.

15. Letter from Tim Wallace to Kim Rodrigues, Cooperative Extension Forest Advisor, Humboldt County, March 19, 1992.

16. The Del Norte group separated from the Northern Klamath Bioregional Council in late 1992, primarily due to the five-hour round-trip drive from Crescent City in Del Norte County to Eureka in Humboldt County. The Del Norte group soon folded, however, in part because most agency staff were based in Humboldt County.

17. There were nevertheless many local groups in the Sonoma-Garberville subregion. One state ecologist reported working with 13 different cooperative groups, most organized around watersheds, and all working on biodiversity issues. (Interview with Greg Giusti, Wildlands Ecology Advisor, Cooperative Extension, Division of Agriculture and Natural Resources, University of California; Ukiah, California, October 14, 1994.)

18. Interview with Gary Nakamura, Area Forest Specialist, Cooperative Extension, Division of Agriculture and Natural Resources, University of California; Redding, California, August 24, 1994.

19. The initial one-year contract included a consultant fee of $62,238 for Moles, plus expenses. (Interagency Agreement #6CA27632, between the Regents of the University of California and the California Department of Forestry and Fire Protection, July 1, 1992.) The second one-year contract included a $70,164 fee for Moles, $40,000 for a GIS consultant in Trinity County, and other expenses. (Interagency Agreement #6CA27632 #2, between the Regents of the University of California and the California Department of Forestry and Fire Protection, December 21, 1993.) In both contracts, Wallace was identified as the project leader for the University of California, but did not receive a fee for his services.

20. For example, subregional leaders participated in a loose-knit organization, known as the Lead Partnership Group, which spanned the Klamath Bioregion, southern Oregon, and the northern portion of the Sierra Bioregion. The Lead Partnership Group was neither an outgrowth of the Klamath Bioregion Project nor a bioregional council, but rather a forum for representatives from many groups to discuss common concerns regarding forest practices and economic development. Jonathan Kusel organized it in 1993 and raised foundation grants to cover expenses.

21. "Minutes of the Trinity River Sub-Region of the Klamath Bioregion," May 29, 1992.

22. Interview with Arnold Whitridge, Trinity County Supervisor; Weaverville, California, June 29, 1994.

23. Figures reported by the TBRG Economic Committee, "Summary of Meeting Notes," Trinity Bio Region Group, July 6, 1993.

24. The Trinity River Basin Fish and Wildlife Act of 1984 (P.L. 98-541) authorized $57 million to execute eleven directives for restoring the Trinity River over a ten-year period. TRRP opened its Weaverville office in 1986, where it was administered by the FWS and the BoR, and coordinated by a fourteen-member task force.

25. See Morris 1994; also see minutes of the South Fork Trinity CRMP.

26. Interview with Chuck Lane, Project Leader, Trinity River Restoration Project, Trinity River Fishery Resource Office, U.S. Fish and Wildlife Service; Weaverville, California, June 28, 1994.

27. Interview #75.

28. Quoted in Smith 1994.

29. President William J. Clinton and Vice President Albert Gore, Jr., *The Forest Plan for a Sustainable Economy and a Sustainable Environment*, Washington, D.C., July 1, 1993.

30. California, Oregon, and Washington were each guaranteed 15 percent of the federal funds. In Northern California, the eight affected counties competed for the state's share of the $270 million allotted for 1994, or about $40 million. The remaining funds were doled out on a competitive basis among the states. A tri-state Regional CERT in Portland ensured that the money was doled out equitably among the three states.

31. "A Proposal for the Trinity National Forest," attached to a letter addressed to President Clinton from Nadine Bailey, representing the Trinity Bio Region Group, May 19, 1993. This proposal was prepared by the so-called "gang of six," an alliance of environmentalists, miners, and loggers. Though it was submitted on behalf of TBRG soon after the Forest Conference, it was not developed through consensus within TBRG.

32. "Summary of Meeting Notes," Trinity Bio Region Group, June 15, 1993.

33. Minutes from the Trinity Bio Region Group meeting on November 30, 1993. TBRG coordinator Patrick Truman subsequently announced in a letter to

Resources Secretary Doug Wheeler and other public officials on May 11, 1994, that TBRG and the Watershed Center "are distinct and separate organizations which have no formal relationship with each other, although they have some members in common."

34. Tim Wallace, *Report on the Resources and Activities of the Biodiversity Project within the Klamath Province*, January 31, 1994.

35. Interagency Agreement #6CA27632 #2, between the Regents of the University of California and the California Department of Forestry and Fire Protection, December 21, 1993. The Klamath Province GIS Project, like the Klamath Bioregion Project, was administered through CDF's ecological analysis unit (FRRAP), which had the most sophisticated GIS capabilities among the agencies.

36. Bailey, Truman & Associates consisted primarily of two people: Nadine Bailey and Patrick Truman. Many considered Truman, who chaired the Trinity County Resource Conservation District, to be acting on behalf of Joseph Bower, a prominent environmental activist, so that Bower would not have to work closely with Nadine Bailey, his long-time adversary in the timber industry. The initial TRRP contract covered $24,000 in personnel costs for the TBRG coordinator and $8100 for related expenses over six months. (Agreement No. TFG 93-04, managed by the Division of Natural Resources, Trinity County Planning Department, September 7, 1993.) Bailey, Truman & Associates also received TRRP money to coordinate the South Fork Trinity CRMP.

37. Letter to Betsy Watson from Jerry Moles, February 1, 1994. Watson was acting director of the Center for Resolution of Environmental Disputes at Humboldt State University in Arcata, in which capacity she served as mediator of this dispute.

38. Letter to Secretary of Resources Doug Wheeler from the Northern Klamath Bioregion Steering Committee, December 8, 1993.

39. Typed transcript from a tape-recorded meeting of the Trinity Bio Region Group, April 5, 1994. Letter from Betsy Watson to Kim Rodrigues, April 28, 1994.

40. Trust in Moles's intentions began to fade shortly after he moved to Trinity County, when he promised TBRG that certain materials would be presented at a quarterly meeting of the Executive Council, and TBRG participants later learned that other materials appeared in the agenda packets instead (Interview #84). Because Moles was the official liaison between the subregional groups and the Executive Council, this event led some to believe that he was manipulating the subregional process, favoring one faction over others. Subsequent events simply confirmed their suspicions.

41. This dual-track strategy has been noted by others, including State Senator Barry Keene, who introduced the bill authorizing the TTF. At a Senate hearing in late 1991, Keene chided the Wilson administration following testimony by Wheeler, stating that "it seems to me that there are a number of areas in which the Administration is growing increasingly schizophrenic. The Governor is trying to retain his environmental credentials; sends you riding off, Secretary Wheeler,

in one direction. Sends another horse off, the business horse—whoever happens to be riding that at a particular time, off in another direction." After allowing Wheeler to respond, Keene suggested that Governor Wilson should be reminded, "when you're riding two horses you can get a hell of a political hernia." (*The Third Annual Natural Diversity Forum: Natural Diversity and Habitat Planning*, Senate Committee on Natural Resources and Wildlife, California Legislature, Sacramento, California, November 20, 1991, p. 20.)

42. Interview with Joan Reiss, former Regional Director of the Wilderness Society; San Francisco, April 11, 1994.

43. For background on the leadership tension in the CDFG, see Woody 1993.

44. Interview with Kim Rodrigues, Forest Advisor and County Director, Cooperative Extension, Division of Agriculture and Natural Resources, University of California; Eureka, California, July 7, 1994.

45. The increases resulted from changing the rule on logging rotation in California, from a 180-year hiatus between cuttings to rotations of 90 to 180 years. See Bowman 1994.

46. In October 1995, approximately 40 line managers, and several county supervisors, attended the first regional managers' forum, in Redding, California.

47. Memo to Jerry Moles from Lloyd I. Keefer, Chief, CDF Region I, January 29, 1993.

48. Interview with Herbert J. Pierce, Wildlife Biologist, California Department of Fish and Game; Eureka, California, August 29, 1994. Letter from Herbert J. Pierce to Klamath Bioregion Project Coordinator Tim Wallace, April 14, 1994.

49. Interview with Kim Rodrigues, Forest Advisor and County Director, Cooperative Extension, Division of Agriculture and Natural Resources, University of California; Eureka, California, July 7, 1994.

50. Interview #79.

51. Interview #75.

52. Interview with Al Wright, Associate State Director, Bureau of Land Management; Sacramento, California, November 16, 1993.

53. Memo from Regional Forester Ron Stewart to Forest Supervisors and Staff Directors, October 20, 1992.

54. Interview #80.

Chapter 6

1. Noss, LaRoe, and Scott (1995: 57–58) cite several estimates of loss, including 70 to 90 percent loss of presettlement CSS in Southern California, and 91.6 percent loss of maritime sage scrub in San Diego County. Ricketts et al. (1999: 321) state that more than 85 percent of CSS has been lost to urban and agricultural development in California and Baja California, Mexico.

2. As Beatley (1994: 58) noted, the biological study undergirding the HCP was "subjected to a peer review process by noted conservation biologists, including Paul Ehrlich, which served to enhance the credibility of the findings." For more background on this HCP, see Marsh and Thornton 1987.

3. The National Marine Fisheries Service (NMFS) reviews and approves HCPs for marine species, including anadromous fish. NMFS is relegated to footnotes in this chapter because most HCPs are land based.

4. Section 10(a)(1)(B) of the Endangered Species Act, as amended in 1982.

5. Section 10(a)(2)(A) of the Endangered Species Act, as amended in 1982. The FWS and NMFS subsequently elaborated these requirements in the *Endangered Species Habitat Conservation Planning Handbook* (U.S. Fish and Wildlife Service and National Marine Fisheries Service, 1996).

6. Scientists have not been completely satisfied by the use of scientific knowledge in HCPs. For some assessments, see Kareiva et al. 1999; Noss, O'Connell, and Murphy 1997.

7. HCPs vary widely in the number of acres and species they cover, the number of participating jurisdictions, and the scope of public participation. For assessments of this variation see Yaffee et al. 1998; Thomas 2001.

8. The most important regulatory assurance was the No Surprises Policy, first implemented in 1994, which assures applicants that no additional land-use restrictions or financial compensation will be required with respect to species covered by an incidental take permit if unforeseen circumstances arise indicating that additional mitigation is needed. See "U.S. Fish and Wildlife Service and National Marine Fisheries Service," 1998.

9. Current statistics can be found on the FWS Web site, at <http://ecos.fws.gov>.

10. This figure is from Beatley 1994: 109. BioSystems Analysis, Inc. (1994: 216) cites a U.S. population of 400 pairs. Except where noted, background on this HCP comes from Beatley 1994.

11. Interview with Kent Smith, McCollum Associates (formerly with the California Department of Fish and Game); Sacramento, California, May 5, 1994.

12. Ibid.

13. Private actors and nonfederal agencies receiving federal funds or permits can pursue Section 7 consultation in lieu of an HCP. Section 7 consultation is usually quicker than producing an HCP because Section 7 includes language stipulating deadlines by which the FWS must respond after a federal agency initiates formal consultation. Federal agencies are not obliged, however, to pursue consultation simply because a nonfederal actor does not want to prepare an HCP. Federal agencies may also participate in a Section 10 HCP and Section 7 consultation simultaneously, allowing the FWS to incorporate the HCP into the biological opinion issued under Section 7.

14. One newspaper article quoted a local city council member as stating: "To me, a rat is a rat. . . . It seems strange the rats are taking a precedent over people and projects." See Johnson 1991.

15. The Coachella Valley fringe-toed lizard HCP was widely considered a successful planning process because it brought together diverse interests, though some questioned whether too much habitat was sacrificed for development (Beatley 1994; Bean, Fitzgerald, and O'Connell 1991). During implementation, the HCP was considered less successful, because new information arose indicating the reserve system did not incorporate crucial habitat for the lizard (Schweik and Thomas 2002).

16. For example, one staff ecologist believed a perverse incentive existed to burn or thin CSS to create grassland for the Stephens' kangaroo rat. Doing so would harm CSS-dependent species, and violate the ESA if these species were listed. (Interview #44.)

17. As Beatley (1994: 46) argues, the threat of enforcement "is often only that" because the FWS "has difficulty even imposing civil and criminal penalties for illegal takes, and such penalties are relatively infrequent."

18. Interview with Monica Florian, Vice President, Irvine Company; Newport Beach, California, May 12, 1994.

19. Ibid.

20. See Frammolino 1990. Frammolino also reported that Irvine Company officials had given at least $19,000 to the Wilson campaign, that Bren hosted a fundraising event at his home netting $86,200, and that Flanigan had taken a month's vacation to help the Republican Party get out the vote.

21. Interview #72.

22. For additional background on NCCP, see Manson 1994; Thompson 1994; McCaull 1994; Welner 1995; Porter 1995; Ebbin 1997; Jasny, Reynolds, and Notthoff 1997.

23. See Frammolino and Newman 1991.

24. Interview #58.

25. Interview #37.

26. *Federal Register*, vol. 58, no. 236, Friday, December 10, 1993; 50 CFR Part 17.

27. San Diego Gas & Electric Company submitted the first approved subregional plan. This subregional plan is not considered significant because it was submitted by a single applicant and covers a relatively small amount of habitat under transmission lines running from southern Orange County to the Mexican border.

28. Interview #37.

29. Interview #37, and interview with Doug Wheeler, Secretary for Resources, California Resources Agency; Sacramento, California, May 3, 1994.

Chapter 7

1. For statistics on the loss of natural communities in the San Joaquin Valley, see Williams, Byrne, and Rado 1992. On the Southern San Joaquin Valley, see

Anderson, Spiegel, and Kakiba-Russell 1991. On the Central Valley more generally, see Noss, LaRoe, and Scott 1995: 58; Ricketts et al. 1999: 278; U.S. Geological Survey, 1998: 603–604. Noss, LaRoe, and Scott (1995: 58) cite the following estimates of habitat loss in the Central Valley: 94–96 percent loss of interior wetlands, 99.9 percent loss of riparian oak forests, and 66–88 percent loss of vernal pools.

2. Interview with Larry Saslaw, Wildlife Biologist, Bakersfield District, U.S. Bureau of Land Management; Bakersfield, California, January 6, 1994.

3. California Public Resources Code, Section 21000 et seq. CEQA's procedural requirements are triggered whenever a local or state agency confronts a discretionary decision to approve or disapprove a project, at which point that agency becomes the lead agency.

4. The INACC minutes from March 23–24, 1988, indicate that Anderson reported the cost of the CEC study to be $700,000, and that he was encouraging the CEC to expand the inventory to cover the entire state, at an estimated cost of $15 million over 10 to 15 years.

5. The CEC study area for the natural lands inventory included the Carrizo Plain, but limited surveys were conducted there due to time constraints, and the Carrizo findings were not included in the final report (Anderson, Spiegel, and Kakiba-Russell 1991: 1).

6. The INACC minutes from March 23–24, 1988, note that BLM ecologist Glen Carpenter stated the Carrizo Plain was targeted to set aside a large portion of habitat for listed species in the Southern San Joaquin Valley, where Kern County had four of the top ten oil-producing wells in the country. The FWS ranked Soda Lake, in the Carrizo Plain, as the fourth most important habitat in California.

7. Interview with Ed Hastey, California State Director, U.S. Bureau of Land Management; Sacramento, California, November 8, 1993.

8. Comments attributed to Leslie Friedman of TNC in the INACC minutes of July 20, 1988.

9. For a list of participants, see the draft *A Biological Framework for Natural Lands and Endangered Species in the Southern San Joaquin Valley*, San Joaquin Valley Biological Technical Committee, May 1993, p. 25.

10. HCPs then underway included the Metropolitan Bakersfield HCP, the Kern County Valley Floor HCP, the Pleasant Valley HCP (near Coalinga in western Fresno County), the Tulare County Valley Floor HCP, and the Kern Water Bank–Kern Fan Element HCP.

11. This was the standard compensation ratio used by the CDFG, FWS, and BLM. The *Biological Framework* suggested other compensation ratios for temporary loss of natural habitat (1.1 : 1) and nonnative grassland (0.3 : 1). It also argued that compensation should not be the only permit requirement because even a high compensation ratio of 3 : 1 necessarily leads to the loss of at least 25 percent of remaining natural habitat.

12. Interview with Larry Saslaw, Wildlife Biologist, Bakersfield District, U.S. Bureau of Land Management; Bakersfield, California, January 6, 1994.

13. Interview with Patty Gradek, Project Manager, San Joaquin Valley Regional Planning, U.S. Bureau of Land Management; Bakersfield, California, January 6, 1994.

14. Draft *A Biological Framework for Natural Lands and Endangered Species in the Southern San Joaquin Valley*, San Joaquin Valley Biological Technical Committee, May 1993, p. 20.

15. Interview #30. Though agency staff and line managers avoided the term *bioregion* in the San Joaquin Valley, it was still used at the state level. In a letter to Executive Council members, Hastey referred to the project as "our San Joaquin Valley bio-regional planning effort." (Letter dated February 4, 1994; reference #1736 CA-910.)

16. Interview with Nick Kinney, Kings County Supervisor; Rio Bravo Country Club, Bakersfield, California, March 24, 1994.

17. Interviews #44 and #54.

18. Interview with Ed Hastey, California State Director, U.S. Bureau of Land Management; Sacramento, California, November 8, 1993.

19. BLM Area Manager Jim Abbott regularly attended Interagency Team meetings and assisted with the outreach effort. Abbott managed the BLM Caliente Resource Area, an administrative jurisdiction encompassing the Southern San Joaquin Valley and the Carrizo Plain, within the Bakersfield District.

20. Interview #30, and interview with Peter Cross, Central Valley Branch Chief, Ecological Services, U.S. Fish and Wildlife Service; Sacramento, California, May 4, 1994.

21. Interview with Ed Hastey, California State Director, U.S. Bureau of Land Management; Sacramento, California, November 8, 1993.

22. Interview with Patty Gradek, Project Manager, San Joaquin Valley Regional Planning, U.S. Bureau of Land Management; Bakersfield, California, January 6, 1994.

23. Ibid.

24. Interview with Ed Hastey, California State Director, U.S. Bureau of Land Management; Sacramento, California, November 8, 1993.

25. "Statement of Intent to Support the San Joaquin Valley Regional Consortium," October 1993. The five signatory agencies were the BLM, CDFG, FWS, BoR, and California Resources Agency. Other members of the Executive Council did not sign because some agencies, like the FS and NPS, had no jurisdiction in the bioregion.

26. See, for example, "Steps to Simplify, Standardize and Expedite CESA and FESA Compliance," San Joaquin Valley Interagency Team, January 7, 1994.

27. See La Ganga 1993.

28. Interview with Patty Gradek, Project Manager, San Joaquin Valley Regional Planning, U.S. Bureau of Land Management; Bakersfield, California, January 6, 1994.

29. Interview with Peter Cross, Central Valley Branch Chief, Ecological Services, U.S. Fish and Wildlife Service; Sacramento, California, May 4, 1994.

30. Interview with Nick Kinney, Kings County Supervisor; Rio Bravo Country Club, Bakersfield, California, March 24, 1994.

31. Kinney attended four of eight Executive Council meetings during the two years (1993–1994) he represented the San Joaquin Valley Regional Association of County Supervisors. Among county supervisors sitting on the Executive Council at that time, his attendance record was average.

32. Kings County Supervisor Nick Kinney chaired these meetings, accompanied by a professional facilitator. The draft agreement was originally entitled "San Joaquin Valley Biological Partnership Memorandum of Understanding" (June 27, 1994). After editing, it emerged as the "San Joaquin Valley Resource Conservation Partnership Agreement" (August 30, 1994).

33. Kinney invited me to the meeting because I had been trying to schedule an interview with him. The interview was cut short, however, when Kern County Supervisor Ashburn approached our table. Kinney quickly walked away with Ashburn rather than introduce me to him. The Kern County General Counsel subsequently approached me, asking who I was and why I was at the meeting. This sequence of events indicated that the supervisors did not expect the public to be present. Other meetings of the San Joaquin Valley Regional Association of County Supervisors in 1993 and 1994 were held in Yosemite National Park, in the Sierra Bioregion, far from their constituents.

34. Transcribed from a tape recording of the semiannual meeting of the San Joaquin Valley Regional Association of County Supervisors, Rio Bravo Country Club, Bakersfield, California, March 24, 1994.

35. Interview with Nick Kinney, Kings County Supervisor; Rio Bravo Country Club, Bakersfield, California, March 24, 1994.

36. Letter from Roy Ashburn to the San Joaquin Valley Supervisors Association, September 19, 1994.

37. Resolution No. 94-547, as quoted in a letter from BLM District Manager Ron Fellows to Kern County Supervisor Roy Ashburn, undated (BLM reference #1120 CA-010.2). This three-page letter rebutted several passages in the resolution, including the quoted passage.

38. San Joaquin Valley Interagency Team, "Steps to Simplify, Standardize and Expedite CESA and FESA Compliance," January 7, 1994.

39. The Interagency Team believed that Kinney was sincere. At one Interagency Team meeting, Gradek stated that Kinney was amenable to their efforts to form a Consortium, but that Ashburn wanted to stall the process until the ESA could be repealed. (Author's notes, Interagency Team meeting, January 7, 1994.)

40. Interview with Larry Saslaw, Wildlife Biologist, Bakersfield District, U.S. Bureau of Land Management; Bakersfield, California, January 6, 1994.

Chapter 8

1. La Porte and Thomas (1995) similarly describe the motivating effects of another regulatory hammer: the ability of Nuclear Regulatory Commission inspectors to shut down nuclear power plants for noncompliance with federal safety standards. In the case of Diablo Canyon Nuclear Power Plant, this meant the potential loss of $2.4 million per day for each of its two reactor units. The threat of the regulatory hammer motivated plant managers to impose much more stringent standards on themselves than strictly required to ensure this revenue stream.

2. The NMFS has authority for marine species under the ESA. On the relative power of the FWS, see Clarke and McCool 1996.

3. For example, when the MOU on Biodiversity was signed, the Bush administration had temporarily halted the listing process. On December 15, 1992, the Bush administration settled a lawsuit with environmental groups, agreeing to list 400 species of plants and animals over the next four years, doubling the rate at which species had been listed since the ESA was enacted in 1973. See Schneider 1992: A1.

4. Interview #47.

5. Interview #77.

6. The BLM received expanded authority to exchange lands under the Federal Land Policy and Management Act (43 U.S.C.A. §1716).

7. Interview with Ed Hastey, California State Director, U.S. Bureau of Land Management; Sacramento, California, November 8, 1993.

8. Interview with Ron Stewart, Regional Forester, Pacific Southwest Region, U.S. Forest Service; San Francisco, March 16, 1994.

9. Interview #26.

10. Interview with Ron Stewart, Regional Forester, Pacific Southwest Region, U.S. Forest Service; San Francisco, March 16, 1994. A U.S. General Accounting Office (1997: 9) study found the FS spent $36,863, on average, to rotate line managers between 1991 and 1996—significantly higher than the average cost of moving line managers in the NPS ($15,156), FWS ($24,502), and BLM ($18,112).

11. Interview #2.

12. For additional findings on distance as a barrier to interagency cooperation, see Hoban 1987.

13. One BLM official in the California State Office, for example, covered his office with the trappings of a left-wing ecologist, including a prominently displayed copy of *Deep Ecology* (Devall and Sessions 1985) on his coffee table.

References

Abbey, Edward. 1968. *Desert Solitaire*. New York: Simon & Schuster.

Aberbach, Joel D. 1990. *Keeping a Watchful Eye: The Politics of Congressional Oversight*. Washington, DC: Brookings Institution.

Adler, Emanuel. 1992. "The Emergence of Cooperation: National Epistemic Communities and the International Evolution of the Idea of Nuclear Arms Control." *International Organization* 46: 101–145.

Agee, James K., and Darryll R. Johnson. 1988. *Ecosystem Management for Parks and Wilderness*. Seattle: University of Washington Press.

Agranoff, Robert. 1986. *Intergovernmental Management: Human Services Problem-Solving in Six Metropolitan Areas*. Albany: State University of New York Press.

Anderson, E. William, and Robert C. Baum. 1987. "Coordinated Resource Management Planning: Does It Work?" *Journal of Soil and Water Conservation* 42: 161–166.

Anderson, Richard L., Linda K. Spiegel, and Karyn M. Kakiba-Russell. 1991. *Southern San Joaquin Valley Ecosystems Protection Program: Natural Lands Inventory and Maps*. Sacramento: California Energy Commission, Energy Facilities Siting and Environmental Protection Division.

Andrews, Richard N. L. 1999. *Managing the Environment, Managing Ourselves: A History of American Environmental Policy*. New Haven, CT: Yale University Press.

Andruss, Van, Christopher Plant, Judith Plant, and Eleanor Wright, eds. 1990. *Home! A Bioregional Reader*. Philadelphia: New Society Publishers.

Argyris, Chris, and Donald A. Schön. 1978. *Organizational Learning: A Theory of Action Perspective*. Reading, MA: Addison-Wesley.

Axelrod, Robert M. 1984. *The Evolution of Cooperation*. New York: Basic Books.

Bakker, Elna. 1984. *An Island Called California: An Ecological Introduction to Its Natural Communities*. 2nd ed. Berkeley: University of California Press.

Bancroft, Ann. 1992. "'Summit' Lists Goals for Sierra: Ideas to Balance and Manage Economic, Environmental Interests." *San Francisco Chronicle*, July 30, p. A19.

Bardach, Eugene. 1996. "Turf Barriers to Interagency Collaboration." In Donald F. Kettl and H. Brinton Milward, eds., *The State of Public Management*. Baltimore: Johns Hopkins University Press.

Bardach, Eugene. 1998. *Getting Agencies to Work Together: The Practice and Theory of Managerial Craftsmanship*. Washington, DC: Brookings Institution.

Bardach, Eugene, and Robert A. Kagan. 1982. *Going by the Book: The Problem of Regulatory Unreasonableness*. Philadelphia: Temple University Press.

Barnard, Chester I. 1938. *The Functions of the Executive*. Cambridge, MA: Harvard University Press.

Bean, Michael J., Sarah G. Fitzgerald, and Michael A. O'Connell. 1991. *Reconciling Conflicts under the Endangered Species Act: The Habitat Conservation Planning Experience*. Washington, DC: World Wildlife Fund.

Beatley, Timothy. 1994. *Habitat Conservation Planning: Endangered Species and Urban Growth*. Austin: University of Texas Press.

Belcher, Elizabeth H., and J. Douglas Wellman. 1991. "Confronting the Challenge of Greenline Parks: Limits of the Traditional Administrative Approach." *Environmental Management* 15: 321–328.

Bell, Robert. 1985. "Professional Values and Organizational Decision Making." *Administration and Society* 17: 21–60.

Bella, David A. 1987. "Engineering and Erosion of Trust." *Journal of Professional Issues in Engineering* 113: 117–129.

Bendor, Jonathan. 1990. "Formal Models of Bureaucracy: A Review." In Naomi B. Lynn and Aaron Wildavsky, eds., *Public Administration: The State of the Discipline*. Chatham, NJ: Chatham House.

Bendor, Jonathan, Serge Taylor, and Roland Van Gaalen. 1987. "Stacking the Deck: Bureaucratic Missions and Policy Design." *American Political Science Review* 81: 873–896.

BioSystems Analysis, Inc. 1994. *Life on the Edge: A Guide to California's Endangered Natural Resources*. Santa Cruz, CA: BioSystems Books.

Blackburn, Dan. 1992. "Saving the Sierra: Master Plan for the Range of Light." *California Journal*, February.

Blais, André, and Stéphane Dion, eds. 1991. *The Budget-Maximizing Bureaucrat: Appraisals and Evidence*. Pittsburgh: University of Pittsburgh Press.

Bonnett, Mark, and Kurt Zimmerman. 1991. "Politics and Preservation: The Endangered Species Act and the Northern Spotted Owl." *Ecology Law Quarterly* 18: 105–171.

Botkin, Daniel B. 1990. *Discordant Harmonies: A New Ecology for the Twenty-First Century*. New York: Oxford University Press.

Bowman, Chris. 1991a. "Mountain of Data Could Help Shape Range's Future." *Sacramento Bee*, November 19.

Bowman, Chris. 1991b. "New Strategy on Conservation." *Sacramento Bee*, September 9.

Bowman, Chris, 1994. "New U.S. Timber Plan Eases North State Limits." *Sacramento Bee*, February 24, p. A1.

Bradley, Dorotha M., and Helen M. Ingram. 1986. "Science vs. the Grass Roots: Representation in the Bureau of Land Management." *Natural Resources Journal* 26: 493–518.

Brick, Philip, Donald Snow, and Sarah Van de Wetering, eds. 2001. *Across the Great Divide: Explorations in Collaborative Conservation and the American West.* Washington, DC: Island Press.

Brooks, Lea. 1994. "Endangered Rat Frustrates Riverside County." *California County*, July/August, pp. 10–13, 20.

Brown, Greg, and Charles C. Harris. 1992a. "The United States Forest Service: Changing of the Guard." *Natural Resources Journal* 32: 449–466.

Brown, Greg, and Charles C. Harris. 1992b. "The U.S. Forest Service: Toward the New Resource Management Paradigm?" *Society and Natural Resources* 5: 231–245.

Brown, Greg, and Charles C. Harris. 1993. "The Implications of Work Force Diversification in the U.S. Forest Service." *Administration and Society* 25: 85–113.

Bureau of Land Management. 1992. "Profile of Fish and Wildlife Resources Administered by Bureau of Land Management California." Sacramento, CA: Bureau of Land Management.

Calef, Wesley. 1960. *Private Grazing and Public Lands.* Chicago: University of Chicago Press.

California Office of State Controller. 1994. *1992–1993 Annual Report of Financial Transactions Concerning Cities of California.* Sacramento: Office of State Controller.

California Office of State Controller. 1993. *1991–92 Annual Report of Financial Transactions Concerning Special Districts of California.* Sacramento: Office of State Controller.

California State Assembly Office of Research. 1991. *California 2000: Biological Ghettos*. Sacramento, CA: Assembly Office of Research.

Callicott, J. Baird. 1989. *In Defense of the Land Ethic: Essays in Environmental Philosophy.* Albany: State University of New York Press.

Callicott, J. Baird, ed. 1987. *Companion to A Sand County Almanac: Interpretive and Critical Essays*. Madison: University of Wisconsin Press.

Cawley, R. McGreggor, and Sally K. Fairfax. 1991. "Land and Natural Resource Policy, I: Development and Current Status." In Clive S. Thomas, ed., *Politics and Public Policy in the Contemporary American West.* Albuquerque: University of New Mexico Press.

Chase, Alston. 1987. *Playing God in Yellowstone: The Destruction of America's First National Park.* San Diego: Harcourt, Brace, Jovanovich.

Chisholm, Donald. 1989. *Coordination without Hierarchy: Informal Structures in Multiorganizational Systems.* Berkeley: University of California Press.

Clark, Tim W. 1997. *Averting Extinction: Reconstructing Endangered Species Recovery.* New Haven, CT: Yale University Press.

Clark, Tim W., Richard P. Reading, and Alice L. Clarke, eds. 1994. *Endangered Species Recovery: Finding the Lessons, Improving the Process.* Washington, DC: Island Press.

Clark, Tim W., Andrew R. Willard, and Christina M. Cromley, eds. 2000. *Foundations of Natural Resources Policy and Management.* New Haven, CT: Yale University Press.

Clarke, Jeanne Nienaber, and Daniel McCool. 1996. *Staking out the Terrain: Power and Performance among Natural Resource Agencies.* Albany: State University of New York Press.

Cochrane, Susan A. 1986. *Programs for the Preservation of Natural Diversity in California.* Sacramento: California Department of Fish and Game.

Coggins, George Cameron, Charles F. Wilkinson, and John D. Leshy. 1993. *Federal Public Land and Resources Law*, 3rd ed. Westbury, NY: Foundation Press.

Coordinated Resource Management and Planning. 1990. *Handbook.* Sacramento, CA: Coordinated Resource Management and Planning Technical Advisory Council.

Cortner, Hanna J., and Margaret A. Moote. 1994. "Trends and Issues in Land and Water Resources Management: Setting the Agenda for Change." *Environmental Management* 18: 167–173.

Cowart, Richard H., and Sally K. Fairfax. 1988. "Public Lands Federalism: Judicial Theory and Administrative Reality." *Ecology Law Quarterly* 15: 375–476.

Culhane, Paul J. 1981. *Public Land Politics: Interest Group Influence on the Forest Service and the Bureau of Land Management.* Baltimore: Johns Hopkins University Press for Resources for the Future.

Curtin, Charles G. 1993. "The Evolution of the U.S. National Wildlife Refuge System and the Doctrine of Compatibility." *Conservation Biology* 7: 29–38.

Cushman, John H., Jr. 1994. "Judge Drops Logging Ban in Northwest." *New York Times*, June 7, p. 17.

Czech, Brian, and Paul R. Krausman. 2001. *The Endangered Species Act: History, Conservation Biology, and Public Policy.* Baltimore: Johns Hopkins University Press.

Dana, Samuel Trask, and Sally K. Fairfax. 1980. *Forest and Range Policy: Its Development in the United States.* 2nd ed. New York: McGraw-Hill.

Dasmann, Raymond F. 1965. *The Destruction of California.* New York: Macmillan.

Derber, Charles, William A. Schwartz, and Yale Magrass. 1990. *Power in the Highest Degree: Professionals and the Rise of a New Mandarin Order.* New York: Oxford University Press.

Devall, Bill, and George Sessions. 1985. *Deep Ecology: Living as if Nature Mattered.* Salt Lake City: Peregrine Smith Books.

Dexter, Lewis Anthony. 1970. *Elite and Specialized Interviewing.* Evanston, IL: Northwestern University Press.

Dietrich, William. 1992. *The Final Forest: The Battle for the Last Great Trees of the Pacific Northwest.* New York: Penguin Books.

Dimock, Marshall E. 1952. "Expanding Jurisdictions: A Case Study in Bureaucratic Conflict." In Robert K. Merton, Alisa P. Gray, Barbara Hockey, and Hanan C. Selvin, eds., *Reader in Bureaucracy.* Glencoe, IL: Free Press.

Doremus, Holly. 1991. "Patching the Ark: Improving Legal Protection of Biological Diversity." *Ecology Law Quarterly* 18: 265–333.

Dottavio, F. Dominic, Peter F. Brussard, and John D. McCrone. 1990. *Protecting Biological Diversity in the National Parks: Workshop Recommendations.* Washington, DC: National Park Service.

Downs, Anthony. 1967. *Inside Bureaucracy.* Boston: Little, Brown.

Duane, Timothy P. 1998. *Shaping the Sierra: Nature, Culture, and Conflict in the Changing West.* Berkeley: University of California Press.

Dunlap, Thomas R. 1988. *Saving America's Wildlife: Ecology and the American Mind, 1850–1990.* Princeton, NJ: Princeton University Press.

Durant, Robert F. 1985. *When Government Regulates Itself: EPA, TVA, and Pollution Control in the 1970s.* Knoxville: University of Tennessee Press.

Durant, Robert F. 1992. *The Administrative Presidency Revisited: Public Lands, the BLM, and the Reagan Revolution.* Albany: State University of New York Press.

Durden, Garey. 1991. "Determining the Classics in Social Choice." *Public Choice* 69: 265–277.

Ebbin, Marc J. 1997. "Is the Southern California Approach to Conservation Succeeding?" *Ecology Law Quarterly* 24: 695–706.

Eckersley, Robyn. 1992. *Environmentalism and Political Theory: Toward an Ecocentric Approach.* Albany: State University of New York Press.

Ehrenfeld, David. 1978. *The Arrogance of Humanism.* New York: Oxford University Press.

Ellison, Brian A. 1995. "Autonomy in Action: Bureaucratic Competition among Functional Rivals in Denver Water Politics." *Policy Studies Review* 14: 25–48.

Emerson, Richard. 1962. "Power-Dependence Relations." *American Sociological Review* 27: 31–40.

Emmerich, Herbert. 1971. *Federal Organization and Administrative Management.* University, AL: University of Alabama Press.

Fairfax, Sally K. 1984. "Beyond the Sagebrush Rebellion: The BLM as Neighbor and Manager in the Western States." In John G. Francis and Richard Ganzel, eds., *Western Public Lands: The Management of Natural Resources in a Time of Declining Federalism.* Totowa, NJ: Rowman and Allanheld.

Fiorina, Morris P. 1989. *Congress: Keystone of the Washington Establishment.* 2nd ed. New Haven, CT: Yale University Press.

Foresta, Ronald A. 1984. *America's National Parks and Their Keepers.* Baltimore: Johns Hopkins University Press for Resources for the Future.

Fortmann, Louise. 1990. "The Role of Professional Norms and Beliefs in the Agency-Client Relations of Natural Resource Bureaucracies." *Natural Resources Journal* 30: 361–380.

Foss, Phillip O. 1960. *Politics and Grass.* Seattle: University of Washington Press.

Frammolino, Ralph. 1990. "Observers Say Irvine Co. Could Land Direct Line to Wilson." *Los Angeles Times*, Orange County Edition, November 3, p. B1.

Frammolino, Ralph, and Maria Newman. 1991. "Songbird Rouses a Chorus of Discord." *Los Angeles Times*, June 2, p. A3.

Freemuth, John C. 1991. *Islands under Siege: National Parks and the Politics of External Threats.* Lawrence: University of Kansas Press.

Freidson, Eliot. 1970. "Dominant Professions, Bureaucracy, and Client Services." In William R. Rosengren and Mark Lefton, eds., *Organizations and Clients: Essays in the Sociology of Service.* Columbus, OH: Merrill.

Galaskiewicz, Joseph. 1985. "Professional Networks and the Institutionalization of a Single Mind Set." *American Sociological Review* 50: 639–658.

Galaskiewicz, Joseph, and Ronald S. Burt. 1991. "Interorganization Contagion in Corporate Philanthropy." *Administrative Science Quarterly* 36: 88–105.

Gilpin, Michael E., and Michael E. Soulé. 1986. "Minimum Viable Populations: Processes of Species Extinction." In Michael E. Soulé, ed., *Conservation Biology: The Science of Scarcity and Diversity.* Sunderland, MA: Sinauer Associates.

Goggin, Malcolm L., Ann Bowman, James Lester, and Laurence O'Toole. 1990. *Implementation Theory and Practice: Toward a Third Generation.* Glenview, IL: Scott Foresman/Little Brown.

Golden, Marissa Martino. 2000. *What Motivates Bureaucrats? Politics and Administration during the Reagan Years.* New York: Columbia University Press.

Golley, Frank Benjamin. 1993. *A History of the Ecosystem Concept in Ecology: More Than the Sum of the Parts.* New Haven: Yale University Press.

Gorman, Tom. 1989. "The Pendleton Preserve: Marines and Environmentalists Are Unlikely Allies in Protecting the Largest Undeveloped Chunk of Southern California Coast." *Los Angeles Times*, December 10, p. 34.

Gray, Barbara. 1989. *Collaborating: Finding Common Ground for Multiparty Problems.* San Francisco: Jossey-Bass.

Griggs, F. Thomas. 1992. "The Remaining Biological Diversity of the San Joaquin Valley, California." In Daniel F. Williams, Sheila Byrne, and Theodore

A. Rado, eds., *Endangered and Sensitive Species of the San Joaquin Valley, California: Their Biology, Management, and Conservation.* Sacramento: California Energy Commission.

Gruber, Judith E. 1994. "Coordinating Growth Management through Consensus-Building: Incentives and the Generation of Social, Intellectual, and Political Capital." Berkeley, CA: Institute of Urban and Regional Development, Working Paper 617.

Grumbine, Edward. 1991. "Cooperation or Conflict?: Interagency Relationships and the Future of Biodiversity for U.S. Parks and Forests." *Environmental Management* 15: 27–37.

Grumbine, Edward. 1992. *Ghost Bears: Exploring the Biodiversity Crisis.* Washington, DC: Island Press.

Grumbine, Edward. 1994. "What Is Ecosystem Management?" *Conservation Biology* 8: 27–38.

Gulick, Luther. 1938. "Notes on the Theory of Organization." In Luther Gulick and L. Urwick, eds., *Papers on the Science of Administration.* Concord, NH: Rumford Press.

Gunnison, Robert B., and Greg Lucas. 1995. "Wilson Tries for All New Species Act: Environmentalists Livid at Proposal." *San Francisco Chronicle*, March 25, p. A1.

Haas, Ernst B. 1990. *When Knowledge Is Power: Three Models of Change in International Organizations*. Berkeley: University of California Press.

Haas, Peter M. 1990. *Saving the Mediterranean: The Politics of International Environmental Cooperation.* New York: Columbia University Press.

Haas, Peter M. 1992. "Introduction: Epistemic Communities and International Policy Coordination." *International Organization* 46: 1–35.

Hagstrom, Jerry. 1994. "Ecomanaging the Forest." *Government Executive*, June.

Hall, Douglas T., Benjamin Schneider, and Harold T. Nygren. 1970. "Personal Factors in Organizational Identification." *Administrative Science Quarterly* 15: 176–190.

Halperin, Morton H. 1974. *Bureaucratic Politics and Foreign Policy.* Washington, DC: Brookings Institution.

Hamilton, Michael S. 1990. "Introduction." In Michael S. Hamilton, ed., *Regulatory Federalism, Natural Resources, and Environmental Management.* Washington, DC: American Society for Public Administration.

Hammond, Thomas H. 1979. *Jurisdictional Preferences and the Choice of Tasks: Political Adaptation by Two State Wildlife Agencies*. Unpublished doctoral dissertation. Berkeley: University of California.

Harris, Larry D. 1984. *The Fragmented Forest: Island Biogeography Theory and the Preservation of Biotic Diversity.* Chicago: University of Chicago Press.

Hasenfeld, Yeheskel, and Thomas Brock. 1991. "Implementation of Social Policy Revisited." *Administration and Society* 22: 451–479.

Hastey, Ed. 1990. *Is There a Role for the Timberland Task Force in Developing a Coordinated Forest Management Strategy for the Klamath Province?: A Status Report to the Timberland Task Force from the Ad Hoc Committee*. Sacramento, CA: Ad Hoc Committee to the Timberland Task Force.

Hays, Samuel P. 1959. *Conservation and the Gospel of Efficiency: The Progressive Conservation Movement, 1890–1920*. Cambridge, MA: Harvard University Press.

Heal, Geoffrey. 2001. *Nature and the Marketplace: Capturing the Value of Ecosystem Services*. Washington, DC: Island Press.

Heclo, Hugh. 1977. *A Government of Strangers: Executive Politics in Washington*. Washington, DC: Brookings Institution.

Hess, Karl, Jr. 1993. *Rocky Times in Rocky Mountain National Park: An Unnatural History*. Niwot, CO: University Press of Colorado.

Himmelspach, Darlene. 1994. "Marines Gird to Guard the Environment: Preserving the Habitat a Key Pendleton Mission." *San Diego Union-Tribune*, April 22, p. A-1.

Hoban, Thomas J. 1987. "Barriers to Interagency Cooperation." *Journal of Applied Sociology* 4: 13–29.

Hodges, Donald G., and Robert F. Durant. 1989. "The Professional State Revisited: Twixt Scylla and Charybdis?" *Public Administration Review* 49: 474–485.

Holden, Matthew. 1966. "'Imperialism' in Bureaucracy." *American Political Science Review* 60: 943–951.

Hood, Laura C. 1998. *Frayed Safety Nets: Conservation Planning under the Endangered Species Act*. Washington, DC: Defenders of Wildlife.

Hoshovsky, Marc. 1990. *Sites Important to California's Natural Diversity*. Administrative Report 90-3. Sacramento: California Department of Fish and Game, Lands and Natural Areas Program.

Hoshovsky, Marc. 1992. "Developing Partnerships in Conserving California's Biological Diversity." *Fremontia* 20: 19–23.

Ingram, Helen. 1990. "Implementation: A Review and Suggested Framework." In Naomi B. Lynn and Aaron Wildavsky, eds., *Public Administration: The State of the Discipline*. Chatham, NJ: Chatham House.

Innes, Judith, Judith Gruber, Michael Neuman, and Robert Thompson. 1994. *Coordinating Growth and Environmental Management through Consensus Building*. Berkeley: California Policy Seminar, University of California.

Jacobsen, John Kurt. 1995. "Much Ado about Ideas: The Cognitive Factor in Economic Policy." *World Politics* 47: 283–310.

Jasny, Michael, Joel Reynolds, and Ann Notthoff. 1997. *Leap of Faith: Southern California's Experiment in Natural Community Conservation Planning*. Washington, DC: Natural Resources Defense Council.

Jenkins, Robert E., Jr. 1988. "Information Management for the Conservation of Biodiversity." In E. O. Wilson, ed., *Biodiversity.* Washington, DC: National Academy Press.

Jensen, Deborah B. 1994. "Conservation through Coordination: California's Experiment in Bioregional Councils." In R. Edward Grumbine, ed., *Environmental Policy and Biodiversity.* Washington, DC: Island Press.

Jensen, Deborah B., Margaret Torn, and John Harte. 1993. *In Our Own Hands: A Strategy for Conserving California's Biological Diversity.* Berkeley: University of California Press.

John, DeWitt. 1994. *Civic Environmentalism: Alternatives to Regulation in States and Communities.* Washington, DC: Congressional Quarterly.

Johnson, Ted. 1991. "Rats! Endangered Rodents Get the Jump on Developers in Norco." *Los Angeles Times*, Orange Country Edition, February 10, p. A3.

Jones & Stokes Associates. 1992. *Reassembling the Pieces: A Strategy for Maintaining Biological Diversity in California.* Sacramento, CA: Jones & Stokes Associates.

Kareiva, Peter, Sandy Andelman, Daniel Doak, Bret Elderd, Martha Groom, Jonathan Hoekstra, Laura Hood, Frances James, John Lamoreux, Gretchen LeBuhn, Charles McCulloch, James Regetz, Lisa Savage, Mary Ruckelshaus, David Skelly, Henry Wilbur, Kelly Zamudio. 1999. *Using Science in Habitat Conservation Plans.* Washington, DC: American Institute of Biological Sciences.

Kaufman, Herbert. 1956. "Emerging Conflicts in the Doctrines of Public Administration." *American Political Science Review* 50: 1057–1073.

Kaufman, Herbert. 1960. *The Forest Ranger.* Baltimore: Johns Hopkins University Press for Resources for the Future.

Kelman, Steven. 1992. "Adversary and Cooperationist Institutions for Conflict Resolution in Public Policymaking." *Journal of Policy Analysis and Management* 11: 178–206.

Kemmis, Daniel. 1990. *Community and the Politics of Place.* Norman: University of Oklahoma Press.

Kenney, Douglas S. 1997. *Resource Management at the Watershed Level: An Assessment of the Changing Federal Role in the Emerging Era of Community-Based Watershed Management.* Boulder: Natural Resources Law Center, University of Colorado Law School.

Kettl, Donald F. 1993. "Public Administration: The State of the Field." In Ada W. Finifter, ed., *Political Science: The State of the Discipline.* Washington, DC: American Political Science Association.

Keystone Center. 1991. *Biological Diversity on Federal Lands: Report of a Keystone Policy Dialogue.* Keystone, CO: Keystone Center.

King, Gary, Robert O. Keohane, and Sidney Verba. 1994. *Designing Social Inquiry: Scientific Inference in Qualitative Research.* Princeton: Princeton University Press.

Kingdon, John W. 1984. *Agendas, Alternatives, and Public Policies*. New York: HarperCollins.

Knight, Richard L., and Peter B. Landres, eds. 1998. *Stewardship across Boundaries*. Washington, DC: Island Press.

Knott, Jack H., and Gary J. Miller. 1987. *Reforming Bureaucracy: The Politics of Institutional Choice*. Englewood Cliffs, NJ: Prentice-Hall.

Knudson, Tom. 1991. "Hopes for Reform Improve." *Sacramento Bee*, June 13, p. A1.

Kreissman, Bern. 1991. *California: An Environmental Atlas & Guide*. Davis, CA: Bear Klaw Press.

Kuhn, Thomas S. 1970. *The Structure of Scientific Revolutions*. 2nd ed. Chicago: University of Chicago Press.

Kunioka, Todd, and Lawrence S. Rothenberg. 1993. "The Politics of Bureaucratic Competition: The Case of Natural Resource Policy." *Journal of Policy Analysis and Management* 12: 700–725.

Lacy, Robert C. 1992. "The Effects of Inbreeding on Isolated Populations: Are Minimum Viable Population Sizes Predictable?" In Peggy L. Fiedler and Subodh K. Jain, eds., *Conservation Biology: The Theory and Practice of Nature Conservation, Preservation, and Management*. New York: Chapman and Hall.

La Ganga, Maria L. 1993. "Panel Takes Desert Squirrel off State's Endangered List." *Los Angeles Times*, May 15, p. A1.

Landau, Martin. 1991. "On Multiorganizational Systems in Public Administration." *Journal of Public Administration Research and Theory* 1: 5–18.

La Porte, Todd R. 1994. "A State of the Field: Increasing Relative Ignorance." *Journal of Public Administration Research and Theory* 4: 5–15.

La Porte, Todd R., and Paula M. Consolini. 1991. "Working in Practice But Not in Theory: Theoretical Challenges of 'High-Reliability Organizations.'" *Journal of Public Administration Research and Theory* 1: 19–47.

La Porte, Todd R., and Craig W. Thomas. 1995. "Regulatory Compliance and the Ethos of Quality Enhancement: Surprises in Nuclear Power Plant Operations." *Journal of Public Administration Research and Theory* 5: 109–137.

Larson, Andrea. 1992. "Network Dyads in Entrepreneurial Settings: A Study of the Governance of Exchange Relationships." *Administrative Science Quarterly* 37: 76–104.

Lax, David A., and James K. Sebenius. 1986. *The Manager as Negotiator: Bargaining for Cooperation and Competitive Gain*. New York: Free Press.

Lebovic, James H. 1994. "Riding Waves or Making Waves? The Services and the U.S. Defense Budget, 1981–1993." *American Political Science Review* 88: 839–852.

Leopold, Aldo. 1933. *Game Management*. New York: Scribner.

Leopold, Aldo. 1949. *A Sand County Almanac*. London: Oxford University Press.

Leopold, Luna. 1953. *Round River: From the Journals of Aldo Leopold*. New York: Oxford University Press.

Lipschutz, Ronnie D. 1996. *Global Civil Society and Global Environmental Governance: The Politics of Nature from Place to Planet*. Albany: SUNY Press.

Lipsky, Michael. 1980. *Street-Level Bureaucracy*. New York: Russell Sage Foundation.

Lowry, William R. 1994. *The Capacity for Wonder: Preserving National Parks*. Washington, DC: Brookings Institution.

Maass, Arthur. 1951. *Muddy Waters: The Army Engineers and the Nation's Rivers*. Cambridge, MA: Harvard University Press.

Manson, Craig. 1994. "Natural Communities Conservation Planning: California's New Ecosystem Approach to Biodiversity." *Environmental Law* 24: 603–615.

March, James G., and Johan P. Olsen. 1983. "Organizing Political Life: What Administrative Reorganization Tells Us about Government." *American Political Science Review* 77: 281–296.

Marsh, Lindell L., and Robert D. Thornton. 1987. "San Bruno Mountain Habitat Conservation Plan." In David J. Brower and Daniel S. Carol, eds., *Managing Land-Use Conflicts: Case Studies in Special Area Management*. Durham, NC: Duke University Press.

Mayhew, David R. 1974. *Congress: The Electoral Connection*. New Haven: Yale University Press.

McCaull, John. 1994. "The Natural Community Conservation Planning Program and the Coastal Sage Scrub Ecosystem of Southern California." In R. Edward Grumbine, ed., *Environmental Policy and Biodiversity*. Washington, DC: Island Press.

McConnell, Grant. 1966. *Private Power and American Democracy*. New York: Knopf.

McGinnis, Michael Vincent, ed. 1999. *Bioregionalism*. New York: Routledge.

McHugh, Paul. 1994. "Endangered Species Bill Blasted: Conservationists Say Plan Leaves Wildlife at Mercy of Developers." *San Francisco Chronicle*, April 21, p. A19.

McHugh, Paul. 1995. "Wilson Order on Wildlife Is Long-Term: Protections Eased through Year 2000." *San Francisco Chronicle*, March 22, p. A1.

Mendeloff, John. 1979. *Regulating Safety: An Economic and Political Analysis of Occupational Safety and Health Policy*. Cambridge, MA: MIT Press.

Moe, Terry M. 1989. "The Politics of Bureaucratic Structure." In John E. Chubb and Paul E. Peterson, eds., *Can the Government Govern?* Washington, DC: Brookings Institution.

Moore, Mark H. 1995. *Creating Public Value: Strategic Management in Government*. Cambridge, MA: Harvard University Press.

Morris, Sally. 1994. "Help Planned for South Fork." *Trinity Journal*, June 1.

Morrison, Michael L., Bruce G. Marcot, and R. William Mannan. 1992. *Wildlife-Habitat Relationships: Concepts and Applications*. Madison: University of Wisconsin Press.

Naess, Arne. 1973. "The Shallow and the Deep, Long-Range Ecology Movements: A Summary." *Inquiry* 16: 95–100.

Nash, Roderick Frazier. 1989. *The Rights of Nature: A History of Environmental Ethics*. Madison: University of Wisconsin Press.

National Biological Service. 1995. *Our Living Resources: A Report to the Nation on the Distribution, Abundance, and Health of U.S. Plants, Animals, and Ecosystems*. Washington, DC: U.S. Government Printing Office.

National Park Service. 1980. *State of the Parks 1980: A Report to the Congress*. Washington, DC: Office of Science and Technology, National Park Service.

National Research Council. 1992. *Science and the National Parks*. Washington, DC: National Academy Press.

National Research Council. 1995. *Science and the Endangered Species Act*. Washington, DC: National Academy Press.

Natural Resources Law Center. 1996. *The Watershed Source Book: Watershed-Based Solutions to Natural Resource Problems*. Boulder: University of Colorado Law School.

Nechodom, Mark. 1994. "The Slow Boring of Hard Boards." Typescript. Santa Cruz: University of California, Santa Cruz.

Needleman, Martin L., and Carolyn Emerson Needleman. 1974. *Guerrillas in the Bureaucracy: The Community Planning Experiment in the United States*. New York: Wiley.

Newmark, William D. 1985. "Legal and Biotic Boundaries of Western North American National Parks: A Problem of Congruence." *Biological Conservation* 33: 197–208.

Niskanen, William A., Jr. 1971. *Bureaucracy and Representative Government*. Chicago: Aldine-Atherton.

Norse, Elliott A. 1990. *Ancient Forests of the Pacific Northwest*. Washington, DC: Island Press.

Norton, Bryan G. 1987. *Why Preserve Natural Variety?* Princeton: Princeton University Press.

Norton, Bryan G., ed. 1986. *The Preservation of Species*. Princeton: Princeton University Press.

Noss, Reed F., and Allen Y. Cooperrider. 1994. *Saving Nature's Legacy: Protecting and Restoring Biodiversity*. Washington, DC: Island Press.

Noss, Reed F., and Larry D. Harris. 1986. "Nodes, Networks, and MUMs: Preserving Diversity at All Scales." *Environmental Management* 10: 299–309.

Noss, Reed F., Edward T. LaRoe III, and J. Michael Scott. 1995. *Endangered Ecosystems of the United States: A Preliminary Assessment of Loss and Degradation*. Washington, DC: National Biological Service, U.S. Department of the Interior.

Noss, Reed F., Michael A. O'Connell, and Dennis D. Murphy. 1997. *The Science of Conservation Planning: Habitat Conservation under the Endangered Species Act*. Washington, DC: Island Press.

Ostrom, Elinor. 1990. *Governing the Commons: The Evolution of Institutions for Collective Action*. New York: Cambridge University Press.

O'Toole, Randal. 1988. *Reforming the Forest Service*. Washington, DC: Island Press.

Ouchi, William C. 1980. "Markets, Bureaucracies, and Clans." *Administrative Science Quarterly* 25: 129–141.

Pearsall, Sam. 1984. "Multi-Agency Planning for Natural Areas in Tennessee." *Public Administration Review* 44: 43–48.

Pfeffer, Jeffrey, and Gerald R. Salancik. 1978. *The External Control of Organizations: A Resource Dependence Perspective*. New York: Harper and Row.

Pickett, Steward T. A., V. Thomas Parker, and Peggy L. Fiedler. 1992. "The New Paradigm in Ecology: Implications for Conservation Biology above the Species Level." In Peggy L. Fiedler and Subodh K. Jain, eds., *Conservation Biology: The Theory and Practice of Nature Conservation, Preservation, and Management*. New York: Chapman and Hall.

Pinchot, Gifford. 1910. *The Fight for Conservation*. Seattle: University of Washington Press.

Pister, Edwin P. 1987. "A Pilgrim's Progress from Group A to Group B." In J. Baird Callicott, ed., *Companion to A Sand County Almanac*. Madison: University of Wisconsin Press.

Porter, Douglas R. 1995. Southern California's Multispecies Planning. In Douglas R. Porter and David A. Salvesen, eds., *Collaborative Planning for Wetlands and Wildlife: Issues and Examples*. Washington, DC: Island Press.

Pressman, Jeffrey L., and Aaron Wildavsky. 1984. *Implementation*. 3rd ed. Berkeley: University of California Press.

Putnam, Robert D. 1993. *Making Democracy Work: Civic Traditions in Modern Italy*. Princeton, NJ: Princeton University Press.

Quammen, David. 1996. *The Song of the Dodo: Island Biogeography in an Age of Extinctions*. New York: Scribner.

Raphael, Ray. 1994. *More Tree Talk: The People, Politics, and Economics of Timber*. Washington, DC: Island Press.

Reisner, Marc. 1993. *Cadillac Desert: The American West and Its Disappearing Water*. 2nd ed. New York: Penguin Books.

Ricketts, Taylor H., Eric Dinerstein, David M. Olson, Colby J. Loucks, William Eichbaum, Dominick DellaSala, Kevin Kavanagh, Prashant Hedao, Patrick T. Hurley, Karen M. Carney, Robin Abell, and Steven Walters. 1999. *Terrestrial Ecoregions of North America: A Conservation Assessment*. Washington, DC: Island Press.

Robinson, W. W. 1948. *Land in California: The Story of Mission Lands, Ranchos, Squatters, Mining Claims, Railroad Grants, Land Scrip, Homesteads.* Berkeley: University of California Press.

Rohlf, Daniel J. 1989. *The Endangered Species Act: A Guide to Its Protections and Implementation.* Stanford, CA: Stanford Environmental Law Society.

Rohlf, Daniel J. 1991. "Six Biological Reasons Why the Endangered Species Act Doesn't Work—And What to Do about It." *Conservation Biology* 5: 273–282.

Rose, Ursula. 1992. "Leslie Fears Sierra Summit." *Sacramento Bee*, June 11, p. N1.

Rourke, Francis E. 1969. *Bureaucracy, Politics, and Public Policy.* Boston: Little, Brown.

Runte, Alfred. 1987. *Our National Parks: The American Experience.* 2nd ed. Lincoln: University of Nebraska Press.

Sabatier, Paul A., John Loomis, and Catherine McCarthy. 1995. "Hierarchical Controls, Professional Norms, Local Constituencies, and Budget Maximization: An Analysis of U.S. Forest Service Planning Decisions." *American Journal of Political Science* 39: 204–242.

Sahagun, Louis, and Mark A. Stein. 1989. "Managing the Mojave: A Struggle for the BLM." *Los Angeles Times*, May 24, p. A1.

Salwasser, Hal. 1988. "Managing Ecosystems for Viable Populations of Vertebrates: A Focus for Biodiversity." In James K. Agee and Darryll R. Johnson, eds., *Ecosystem Management for Parks and Wilderness.* Seattle: University of Washington Press.

Salwasser, Hal, Christine Schonewald-Cox, and Richard Baker. 1987. "The Role of Interagency Cooperation in Managing for Viable Populations." In Michael E. Soulé, ed., *Viable Populations for Conservation.* New York: Cambridge University Press.

Sax, Joseph L., and Robert B. Keiter. 1987. "Glacier National Park and Its Neighbors: A Study of Federal Interagency Relations." *Ecology Law Quarterly* 14: 207–263.

Schiff, Ashley L. 1962. *Fire and Water: Scientific Heresy in the Forest Service.* Cambridge, MA: Harvard University Press.

Schiff, Ashley L. 1966. "Innovation and Administrative Decision Making: The Conservation of Land Resources." *Administrative Sciences Quarterly* 11: 1–30.

Schneider, Keith. 1992. "U.S. to Speed Up Saving of Species." *New York Times*, December 16, pp. A1, A13.

Schoenherr, Allan A. 1992. *A Natural History of California.* Berkeley: University of California Press.

Schonewald-Cox, Christine, and Marybeth Buechner. 1992. "Park Protection and Public Roads." In Peggy L. Fiedler and Subodh K. Jain, eds., *Conservation Biology: The Theory and Practice of Nature Conservation, Preservation, and Management.* New York: Chapman and Hall.

Schweik, Charles, and Craig W. Thomas. 2002. "Using Remote Sensing to Evaluate Environmental Institutional Designs: A Habitat Conservation Planning Example." *Social Science Quarterly* 83: 244–262.

Scicchitano, Michael J., and David M. Hedge. 1993. "From Coercion to Partnership in Federal Partial Preemption: SMCRA, RCRA, and OSH Act." *Publius* 23: 107–121.

Seidman, Harold, and Robert Gilmour. 1986. *Politics, Position, and Power: From the Positive to the Regulatory State*. 4th ed. New York: Oxford University Press.

Seligsohn-Bennett, Kyla. 1990. "Mismanaging Endangered and 'Exotic' Species in the National Parks." *Environmental Law* 20: 415–440.

Sellars, Richard West. 1997. *Preserving Nature in the National Parks: A History*. New Haven, CT: Yale University Press.

Selznick, Phillip. 1949. *TVA and the Grassroots: A Study of Politics and Organization*. Berkeley: University of California Press.

Serfis, Jim. 1991. "Pesticide Regulation." In Kathryn A. Kohm, ed., *Balancing on the Brink of Extinction: The Endangered Species Act and Lessons for the Future*. Washington, DC: Island Press.

Sheppard, George. 1992. "Endangered Species Management and Oil and Gas Development in a Multiple-use Agency." In Daniel F. Williams, Sheila Byrne, and Theodore A. Rado, eds., *Endangered and Sensitive Species of the San Joaquin Valley, California: Their Biology, Management, and Conservation*. Sacramento: California Energy Commission.

Sher, Victor M. 1993. "Travels with Strix: The Spotted Owl's Journey through the Federal Courts." *Public Land Law Review* 14: 41–79.

Sigelman, Lee. 1986. "The Bureaucrat as Budget Maximizer: An Assumption Examined." *Public Budgeting and Finance* 6: 50–59.

Simon, Herbert A. 1976. *Administrative Behavior*. 3rd ed. New York: Free Press.

Smith, Gordon. 1994. "At Loggerheads, But They Try: An Unlikely Pair Work to Save a Lumber Town." *San Diego Union-Tribune*, March 1, p. A1.

Soulé, Michael E. 1985. "What Is Conservation Biology?" *BioScience* 35: 727–734.

Soulé, Michael E. 1987. *Viable Populations for Conservation*. New York: Cambridge University Press.

Soulé, Michael E., ed. 1986. *Conservation Biology: The Science of Scarcity and Diversity*. Sunderland, MA: Sinauer Associates.

Soulé, Michael E., and John Terborgh. 1999. *Continental Conservation: Scientific Foundations of Regional Reserve Networks*. Washington, DC: Island Press.

"State, Federal Land Managers to Cooperate in Regional Goals." 1991. *San Diego Union-Tribune*, September 21, p. A32.

"State, Federal Officials Announce New Conservation Strategy." 1991. Distributed by United Press International, September 19.

Stegner, Wallace. 1954. *Beyond the Hundredth Meridian: John Wesley Powell and the Second Opening of the West*. New York: Penguin Books.

Steinbruner, John D. 1974. *The Cybernetic Theory of Decision*. Princeton, NJ: Princeton University Press.

Stevens, William K. 1996. "Wildlife Finds Odd Sanctuary on Military Bases." *New York Times*, January 2, p. B5.

Stoker, Robert P. 1991. *Reluctant Partners: Implementing Federal Policy*. Pittsburgh: University of Pittsburgh Press.

Susskind, Lawrence E. 1994. *Environmental Diplomacy: Negotiating More Effective Global Agreements*. New York: Oxford University Press.

Takacs, David. 1996. *The Idea of Biodiversity: Philosophies of Paradise*. Baltimore: Johns Hopkins University Press.

Taylor, Frederick W. 1911. *The Principles of Scientific Management*. New York: Harper.

Taylor, Serge. 1984. *Making Bureaucracies Think: The Environmental Impact Strategy of Administrative Reform*. Stanford, CA: Stanford University Press.

Thomas, Craig W. 1993. "Reorganizing Public Organizations: Alternatives, Objectives, and Evidence." *Journal of Public Administration Research and Theory* 3: 457–486.

Thomas, Craig W. 1997. "Public Management as Interagency Cooperation: Testing Epistemic Community Theory at the Domestic Level." *Journal of Public Administration Research and Theory* 7: 221–246.

Thomas, Craig W. 1998. "Maintaining and Restoring Public Trust in Government Agencies and Their Employees." *Administration and Society* 30: 166–193.

Thomas, Craig W. 1999. "Linking Public Agencies with Community-Based Watershed Organizations: Lessons From California." *Policy Studies Journal* 27: 544–564.

Thomas, Craig W. 2001. "Habitat Conservation Plans: Certainly Empowered, Somewhat Deliberative, Questionably Democratic." *Politics and Society* 29: 105–130.

Thomas, Jack Ward, and Hal Salwasser. 1989. "Bringing Conservation Biology into a Position of Influence in Natural Resource Management." *Conservation Biology* 3: 123–127.

Thompson, James D. 1967. *Organizations in Action*. New York: McGraw-Hill.

Thompson, Robert. 1994. "Natural Communities Conservation Planning." In Judith Innes, Judith Gruber, Michael Neuman, and Robert Thompson, eds., *Coordinating Growth and Environmental Management through Consensus Building*. Berkeley: California Policy Seminar, University of California.

Thornton, Robert D. 1991. "Searching for Consensus and Predictability: Habitat Conservation Planning Under the Endangered Species Act of 1973." *Environmental Law* 21: 604–656.

Timberland Task Force. 1993. *The Report of the California Timberland Task Force*. Sacramento: California Resources Agency.

Tobin, Richard J. 1990. *The Expendable Future: U.S. Politics and the Protection of Biological Diversity*. Durham, NC: Duke University Press.

Tullock, Gordon. 1965. *The Politics of Bureaucracy*. Washington, DC: Public Affairs Press.

Tversky, Amos, and Daniel Kahneman. 1991. "Loss Aversion in Riskless Choice: A Reference-Dependent Model." *Quarterly Journal of Economics* 106: 1039–1061.

Twight, Ben W. 1983. *Organizational Values and Political Power: The United States Forest Service versus the Olympic National Park*. University Park: Pennsylvania State University Press.

Twight, Ben W., Fremont J. Lyden, and E. Thomas Tuchmann. 1990. "Constituency Bias in a Federal Career System?: A Study of District Rangers of the U.S. Forest Service." *Administration and Society* 22: 358–389.

U.S. Fish and Wildlife Service and National Marine Fisheries Service. 1996. *Endangered Species Habitat Conservation Planning Handbook*. Washington, DC: U.S. Fish and Wildlife Service and National Marine Fisheries Service.

U.S. Fish and Wildlife Service and National Marine Fisheries Service. 1998. "Habitat Conservation Plan Assurances Rule." *Federal Register* 63, no. 35 (February 23): 8859–8873.

U.S. General Accounting Office. 1987. *Endangered Species: Limited Effect of Consultation Requirements on Western Water Projects*. GAO/RCED-87-78. Washington, DC: U.S. General Accounting Office.

U.S. General Accounting Office. 1989 *Endangered Species: Spotted Owl Petition Evaluation Beset by Problems*. GAO/RCED-89-79. Washington, DC: U.S. General Accounting Office.

U.S. General Accounting Office. 1991. *Public Land Management: Attention to Wildlife Is Limited*. GAO/RCED-91-64. Washington, DC: U.S. General Accounting Office.

U.S. General Accounting Office. 1993. *Endangered Species: Factors Associated with Delayed Listing Decisions*. GAO/RCED-93-152. Washington, DC: U.S. General Accounting Office.

U.S. General Accounting Office. 1995. *National Wildlife Refuge System: Contributions Being Made to Endangered Species Recovery*. GAO/RCED-95-7. Washington, DC: U.S. General Accounting Office.

U.S. General Accounting Office. 1996. *National Park Service: Activities within Park Borders Have Caused Damage to Resources*. GAO/RCED-96-202. Washington, DC: U.S. General Accounting Office.

U.S. General Accounting Office. 1997. *Land Management Agencies: Information on Selected Administrative Policies and Practices*. GAO/RCED-97-40. Washington, DC: U.S. General Accounting Office.

U.S. Geological Survey. 1998. *Status and Trends of the Nation's Biological Resources*. 2 vols. Washington, DC: U.S. Government Printing Office.

U.S. Office of Technology Assessment. 1987. *Technologies to Maintain Biological Diversity*. OTA-F-330. Washington, DC: U.S. Office of Technology Assessment.

U.S. Office of Technology Assessment. 1992. *Forest Service Planning*. Washington, DC: U.S. Office of Technology Assessment.

Wagner, Frederic H., Ronald Foresta, and R. Bruce Gill. 1995. *Wildlife Policies in the U.S. National Parks*. Washington, DC: Island Press.

Wallace, David Rains. 1983. *The Klamath Knot*. San Francisco: Sierra Club Books.

Weatherley, Richard, and Michael Lipsky. 1977. "Street-Level Bureaucrats and Institutional Innovation: Implementing Special-Education Reform." *Harvard Educational Review* 47: 171–197.

Weiss, Janet A. 1987. "Pathways to Cooperation among Public Agencies." *Journal of Policy Analysis and Management* 7: 94–117.

Welner, Jon. 1995. "Natural Communities Conservation Planning: An Ecosystem Approach to Protecting Endangered Species." *Stanford Law Review* 47: 319–361.

Wilcove, David S., Margaret McMillan, and Keith C. Winston. 1993. "What Exactly Is an Endangered Species? An Analysis of the U.S. Endangered Species List: 1985–1991." *Conservation Biology* 7: 87–93.

Wilensky, Harold. 1964. "The Professionalization of Everyone?" *American Journal of Sociology* 70: 137–158.

Williams, Daniel F., Sheila Byrne, and Theodore A. Rado. 1992. *Endangered and Sensitive Species of the San Joaquin Valley, California: Their Biology, Management and Conservation*. Sacramento: California Energy Commission.

Williamson, Oliver E. 1985. *The Economic Institutions of Capitalism*. New York: Free Press.

Wilson, Edward O. 1992. *The Diversity of Life*. Cambridge, MA: Harvard University Press.

Wilson, James Q. 1989. *Bureaucracy: What Government Agencies Do and Why They Do It*. New York: Basic Books.

Wilson, James Q., and Patricia Rachal. 1977. "Can the Government Regulate Itself?" *Public Interest* 46: 3–14.

Wondolleck, Julia M., and Steven L. Yaffee. 2000. *Making Collaboration Work: Lessons from Innovation in Natural Resource Management*. Washington, DC: Island Press.

Woody, Todd. 1993. "Fishy Deals." *The Recorder*, August 11. Reprinted in the *San Francisco Examiner*, September 19, 1993, pp. 7, 10, 11, 12.

Worster, Donald. 1994. *Nature's Economy: A History of Ecological Ideas*. 2nd ed. New York: Cambridge University Press.

Yaffee, Steven L. 1994. *The Wisdom of the Spotted Owl: Policy Lessons for a New Century*. Washington, DC: Island Press.

Yaffee, Steven L., editor. 1996. *Ecosystem Management in the United States: An Assessment of Current Experience*. Washington, DC: Island Press.

Yaffee, Steven L., Peter Aengst, Jeremy Anderson, Jay Chamberlin, Christopher Grunewald, Susan Loucks, and Elizabeth Wheatley. 1998. *Balancing Public Trust and Private Interest: Public Participation in Habitat Conservation Planning*. Ann Arbor: School of Natural Resources and Environment, University of Michigan.

Yates, Douglas. 1982. *Bureaucratic Democracy: The Search for Democracy and Efficiency in American Government*. Cambridge, MA: Harvard University Press.

Zea, Donn. 1991. "California Forestry Association Opinion Piece." *Business Wire*, October 1.

Index